THE GEOGRAPHY BOOK

BIG IDEAS

THE ANTHROPOLOGY BOOK
THE ARCHITECTURE BOOK
THE ART BOOK
THE ASTRONOMY BOOK
THE BIBLE BOOK
THE BIOLOGY BOOK
THE BLACK HISTORY BOOK
THE BUSINESS BOOK
THE CHEMISTRY BOOK
THE CLASSICAL MUSIC BOOK
THE CRIME BOOK
THE DESIGN BOOK
THE ECOLOGY BOOK
THE ECONOMICS BOOK
THE FEMINISM BOOK
THE GEOGRAPHY BOOK
THE HISTORY BOOK
THE ISLAM BOOK
THE LAW BOOK

THE LGBTQ+ HISTORY BOOK
THE LITERATURE BOOK
THE MATHS BOOK
THE MEDICINE BOOK
THE MILITARY HISTORY BOOK
THE MOVIE BOOK
THE MYTHOLOGY BOOK
THE PHILOSOPHY BOOK
THE PHYSICS BOOK
THE POETRY BOOK
THE POLITICS BOOK
THE PSYCHOLOGY BOOK
THE RELIGIONS BOOK
THE SCIENCE BOOK
THE SHAKESPEARE BOOK
THE SHERLOCK HOLMES BOOK
THE SOCIOLOGY BOOK
THE WORLD WAR I BOOK
THE WORLD WAR II BOOK

SIMPLY EXPLAINED

THE GEOGRAPHY BOOK

DK LONDON

SENIOR EDITOR
Alison Sturgeon

SENIOR ART EDITOR
Gadi Farfour

EDITORS
John Andrews,
Rachel Warren Chadd, Vicki Murrell

ILLUSTRATIONS
James Graham

SENIOR PRODUCTION EDITOR
Andy Hilliard

SENIOR PRODUCTION CONTROLLER
Laura Andrews

MANAGING EDITOR
Gareth Jones

MANAGING ART EDITOR
Luke Griffin

PUBLISHING DIRECTOR
Georgina Dee

ART DIRECTOR
Maxine Pedliham

This book was made with Forest Stewardship Council™ certified paper – one small step in DK's commitment to a sustainable future. Learn more at www.dk.com/uk/information/sustainability

DK DELHI

SENIOR EDITOR
Janashree Singha

PROJECT ART EDITOR
Shipra Jain

ART EDITOR
Arshti Narang

MANAGING EDITOR
Soma B. Chowdhury

SENIOR MANAGING ART EDITOR
Arunesh Talapatra

SENIOR JACKET DESIGNER
Suhita Dharamjit

SENIOR JACKETS COORDINATOR
Priyanka Sharma Saddi

SENIOR DTP DESIGNER
Harish Aggarwal

ASSISTANT PICTURE RESEARCH ADMINISTRATOR
Samrajkumar S.

PRE-PRODUCTION IMAGE EDITORS
Ashok Kumar, Vikram Singh

PRE PRODUCTION DESIGNER
Umesh Singh Rawat

PRE-PRODUCTION COORDINATOR
Tarun Sharma

PRE-PRODUCTION MANAGER
Balwant Singh

PRODUCTION MANAGER
Pankaj Sharma

CREATIVE HEAD
Malavika Talukder

SANDS PUBLISHING SOLUTIONS

EDITORIAL PARTNERS
David and Silvia Tombesi-Walton

DESIGN PARTNER
Simon Murrell

original styling by
STUDIO 8

First published in Great Britain in 2026 by
Dorling Kindersley Limited
20 Vauxhall Bridge Road,
London, SW1V 2SA

The authorised representative in the EEA is
Dorling Kindersley Verlag GmbH.
Arnulfstr. 124, 80636 Munich, Germany

Copyright © 2026 Dorling Kindersley Limited
A Penguin Random House Company
10 9 8 7 6 5 4 3 2 1
001–355489–April/2026

All rights reserved.
No part of this publication may be reproduced, stored in or introduced into a retrieval system, or transmitted, in any form, or by any means (electronic, mechanical, photocopying, recording, or otherwise), without the prior written permission of the copyright owner.

DK values and supports copyright. Thank you for respecting intellectual property laws by not reproducing, scanning or distributing any part of this publication by any means without permission. By purchasing an authorised edition, you are supporting writers and artists and enabling DK to continue to publish books that inform and inspire readers. No part of this publication may be used or reproduced in any manner for the purpose of training artificial intelligence technologies or systems. In accordance with Article 4(3) of the DSM Directive 2019/790, DK expressly reserves this work from the text and data mining exception.

A CIP catalogue record for this book
is available from the British Library.
ISBN: 978-0-2417-8426-6

Printed and bound in India

www.dk.com

CONSULTANT & CONTRIBUTORS

JERRY T. MITCHELL – CONSULTANT

Jerry Mitchell, PhD, is Professor and Chair of the Department of Geography at the University of South Carolina in the US, where he heads the Center of Excellence for Geographic Education and teaches courses in environmental hazards, geography education, and Latin America. He is the author of *Geography for Dummies* and former Editor of the *Journal of Geography*.

MICHAEL BRIGHT

Michael Bright is a graduate of the University of London, a corporate biologist, and a member of the Royal Society of Biology. He has worked for the BBC Natural History Unit in Bristol, UK, and is now a freelance writer and ghostwriter.

TIM HARRIS

After studying Norwegian glaciers at university, Tim Harris has travelled the world in search of extraordinary landscapes and unusual wildlife. He has explored the dunes of the Namib Desert, climbed Popocatepetl volcano, and camped in the Sumatran rainforest. A former Deputy Editor of *Birdwatch* magazine in the UK, Tim has written many books about nature and geography for adults and children.

TOM JACKSON

Tom Jackson has been a science writer for 30 years and has a particular interest in the history of ideas. He has written more than 100 non-fiction books for adults and children and contributed to many more. He studied at Bristol University and has worked variously as a zookeeper, conservationist, travel writer, and tech journalist before turning to a career in non-fiction writing.

ROBERT SNEDDEN

Robert Snedden went from calculating royalties to writing and editing books for children and adults. Over more than 40 years in publishing he has produced works on topics ranging from the environment and space exploration to mathematics and medicine. He contributed to Dorling Kindersley's *The Biology Book*, *The Medicine Book*, *The Physics Book* and *The Chemistry Book*.

MARCUS WEEKS

Marcus weeks studied Music and Philosophy at Sheffield University, and worked as a teacher, piano restorer, and musician before embarking on a career as an author. He has written and contributed to numerous books on music, philosophy, psychology and the arts, including several titles in Dorling Kindersley's "Big Ideas" series.

CONTENTS

10 INTRODUCTION

MAPS AND MAPPING

20 On the measure of the Earth
The development of cartography

22 One representing the length and the other the breadth
A geographical coordinates system

24 To spread on a plane the surface of the sphere
Map projections

32 By geometrical calculation, we determine the distances of places
Triangulation

34 The connection between the Broad Street pump and the outbreak of cholera
Thematic mapping

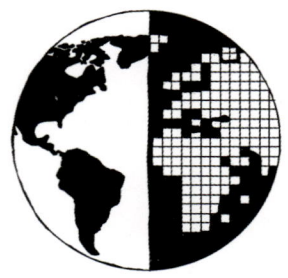

38 The magnetic phenomena of Earth
Geophysical mapping

40 The earth unfolds into an immense carpet without borders
Aerial photography and remote sensing

46 A revolution in geological thinking
Mapping the ocean floor

48 Accuracy, integrity, and availability
The global positioning system

52 The power of geography
Geographic information systems

60 A paradigm shift in Earth system modelling
Digital twins of Earth

PHYSICAL GEOGRAPHY

66 Neither from springs nor from the sea: but from the run-off of rains
The hydrological cycle

70 At the time the lower stratum was being formed, none of the upper strata existed
Steno's law of superposition

72 The action of the Sun is the original cause
Atmospheric circulation and winds

74 The traces of revolutions which have taken place
Catastrophism

75 A continual succession
Uniformitarianism

76 Raised from the deep by a succession of upward movements
Orogeny

80 A tendency to accumulate heat at the surface
The Greenhouse effect

84 The compound centrifugal force
The Coriolis effect

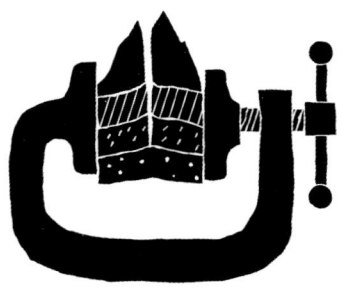

86 This great engine set at work ages ago
Glaciation and ice ages

90 Species in the inorganic kingdom
Mineralogy

94 A solid crust, resting upon a plastic substratum
Isostasy

96 Temperature, precipitation, and seasonal cycle
Climatic zones

102 Structure, process, and time
The cycle of erosion

108 Either mechanically or through chemical corrosion
Karst landscapes

110 The continents must have shifted
Continental drift

112 At that depth, there must be a sudden change of material making up the interior of the Earth
The structure of Earth

114 Forces which displace continents are the same as those which produce great fold-mountain ranges
Plate tectonics

122 A thermal anomaly in the mantle
Volcanic activity and hotspots

126 When you look deeper into the ice, you go back in time
Paleoclimatology

128 Soil has a profile
Pedology

132 A method of segregating large, moderate, and small shocks
Seismology and earthquake magnitude

136 There is nothing large or small in nature
Biomes and ecological zones

140 Great ocean conveyor
Global ocean circulation

142 There was no type of building code that was of any value...
Extreme weather

HUMAN GEOGRAPHY

148 Famine seems to be the last, the most dreadful resource of nature
Malthusian theory

150 The nature of the industry determines its location
Location theory

152 Migration means life and progress
Migration theory

154 The state is an organism
Organic theory

156 Man is a product of the Earth's surface
Environmental determinism

157 There are no necessities, but everywhere possibilities
Possibilism

158 Who rules the Heartland commands the World-Island
Heartland theory

159 Who controls the Rimland rules Eurasia
Rimland theory

160 A relationship between size and rank that is both quite precise and quite simple
Rank-size rule

161 A country's leading city is disproportionately large
Primate city rule

162 Geography bore many children
Geography as human ecology

- **164** Diffusion of customs over enormous areas
  Cultural diffusion

- **172** Culture is the agent, the natural area the medium
  Cultural landscape theory

- **176** We are becoming more and more interested in the relation of our numbers to our welfare
  Demographic transition model

- **182** The range of a good is greater when it is offered in a larger central place
  Central place theory

- **184** The city is the nucleus
  The concentric zone model

- **188** Human capital is the most critical capital for contemporary societies
  Censuses and population geography

- **190** The story of each national economy
  Theories of development

- **192** Near things are more related than distant things
  Spatial interaction theory

- **194** Every citizen has had long associations with some part of his city
  Behavioural geography

- **196** The whole social character of the district is changed
  Gentrification

- **198** The only kind of social system is a world-system
  World-systems theory

- **200** Spatial feelings and ideas in the stream of experience
  Humanistic geography

- **202** Place has power
  Territoriality theory

- **204** The capitalist production and reconstruction of space
  The urbanization of capital

- **206** The accumulation of capital has always been a profoundly geographical affair
  Spatial justice

- **208** Places are not closed or bounded
  Local and global

- **210** Spaces and places are gendered
  Gender and space theory

- **214** The border between real and digital virtual worlds is already porous
  Digital spaces as cultural spaces

ENVIRONMENTAL GEOGRAPHY

- **220** Why is the common itself so bare-worn?
  The tragedy of the commons

- **222** All our living trees will clap their hands
  Natural resource management and conservation

- **228** Man-induced soil erosion
  Desertification

- **230** Floods are "acts of God", but flood losses are largely acts of man
  Floodplain management

- **232** An uncontrolled increase in atmospheric CO_2
  Climate change

- **240** This sudden silencing of the song of the birds...
  Silent Spring

- **241** The guardian of life for all of its existence
  The Gaia hypothesis

242 How do we rediscover where we actually live?
Bioregionalism

244 For the entire planet into the distant future
Sustainable development

246 All people are entitled to equal environmental protection
The environmental justice movement

248 Earth was in a state of overuse
Ecological footprint concept

250 A universalization of hazards
Globalization

254 The best single measure of global environmental decline
Climate migration

256 The central role of mankind in geology and ecology
The Anthropocene

APPLIED GEOGRAPHY

264 Zoning seeks to protect and stabilize what is good
Land use zoning

266 People are not paths but they cannot avoid drawing them in space-time
Time-geography

268 The extreme sensitivity of Earth's climate
Climate modelling

272 We must design with nature
Environmental impact assessment

274 The new magic
Geodemographic analysis

278 Justice has a geography
Mapping social injustice

280 The importance of "where"
Spatial data analysis

282 First we shape the cities – then they shape us
The "Smart Growth" movement

284 It's hard to manage what you can't measure
Monitoring environmental change

290 The sun and wind won't be sending you a bill
Renewable energy geography

292 You cannot live without space
Remote-sensing agriculture

296 There's no such thing as a natural disaster
Early warning systems

304 Everyone, everywhere
Crisis mapping with crowdsourcing

306 Biodiversity can be monitored on a global scale
Monitoring biodiversity

308 To take the pulse of our planet
Global Earth observation programmes

310 Information has to move even faster
Tracking disease

314 DIRECTORY

324 GLOSSARY

328 INDEX

335 QUOTE ATTRIBUTIONS

336 ACKNOWLEDGMENTS

INTRODU

CTION

INTRODUCTION

The word "geography" is derived from the Greek γεωγραφία and literally means "Earth writing". However, such a simple description belies the scope of this subject. As one of the most varied fields of study, geography encompasses a broad range of disciplines from both the natural sciences and social sciences. It encompasses the physical features of Earth, its inhabitants, and the relationships between them, and embraces fields as diverse as geology and politics in the study of our planet and how we live on it.

Branches of geography

Geography as a discipline can broadly be divided into two main branches – physical geography and human geography – with each incorporating different categories and approaches. Cartography, environmental geography, and applied geography straddle both main branches, and form distinct fields of study.

Cartography, or mapmaking, was the primary goal of early geographers in ancient civilizations. With its emphasis on location and navigation, cartography laid the foundations for geography as an academic discipline and has continued to provide vital tools and techniques to the present time. Communicating spatial information about both natural (physical) phenomena, such as landforms and rivers, and human phenomena, such as cities and roads, it serves both main branches of geography.

Physical geography is the study of Earth's natural environment, its features, and processes. A subject with links to all the natural sciences, it examines the atmosphere (air), hydrosphere (water), biosphere (life), and geosphere (land), and the way these elements interact.

The branch known as human geography studies the complex relationship between people and their physical environment. Allied with the social sciences and humanities, it looks at issues such as population and urbanization, as well as the role of culture, politics, and economics in the way people relate to their surroundings. Its focus is therefore on where, how, and why spatial patterns develop (such as migration or urban sprawl).

The field of environmental geography also combines elements of physical and human geography. Drawing on knowledge from various disciplines, including ecology and environmental science, it addresses problems of environmental damage, such as climate change, depletion of resources and biodiversity, and the possibility of sustained environmental management.

Applied geography takes the tools, techniques, and methods used by geographers, and applies these to solve real-world social, economic, and environmental problems, using technologies such as remote sensing and the global positioning system (GPS) that we consider essentials of modern life.

Early geographers

The quest for what we now call geographic knowledge is as old as human civilization. When ancient peoples began to explore the world,

Geography is everywhere.
Denis Cosgrove
"Geography is Everywhere: Culture and Symbolism in Human Landscapes", *Horizons in Human Geography,* **1989**

they recorded their observations of physical landscapes in maps and verbal descriptions, often coupled with mythological or religious explanations. The earliest known maps were created in ancient Babylon, Mesopotamia, in the 9th century BCE. Drawn on clay tablets, these maps portray prominent geographical features, most likely for practical use in trade or agriculture, or for delineating territory.

As civilizations in Babylon, Egypt, Greece, China, and India expanded their horizons, geographical knowledge became key to aiding navigation. The Ancient Greeks in particular became dissatisfied with simply describing natural phenomena and began to seek rational explanations for them. Early Greek philosophers, including Thales of Miletus and his protégé Anaximander, produced the first world maps in the 6th century BCE, depicting Earth as a flat disc divided into three continents and encircled by a vast ocean.

An emerging science

Geography as a distinct area of study emerged as Classical Greek culture flourished. Philosophers such as Pythagoras and Aristotle developed a scientific approach, based on observation and reason, and hypothesized Earth as spherical. Eratosthenes (c. 240 BCE) – credited with coining the word *"geographia"* – later proved this to be true, and calculated Earth's circumference.

Geography continued to thrive in the regions colonized by Alexander the Great. Significant advances were made in cartography, as geographers such as Hipparchus, Marinus of Tyre, and Claudius Ptolemy produced increasingly comprehensive maps and atlases of the known world.

With the rise of Christianity in Europe throughout the Middle Ages (c. 500–1500 CE), religious dogma largely replaced empirical scientific study, and European geographical knowledge was discredited or simply lost. However, the establishment of Islam prompted a Golden Age of scholarship around the 9th century CE in the Persian and Arab world and the rediscovery of Classical Greek texts. Islamic scholars such as Abū Zayd al-Balkhī, Al-Idrisi, and Ibn Battuta built on the Greek texts, making advances not only in cartography, but also the scientific study of landforms and climate.

Scientific revolution

From c. 1500, the cultural shake-up of the European Renaissance and the so-called "Scientific Revolution" that accompanied it inspired a renewed interest in the quest for geographic knowledge. This in turn prompted the "Age of Exploration", with European seafarers sailing the globe in search of goods to trade and territories to conquer. Explorers such as Christopher Columbus and Ferdinand Magellan "discovered" lands unknown to Europeans and, in the process of mapping new territories and seas, gained a more complete spatial understanding of the globe. At the same time, advances were also taking place in the East, with Chinese mariners such as Zheng He exploring and mapping the South Seas. »

> The purpose of Geography is to provide a view of the whole Earth by mapping the location of places.
> **Ptolemy**
> c. 150 CE

INTRODUCTION

Long-range exploration required a world map that represented the curved surface of Earth on a flat sheet, while retaining lines of constant course for navigation. In 1569, Flemish cartographer Gerardus Mercator made the breakthrough, with his eponymous map projection that became the standard for nautical navigation and continues to influence digital mapping in the modern age.

The natural world

By establishing geography as an analytical and empirical science, the "Scientific Revolution" in Europe led to many key advances in the field of physical geography. In 1580, French hydraulics engineer Bernard Palissy proposed the first comprehensive explanation of the water cycle. A century later, Danish scientist Nicolas Steno's rigorous field studies in the mountains and quarries of Tuscany, Italy, led to the understanding that rock layers represent a chronological history. This established fundamental principles for the study of geology and paleontology.

By the 18th century, the Scientific Revolution had spawned the Enlightenment, or Age of Reason. This manifested itself in the establishment of modern scientific disciplines such as chemistry, physics, and other natural sciences, and these were incorporated into the study of physical geography, especially in the study of weather, climate, and ecosystems. In 1735, English physicist George Hadley described global atmospheric circulation as driven by solar heating and the rotation of Earth. A century later, French scientist Gaspard-Gustave de Coriolis observed how Earth's rotation created horizontal deflections of air. Both theories became fundamental to explaining global weather systems.

Building on the work of Nicolas Steno, in the 1820s Scottish geologist Charles Lyell proposed the theory that slow, gradual geological processes had shaped the planet over immense timescales. This radically expanded the scientific understanding of Earth's history and provided the temporal framework for German geologist Alfred Wegener's continental drift theory in 1912. In turn, the latter laid the groundwork for the theory of plate tectonics, which became widely accepted in the 1960s.

People and place

In the late 18th century, the Industrial Revolution soon radically and permanently transformed the environment of the modern world. The human experience in the context of rapid urbanization and change led to human geography as a distinct field of study.

German-English geographer E.G. Ravenstein analysed migration statistics to identify underlying principles in the mass movements of people. German ethnologist Leo Frobenius examined the spread of cultural traits, while Canadian sociologist Ernest Burgess studied the process of urbanization to explain how cities grow and social groups are spatially organized. In the late 20th century, American sociologist Robert D. Bullard and British geographer Doreen Massey challenged traditional concepts of

Geography is the study of the Earth as the home of people.
Yi-Fu Tuan
"A View of Geography",
Geographical Review, 1991

  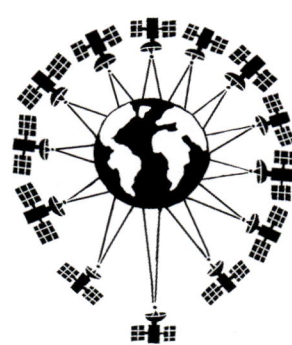

spatial theory to show how race and gender shape people's experiences of space and place.

Environmental challenges

In recent decades, as the effects of human activity on the environment have been brought to the fore, the growing field of environmental geography has sought to quantify the impact of society on biological and physical systems. In 1958, American geochemist Charles David Keeling confirmed the link between the burning of fossil fuels and the continuing rise in the level of atmospheric CO_2. A few years later, with the publication of her book *Silent Spring* (1962), American marine biologist Rachel Carson brought environmental issues firmly into the mainstream.

The UN-sponsored Brundtland Report, "Our Common Future" (1987) also provided a foundational text. By emphasizing the link between environment, social equity, and economic growth, the report laid out a long-term framework for global cooperation on sustainable development.

Technology and data

Throughout the 20th century, major advances in technology revolutionized the way in which geographers acquire geographic knowledge and also how they analyse and make use of that information. Mapping was transformed by aerial photography, satellite imagery, and global positioning system (GPS) technology, moving from static paper maps to interactive, digital formats. In the 1960s, what became known as the Quantitative Revolution introduced the widespread use of statistical techniques, mathematical models, and computer technology to analyse large data sets and identify spatial patterns.

The unique purpose of Geography is to seek comprehension of the variable character of areas in terms of all the interrelated features which together form that variable character.
Richard Hartshorne
Perspective on the Nature of Geography, 1959

The invention of geographic information systems (GIS) by British-Canadian geographer Roger Tomlinson in 1963 enabled an unprecedented amount of data to be gathered and analysed for mapping and data modelling. With the launch of NASA's Landsat-1 satellite in 1974, multispectral scanners began to capture Earth's surface data across multiple wavelengths. This combination of remote sensing (data collection) and GIS (data analysis) has provided detailed information that can be applied across numerous fields, such as public health and urban planning, and has enhanced geography's ability to address complex real-world issues.

From its earliest beginnings in the rudimentary maps of ancient civilizations, geography has evolved into a comprehensive field of study. By incorporating knowledge and methods from the spectrum of natural and social sciences, it offers an expansive understanding of the world and provides practical applications in human society. Combined with the sophistication of modern technology, this ensures geography is an essential tool to help humankind find solutions to the challenges of the world today. ∎

MAPS AND MAPPING

D

18 INTRODUCTION

Anaximander produces an early **scientific map of the world**, encircled by a vast ocean.

6TH CENTURY BCE

Marinus of Tyre **superimposes a grid of squares** onto his maps for reference purposes.

c. 120 CE

Chinese map-makers **create sea charts** based on Zheng He's voyages to East Africa.

15TH CENTURY CE

A 182-sheet map of France – a huge **feat of triangulation** – is finally completed.

1815

3RD CENTURY BCE

Eratosthenes **uses early trigonometry** to calculate Earth's circumference.

c. 150 CE

Ptolemy's treatise *Geography* includes instructions for map projections and describes and plots known places.

1569

Gerardus Mercator produces a world map based on a new **projection that transforms sea navigation**.

1840

Carl Gauss and Wilhelm Weber **publish a book of magnetic maps** and theoretical principles: *Atlas of Geomagnetism*.

A familiar part of everyday life, maps are used to follow a route, find a destination, or check a weather forecast. However, they also play a fundamental role in every branch of geography. By enabling an understanding of spatial relationships and representing data about both natural and human features, they are an invaluable tool for understanding Earth's geography and how humans interact with it.

Mapmaking stretches back into prehistory, born of the basic desire for humans to understand their surroundings. In cave paintings and carved on bones and tusks left by nomadic Stone Age peoples are what appear to be rudimentary maps showing geographical features such as rivers and mountains.

By the 1st millennium BCE, early civilizations were beginning to illustrate their concept of the world. In Mesopotamia – modern-day Iraq – the Babylonians made the first known attempt at a world map. Inscribed on clay, it included Babylon and its empire, surrounded by a river with mountains beyond.

Ancient Greek insight
By the 6th century BCE, Greek scholars had begun to envisage the world as a sphere floating in space. Aristotle confirmed this idea in the 4th century BCE, and a century later Eratosthenes, applying mathematics to mapmaking, calculated Earth's circumference with remarkable accuracy. He also created a world map from the latest contemporary knowledge and plotted locations on it using parallels and meridians.

In the 2nd century BCE, Hipparchus refined the coordinates system, and Marinus of Tyre produced a list of places, or gazetteer, each with assigned coordinates.

Next, scholars tackled the geometrical conundrum of map projections – how to represent a roughly spherical world on a flat surface. Greek polymath Ptolemy calculated two solutions, but it wasn't until 1569 that Flemish geographer Gerardus Mercator developed a map projection that mariners could use to navigate the world's oceans.

Mapping the land
While short distances could be measured – for example, by using a chain – those over large areas remained largely approximate until Dutch mathematician Gemma

MAPS AND MAPPING

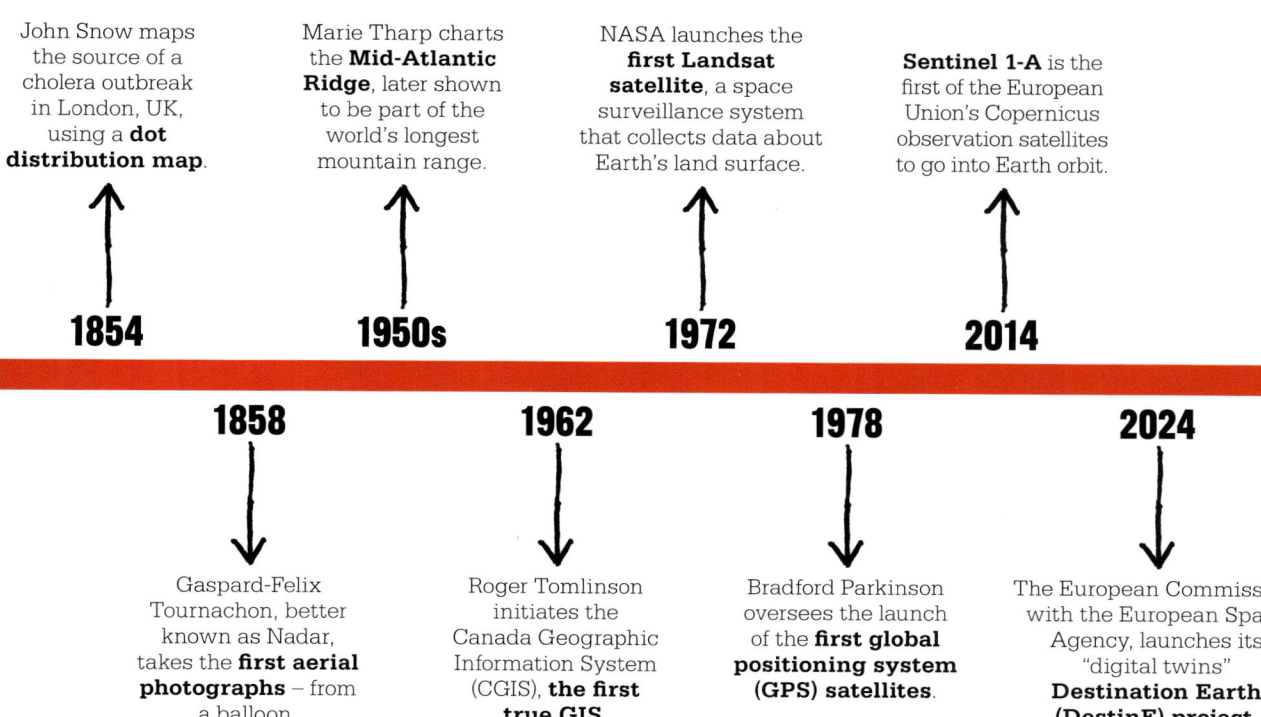

Frisius established the principles of triangulation in the 1530s. This created a method of surveying based on the trigonometric proposition that if one side and two angles of a triangle are known, the remaining sides can be calculated. Frisius's technique of connecting multiple triangles, and ever more sophisticated surveying tools, enabled increasingly accurate regional and national maps.

Tracking other features

Over time, maps became tools for representing more than just physical features. The first thematic maps emerged in the 17th century and by the 19th century were in wide use to illustrate a specific subject or theme within a geographic area. With scientific advances came geophysical maps that use data – seismic waves, gravity, and magnetic fields – to visualize Earth's subsurface features.

Technological breakthroughs in the 19th and 20th centuries revolutionized the surveying of land. The invention of photography and the advent of flight enabled detailed aerial scanning and observation. On the oceans, research ships equipped with seismographs and sonar sounding equipment collected data for the first maps of the sea floor.

Huge digital leaps

Radio-based navigation systems proved their worth in World War II, while the satellite global positioning system (GPS) transformed map usage by making it accessible, real-time, and interactive for all. In turn, the creation of geographical information systems (GIS) has led to dynamic, digital, and multi-layered maps. Each layer represents a specific type of data and insights on topics as diverse as crime statistics, rainfall levels, or voting intentions can all be integrated.

Ever-increasing computer power and the advent of artificial intelligence (AI) have also allowed geographers to create 3-D virtual models of Earth. Launched in 2022, the European Space Agency's Digital Twin Earth (DTE) has been designed to replicate and simulate the behaviour of Earth's systems with remarkable precision. Its ability to monitor, forecast, and analyse makes it an invaluable resource for dealing with climate change impacts. By providing models for alternative realities, it could also hold the key to ensuring a more positive future for the planet. ∎

ON THE MEASURE OF THE EARTH
THE DEVELOPMENT OF CARTOGRAPHY

IN CONTEXT

KEY FIGURES
Aniximander (610–546 BCE),
Eratosthenes
(c. 276–c. 194 BCE)

BEFORE
c. 1300 BCE The Egyptian Turin Papyrus includes a local map, depicting the topography of the Wadi Hammamat valley.

c. 7th century BCE Babylonian world maps on clay tablets feature Babylon as a rectangle surrounded by a circular body of water.

AFTER
9th century CE Arab mathematician Al-Khwarizmi calculates the Earth's circumference within 15 per cent of its true modern value.

1686 British polymath Isaac Newton theorizes the shape of the Earth as an oblate spheroid, meaning it bulges at the equator and is flattened at the poles.

In ancient times, mapmaking was either used to show the topography of a small region or depicted the world as it was conceived at the time – as a large, flat disc surrounded by an ocean. However, this idea came under scrutiny in the 6th century BCE with the emergence of the first philosophers in Greek Asia Minor, and by c. 240 BCE, Greek polymath Eratosthenes had calculated the circumference of a spherical world with remarkable accuracy.

Towards a sphere
In the 6th century BCE, Greek philosopher Anaximander of Miletus proposed the radical notion that the disc imagined by the ancients was instead the surface of a cylindrical world floating freely in infinite space. It was, he believed, on this surface that the known world existed, with its land surrounded by ocean. The world map he produced, considered

Eratosthenes' world map, recreated in the 1870s from his descriptions (quoted by later writers), shows the grid system of parallels and meridians that he introduced.

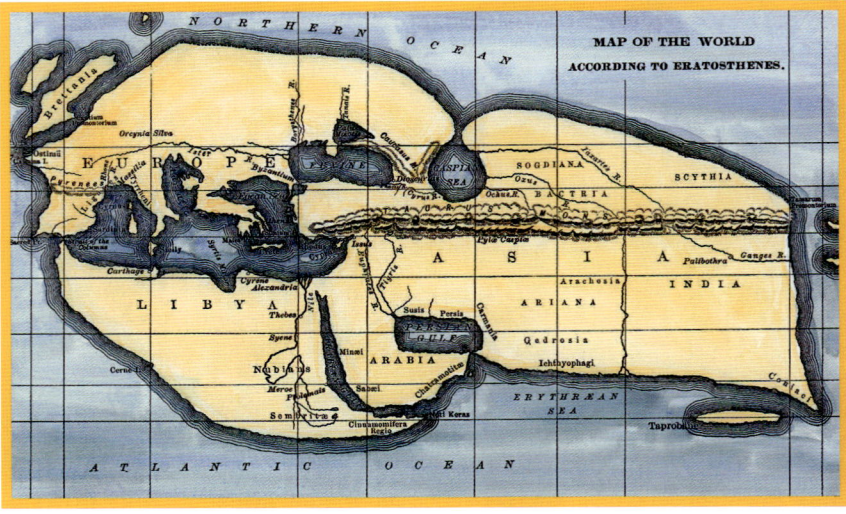

MAPS AND MAPPING 21

See also: Map projections 24–31 ▪ Triangulation 32–33 ▪ The global positioning system 48–51 ▪ Digital twins of Earth 60–61

---

the first of its kind, depicted a circle with the eastern Mediterranean at its centre, and around it the lands of Europe, Asia, and Libya, ringed by an outer ocean. Although his original map has not survived, the concept can be inferred from a version made by Greek geographer Hecataeus of Miletus (c. 550–c. 476 BCE).

By the time of Pythagoras (c. 570–c. 495 BCE), the concept of a world floating in space had taken hold and the idea of a spherical Earth began to emerge. Confirmation came when Aristotle (384–322 BCE) provided evidence from his observation of ships disappearing over the horizon, the shape of lunar eclipses, and the fact that certain stars are not visible in different parts of the world.

Earth's circumference

Following Aristotle's empirical approach, in the 3rd century BCE, Greek polymath Eratosthenes determined to place the study of geography on a solid scientific footing. In his treatise "On the measurement of the Earth", he

Eratosthenes observed, on the summer solstice, the angle of the shadow cast by a vertical rod at Syene (present-day Aswan, Egypt) and Alexandria respectively. In Syene, the rod cast no shadow, while in Alexandria, the shadow revealed the Sun's rays to be at a 7.2-degree angle, which he calculated as being 1/50th of a sphere. By multiplying the distance by 50, he could deduce the circumference measurement.

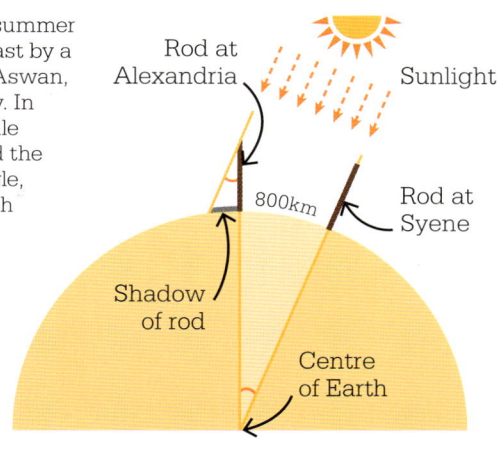

showed how mathematics could be used to achieve greater accuracy in mapmaking. First, he compared the angle of the Sun's shadow at two different locations on the summer solstice. Then, he used the work of professional "bematists", who measured distances by counting their paces, to determine the distance between these locations. With these measurements, and simple trigonometry, he calculated the circumference of Earth to be approximately 252,000 stadia.

This is the equivalent of about 40,000 km (24,855 miles), which is remarkably close to the modern measurement, with an error margin of only 1–2 per cent.

Eratosthenes' calculations paved the way for more advanced mapmaking. Along with knowledge derived from Alexander the Great's military campaigns, he created a world map, and imposed on it a grid of parallels and meridians, which allowed him to plot the locations of more than 400 places. ■

Eratosthenes

Although originally from Cyrene, a Greek city in present-day Libya, Eratosthenes moved to Athens in c. 276 BCE to study with the Stoic philosopher Zeno of Citium, and later at the Academy founded by Plato. As well as philosophy, he studied mathematics, astronomy, and music, and was also an accomplished poet.

His scholarship brought him to the attention of the Egyptian ruler Ptolemy III Euergetes, who in c. 245 BCE invited him to work at the Library of Alexandria, where he soon rose to the position of chief librarian and began work on a treatise, which he titled *Geographica*, from the Greek words *Geo* (Earth) and *graphien* (writing). It was intended as a comprehensive study of the emerging discipline, comparable to those that also existed for medicine or philosophy. His works, now largely lost, were cited by later writers, such as Strabo.

Key works

c. 194 BCE *Geographica*
c. 240 BCE *On the measurement of the Earth*

ONE REPRESENTING THE LENGTH AND THE OTHER THE BREADTH
A GEOGRAPHICAL COORDINATES SYSTEM

IN CONTEXT

KEY FIGURE
Hipparchus
(c. 190–c. 120 BCE)

BEFORE
c. 3000 BCE A standardized system of measurement for use in land surveying and building projects evolves in Ancient Egypt.

3rd century BCE Eratosthenes shows that large distances on Earth's surface can be calculated from observation of the Sun's position.

AFTER
276 CE Chinese cartographer Pei Xiu uses a plotted geometrical grid reference and graduated scale in his maps to improve accuracy.

1569 Flemish geographer Gerardus Mercator's world map presents a novel cylindrical projection with meridians shown as vertical lines, and circles of latitude shown as horizontal lines.

Eratosthenes sets **a grid over his maps** to help him **pinpoint different locations**.

Hipparchus calculates **lateral parallels and vertical meridians** centred on Rhodes to determine and describe positions on Earth's surface.

Marinus of Tyre **moves the prime meridian** (0° longitude) to an island west of Africa – a reference point used until the 18th century.

Ptolemy **measures latitude from the equator** and describes **curved meridians and parallels** that conform to Earth's spherical form.

The earliest maps were often simple visual depictions of landscape until the 3rd century BCE, when the Greek geographer Eratosthenes superimposed a grid of vertical and lateral lines onto his maps. This grid made it possible to specify a location in terms of its position, east or west and north or south of the grid's central point, the city of Rhodes, a Greek island. However, these lines took no account of the curvature of Earth's surface, leading to distortions in measurements.

In the 2nd century BCE, the Greek astronomer Hipparchus established a rigorous scientific

MAPS AND MAPPING 23

See also: The development of cartography 20–21 ▪ Map projections 24–27 ▪ The global positioning system 48–51 ▪ Early warning systems 296–303 ▪ Crisis mapping with crowdsourcing 304–05

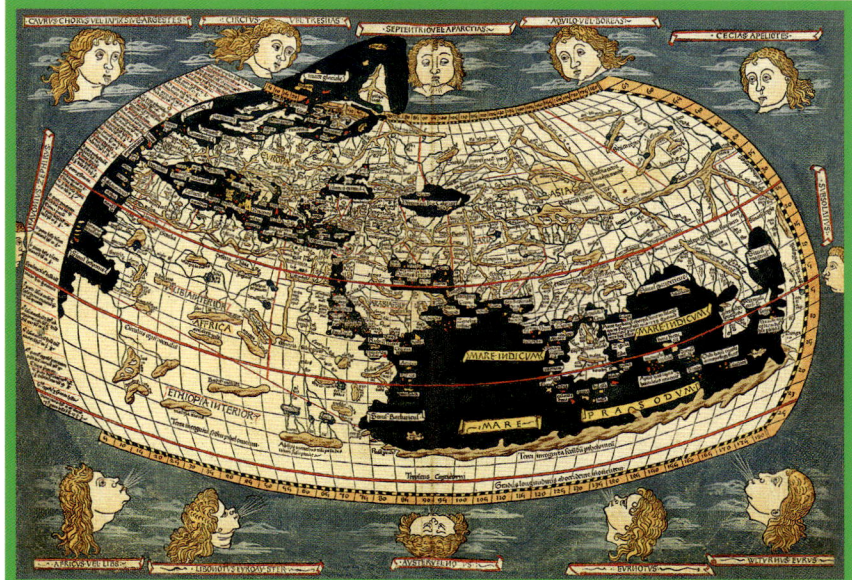

Ptolemy's world map, recreated by Renaissance scholars, includes latitude and longitude coordinates for 8,000 global locations, as recorded in *Geographia*.

mapmaking in his work *Geographia*. A large section of it includes instructions for creating a world map and several regional maps. When *Geographia* was rediscovered in the early 15th century, its scholarship was considered revolutionary and significantly impacted the Age of Exploration.

Latitude and longitude

Indirectly, Ptolemy helped standardize the geographical coordinate terms "latitude" and "longitude". The original text for *Geographia* was written in Ancient Greek, with parallels and meridians represented as Greek symbols. However, when European Renaissance scholars created the first Latin editions, these were translated as *latitudo*, meaning "breadth" or "width", and *longitudo*, meaning "length". ∎

basis for the grid by combining astronomical observations with spherical trigonometry to create a system of geographical coordinates that laid down the foundations for modern cartography.

Astronomical observations

Hipparchus divided Earth into a 360-degree circle, with lateral lines running parallel to the equator and vertical lines running from pole to pole. Taking the city of Rhodes as a reference point, where local time would be considered standard time, he determined the parallels (lines of latitude) by comparing hours of sunlight and star positions in the night sky at different locations. He then calculated the meridians (lines of longitude) by observing the time difference, in various locations, in between the length of a lunar eclipse.

In the 2nd century CE, Greek cartographer Marinus of Tyre expanded on this framework, developing an extensive gazetteer or geographical index of places, and assigning each a mathematically calculated latitude and longitude. This work laid the foundation for the theories of Roman mathematician and astronomer Claudius Ptolemy, who emphasized the importance of using geographical coordinates for

Hipparchus

Little is known of Hipparchus's life, other than what can be inferred from his few surviving works and references by later writers, such as Strabo and Ptolemy. He was born in Nicaea, Bithynia, in the northeast of modern-day Turkey around 190 BCE, and it is likely that he spent most of his adult life on the island of Rhodes. His knowledge of Egyptian and Babylonian astronomy and mathematics suggests contact with Egypt and Mesopotamia.

He is believed to have written more than 14 treatises, detailing his many astronomical discoveries: the gradual precession of the equinoxes, a star catalogue, and a comprehensive overview of trigonometry and its application in mapmaking. Hipparchus is thought to have died in Rhodes, aged about 70, around 120 BCE.

Key work

c. 150–125 BCE *Commentary on the Phaenomena of Eudoxus and Aratus*

TO SPREAD ON A PLANE THE SURFACE OF THE SPHERE

MAP PROJECTIONS

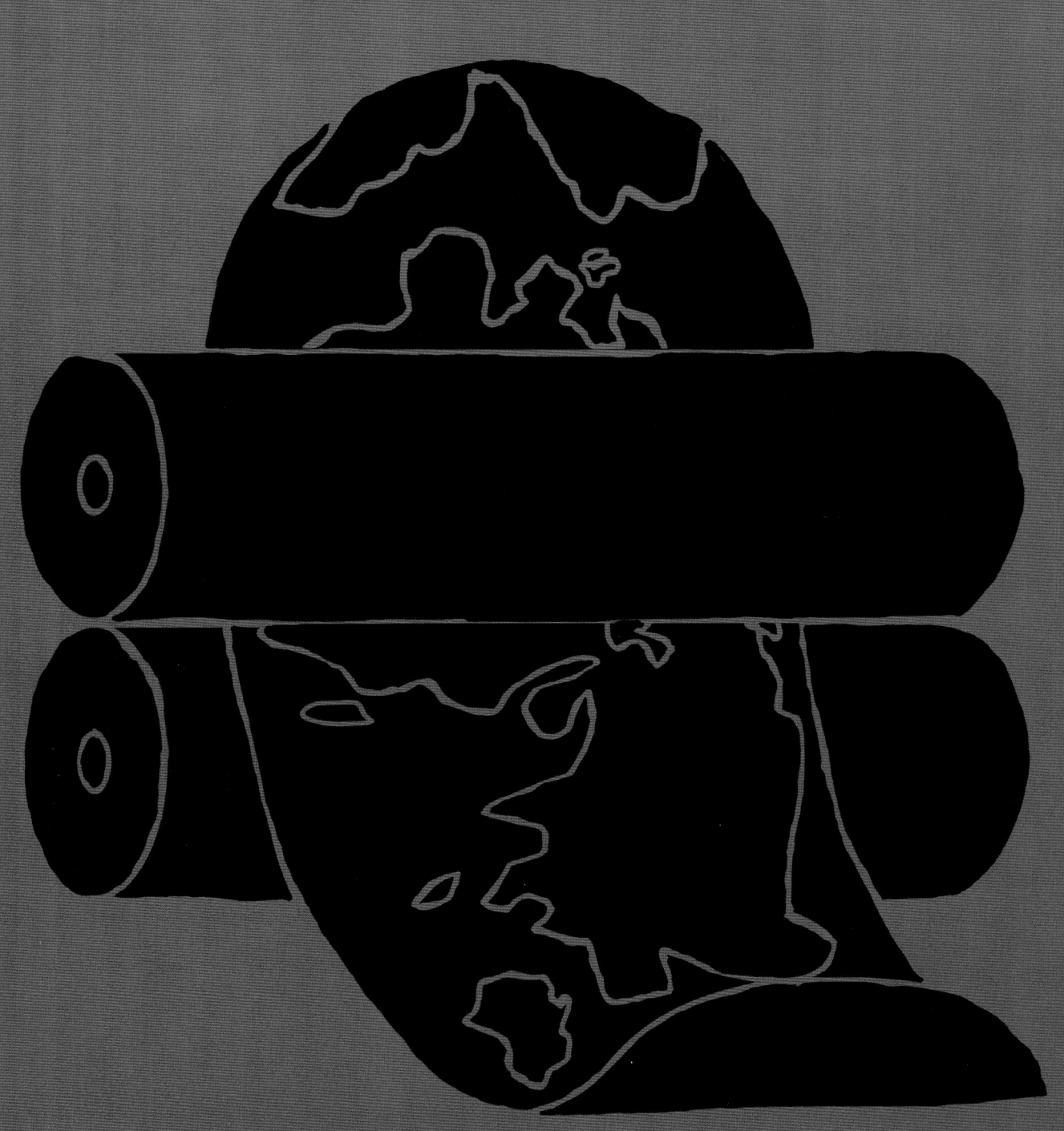

26 MAP PROJECTIONS

IN CONTEXT

KEY FIGURE
Gerardus Mercator
(1512–94)

BEFORE
6th century BCE Greek scholar Anaximander's circular world map is a more realistic depiction than earlier maps.

3rd century BCE Planning to map the world, Greek scholar Eratosthenes measures the circumference of Earth.

13th century European mariners begin to create their own navigational charts based on collective knowledge.

AFTER
1772 Swiss polymath Johann Heinrich Lambert introduces new projections that include the Lambert conformal conic, and the Lambert azimuthal equal-area. In the 21st century, both projections are still used.

1963 In response to an appeal by American map publishers Rand McNally, cartographer Arthur Robinson creates a projection for a world map that balances distortions.

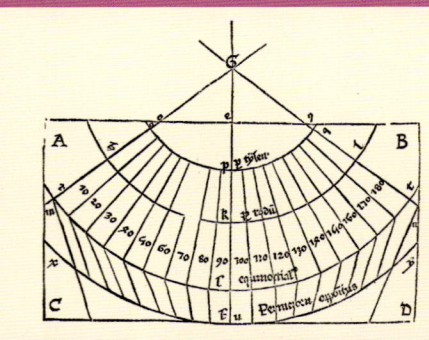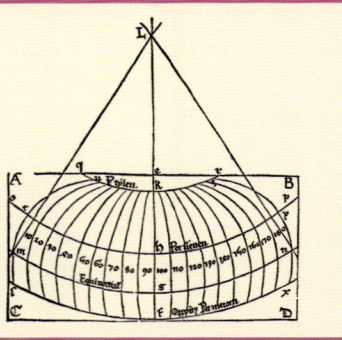

Ptolemy's first projection (left) shows meridians as straight lines angled towards the North Pole. In his second, both meridians and parallels are curved.

Ancient Hellenic astronomy showed that Earth is roughly spherical, which presented cartographers with a problem. Projecting the curvature of Earth – or any spherical surface – onto a flat sheet of paper inevitably involves some degree of distortion.

When mapping larger distances, primarily for sea navigation, early mapmakers drew on accumulated experience and prioritized details such as coastline shape, distance, direction, and landmarks. It was only in the 16th century, when a new age of maritime exploration coincided with the rediscovery of Ancient Greek scholarship, that Flemish geographer Gerardus Mercator created a world map that overcame these problems. His method was a major advance that shaped all future world mapping.

The geometrical task

The surface of a sphere is what is known as a non-developable surface; that is, it cannot be flattened onto a sheet without distorting it in some way. Discrepancies may be negligible when mapping a small area, but over larger areas, Earth's curvature becomes a problem.

In the early 2nd century CE, the Greek geographer Marinus of Tyre attempted the task of mapping a spherical world by projecting the globe onto a rectangular grid – the equirectangular projection method. Using the geographic coordinates system developed by Eratosthenes, Marinus drew straight lines of longitude and latitude (meridians and parallels) that were equally spaced and represented a specific distance on the ground. Inscribed onto a spherical globe, however, vertical north-south lines would not be parallel but would bend and converge at the poles – something that a grid map could not reflect.

Ptolemy's *Geography*

Later in the 2nd century, Greco-Roman scholar Claudius Ptolemy drew on the work of Marinus when compiling his treatise *Geography* but took a quite different approach. Addressing the distortion inherent in older methods, Ptolemy outlined two new projections for displaying spherical Earth on a flat plane.

While Marinus based his maps mainly on information from physical surveys, Ptolemy used astronomical observations to

This method of drawing the map is the better one, yet is less satisfactory in this respect, that it is not as simple as the other.
Ptolemy
Geography of Claudius Ptolemy, 1932

MAPS AND MAPPING 27

See also: The development of cartography 20–21 ▪ A geographical coordinates system 22–23 ▪ Triangulation 32–34
▪ The global positioning system 48–51

determine latitude, and less reliable accounts and physical surveys for longitude. He then employed geometry and a basic form of trigonometry to calculate and develop his map projections.

Ptolemy's first projection method describes straight meridians, each made up of two segments, with the upper segments converging at a hypothetical point north of the North Pole, while the equidistant parallels are circular arcs. His second projection method again describes meridians that converge at the North Pole but, in this model, they are now also curved, somewhat resembling modern polar-centred map projections of the northern hemisphere. In Ptolemy's time the southern hemisphere was virtually unknown. While no diagrams of Ptolemy's projections survive, later scholars were able to draw them from instructions and coordinates given in *Geography*.

After the decline of the Roman Empire, interest in global map projection stagnated in the West, although Islamic scholars preserved its legacy. Ptolemy's *Geography* was translated into Arabic by at least the 9th century, but not into Latin until the early 15th century. From the mid-9th century, several Islamic astronomers refined the calculations and descriptions in *Geography*. Their work influenced later European cartographers.

Portolan charts

In the late 12th century, European mariners began to use the newly developed needle compass, which enabled them to venture further from familiar coastlines into open seas. From their early rough charts, a new style of map specifically designed for seafaring – portolan charts – emerged in the late 13th century. These charts initially mapped the Mediterranean Sea and eastern Atlantic Ocean but later extended further into the Atlantic and also around Africa, as European sailors discovered lands hitherto unknown to them.

What mariners needed to know, above all else, was the bearing they should take to steer a straight course from one destination to another. Portolan charts consisted of networks of rhumb lines – lines »

Depicting coastlines with great accuracy, this 16th-century portolan chart maps the Mediterranean area best known to European mariners, who used its rhumb lines to plot their sea routes.

28 MAP PROJECTIONS

Gerardus Mercator

Gerardus Mercator, born in 1512 in Rupelmonde, Flanders (present-day Belgium), was orphaned at a young age. He became the ward of his uncle, a priest, who ensured he had a good education. At the University of Leuven (Louvain), he studied philosophy, mathematics, astronomy, and cosmography, under the guidance of Gemma Frisius. In Frisius's workshop, Mercator began to produce maps, globes, and scientific instruments. Suspected of heresy for his links to Lutherans, Mercator was arrested in 1544 but was released after the university vouched for him.

In 1552, Mercator moved to Duisburg, Germany, where he continued to produce many maps. He coined the word "atlas" for his last collection, unfinished at his death in 1594.

Key works

1569 *New and more complete representation of the terrestrial globe properly adapted for use in navigation*
1595 *Atlas or cosmographical meditations upon the creation of the universe, and the universe as created*

emanating from compass roses. These circular symbols indicated north, south, east, and west (the principal directions and winds) and the points between them. Rhumb lines subdivided the compass rose into 16 or 32 parts and radiated from its centre across the chart to represent constant compass bearings. Used together with coastline details and the rhumb lines from other roses on the chart, they provided information about a vessel's direction and distance, enabling navigators to plot a workable course between points – though not extensive voyages.

They [the pilots] represent as straight what has been sailed with so many detours.
Pedro Nunes
Treatise on the Sphere, 1537

In early 15th-century China, ruled by the Ming dynasty, the nation's cartographers were producing nautical charts using similar methods. The Mao Kun map, for example, based on the voyages of renowned Chinese navigator Zheng He, used a 24-point compass system and documented the sea route from Nanjing in eastern China to East Africa. This map also included astronomical observations for navigation, such as four stellar diagrams that showed the positions of constellations, helping sailors to determine their latitude. This additional knowledge was especially important when Chinese mariners made longer journeys.

In the late 15th century, as the focus of European exploration turned westwards across the Atlantic, navigators such as Christopher Columbus, Ferdinand Magellan, and Vasco da Gama discovered the inadequacies of the portolan chart. Useful as the charts were for short coastal journeys, their rhumb lines did not take into account the curvature of Earth (nor the position of the magnetic North Pole). As a result, navigation beyond the Mediterranean Sea and eastern Atlantic Ocean was imprecise. When Columbus sailed west in 1492, he calculated speed and time to estimate distance, and followed a compass direction west. Arriving further south than intended, he also believed he had reached Asia, as Europeans at that time were unaware of the existence of the American continent.

Mapmaking flourishes

By the early 14th century, Islamic astronomy and mathematics had largely spread through Europe and its scholars were also rediscovering Ancient Greek texts, as Islamic scholars had before them. Ptolemy's treatise was translated into Latin in around 1406, sparking a fresh interest in scientific mapmaking. In a 1513 edition of *Geography* that was widely circulated, to accompany the 27 Ptolemaic maps, German cartographer Martin Waldseemüller created 20 new maps showing parts of the Americas and the Indian Ocean. In his world map *Universalis Cosmographica* (1507), he had been the first to use the word "America" (named after Italian explorer Amerigo Vespucci) for the lands previously unknown in Europe.

From the 1530s, Pedro Nunes, a Portuguese mathematician and cosmographer, began investigating rhumb lines. Mariners of his time believed that if they followed a constant compass bearing, they could plot their progress as a straight line along a corresponding rhumb line on a portolan chart. However, Nunes saw that rhumb lines took no account of the fact that meridians converge at the poles so were inadequate for long sea voyages. He calculated that, in fact, a rhumb line traces a spiral (later known as a loxidrome) that crosses Earth's meridians at the same fixed angle. His insights provided a mathematical basis for Mercator's innovative projection.

A cylindrical projection

Mercator and Nunes were close contemporaries and, although they never worked together, Mercator was familiar with Nunes' work. As a skillful maker of globes as well as maps, Mercator understood the problems involved in translating the features of a three-dimensional sphere onto a two-dimensional surface. He set himself the task of resolving the problem well enough to create a map that worked for long-distance navigation.

On his renowned 1569 world map – a huge work printed from 18 copperplates – Mercator expressed his goal that it should "spread on a plane the surface of the sphere" in a way that was both of practical use in navigation and presented a useful geography of the world. It was the first ever attempt to integrate the practical tradition of portolan charts with the more scientific process of map projection.

Although he never described it as such, what Mercator developed was a cylindrical projection – as if a globe were encircled in a paper cylinder and the image of the globe projected onto the paper. He preserved equally spaced lines of longitude, but what distinguished this 1569 world map from any of its predecessors was Mercator's decision to ignore the conventional equal spacing of the parallels and, instead, to widen some of them »

Map projection models

Projections can be visualized by imagining a developable surface, such as a cylinder or cone, onto which the image of Earth can be projected. The Mercator projection imagines a cylinder projection (see below) with the cylinder oriented along Earth's equator, although there are types of cylindrical projection that orient the cylinder in other ways in relation to Earth. Alternative methods include conic projections that imagine Earth within a cone, the apex of which is above one of the poles and its interior touching the globe along one or two lines of latitude. Azimuthal or planar map projections visualize Earth touching a plane at a single point and extending outwards; these are not suitable for showing the entire world on one map. All methods require complex mathematical calculations.

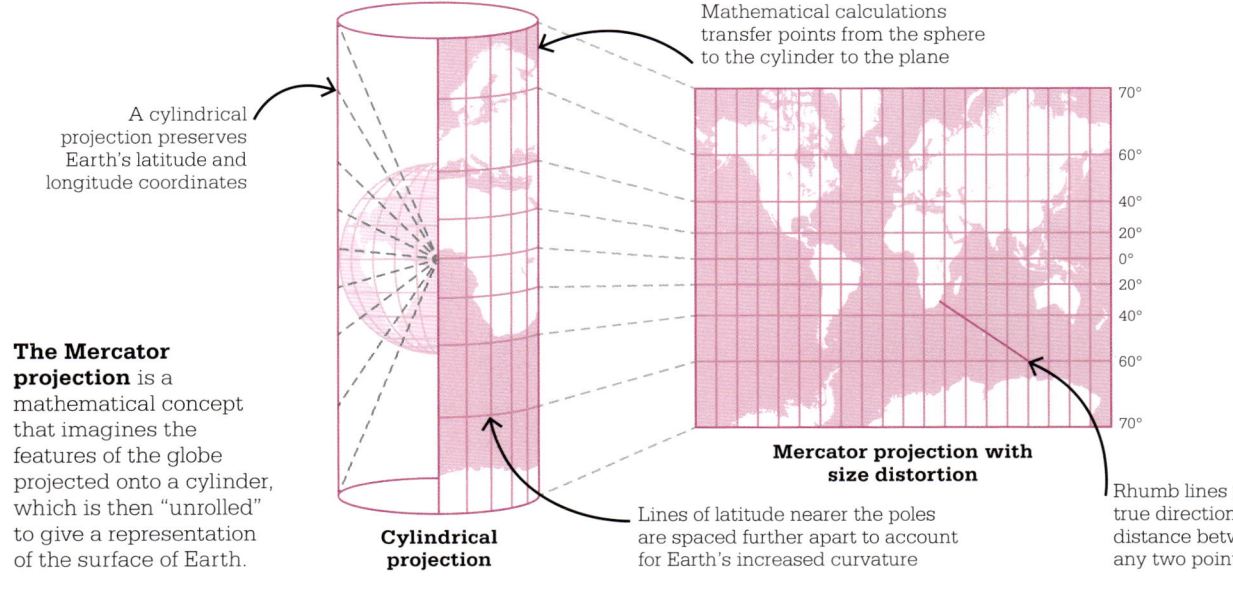

A cylindrical projection preserves Earth's latitude and longitude coordinates

The Mercator projection is a mathematical concept that imagines the features of the globe projected onto a cylinder, which is then "unrolled" to give a representation of the surface of Earth.

Cylindrical projection

Mathematical calculations transfer points from the sphere to the cylinder to the plane

Mercator projection with size distortion

Lines of latitude nearer the poles are spaced further apart to account for Earth's increased curvature

Rhumb lines give a true direction and distance between any two points while preserving the 90-degree angle at which lines of longitude and latitude intersect. To achieve this effect, Mercator enlarged the gap between lines of latitude north and south of the equator as they neared the poles.

Using loxidromic calculations, possibly borrowed from Nunes' tables, Mercator ensured that the spirals of constant compass bearing (rhumb lines) would appear on the map as straight lines. In this way, he cleverly integrated the practical tradition of portolan sea charts into a mathematically derived representation of the world and created a map that was ideally suited to long-distance seafaring.

Mercator's world map became widely adopted as a nautical chart. Sailors could plot a straight course across long distances – a major improvement over earlier maps, where navigating a straight line had required constant adjustments for Earth's curvature. His innovative

Mercator's 1569 world map
represented a milestone in cartography as it allowed sailors to follow a rhumb line with a fixed compass bearing.

cartography improved navigational accuracy and made long sea voyages safer and more efficient. Sailors could venture further with more confidence, leading to increased trade and exploration and the mapping of new territories. By the 19th century, the Mercator projection had become the most widely accepted two-dimensional representation of the world and was used for naval charts and many other practical purposes.

Distortions in size
Like any map projection, the Mercator model has shortcomings. Especially at higher latitudes, a Mercator world map distorts the size of landmasses so that those near the poles, such as Greenland and Antarctica, appear much larger than they are relative to those, such as Africa and South America, that are near the equator.

While it is highly unlikely that Mercator had any political intent, map projections are never neutral. The distortion of scale inherent in the Mercator projection, which makes Europe in the northern hemisphere look much larger than

Great thanks are owed to the Ptolemaic charts, and great thanks to you, Mercator, for having at last surpassed that ancient labour.
Johannes Vivianus
Text around Franz Hogenberg's portrait of Mercator in *Atlas*, 1595

equatorial regions such as Africa and South America, flattered European powers during and after the European Age of Exploration. It underscored the narrative of power and influence that Western rulers used to pursue a project of colonization around the world.

Later developments
The distortions inherent in the Mercator method prompted other geographers to explore different ways of projecting the planet's non-developable surface onto a plane. In a 1730 double-hemisphere world map, German map publisher Matthäus Seutter borrowed from scientific diagrams to present the western and eastern hemispheres in circular form, framed by other projections of different areas of the world. While still including many inaccuracies, the work offers bird's-eye views of Earth's surface from high above it as satellite images of Earth do today. This type of vertical projection is now popular in computer programs.

Several equal-area projections were devised in the 18th century. These overcame some of the

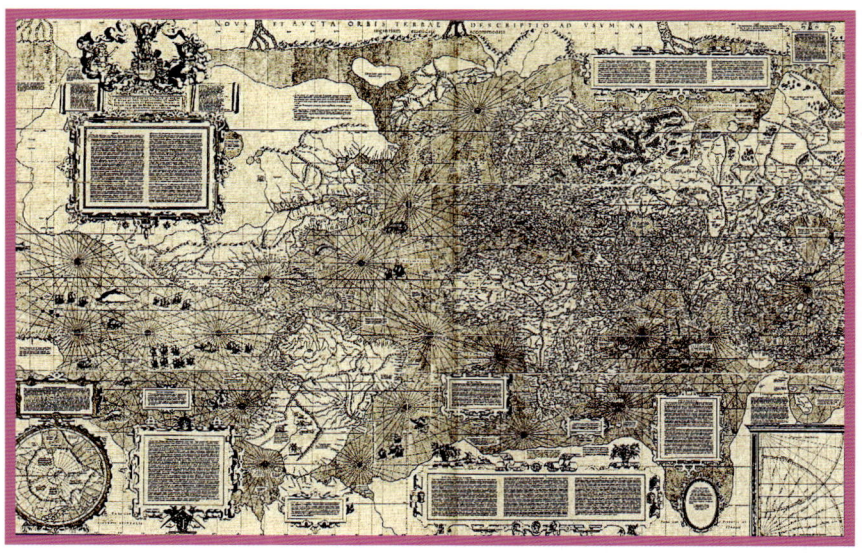

MAPS AND MAPPING

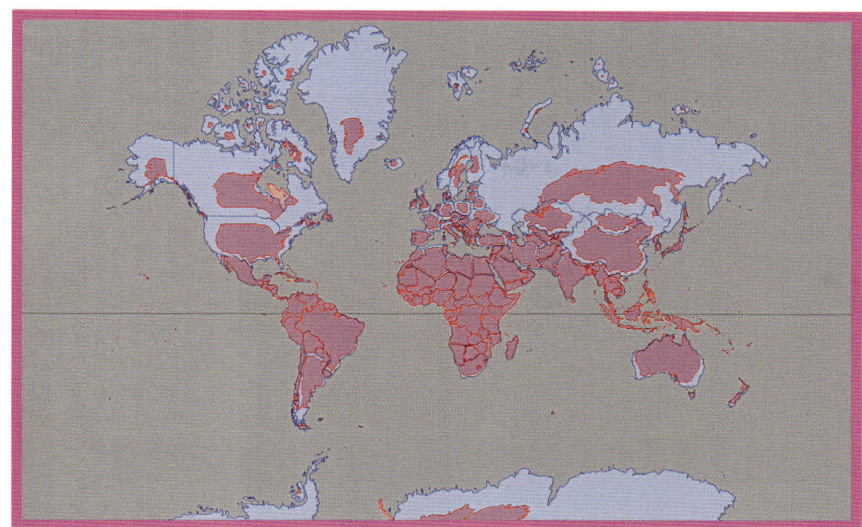

Size distortion near the poles is the price for preserving angles in Mercator's projection (in blue). Here, the red shading indicates true size.

failings of the Mercator projection by showing the correct area of features such as landmasses and oceans, but only by distorting shapes, angles, or directions. Unlike maps using the Mercator projection, straight lines do not correspond to compass bearings.

In 1805, Heinrich Christian Albers developed the Albers equal-area conic projection, which uses two standard parallels of latitude in an attempt to reduce distortion. This resulted in greater accuracy in scale, directions, and distances across landmasses extending east-west at mid-latitudes, such as the US, but distortion increased away from the standard parallels as the map extended north-south.

In 1855, the Scottish clergyman and cartographer James Gall presented an equal-area cylindrical projection, which he modified with standard parallels at 45-degrees north and south latitude. This had the effect of reducing the distortion near the poles (as in the Mercator projection), and offered a more balanced representation of Earth, yet distortion is still present in the shape of countries. Gall's projection remained obscure until it was popularized by German historian Arno Peters in the 1970s. It is still used today but is not as widespread as the Mercator version.

Other map projections include azimuthal projections, which map the globe onto a plane from a chosen central point, and preserve desired properties such as angles and distances from that point, while distorting other features.

Digital projections

Although map projections can be visualized by projecting spherical Earth onto a developable surface, in reality complex geometrical or trigonometrical calculations are always required. These underlying mathematical principles remain the basis for map projections today, but the formulas are much more easily applied using digital technology.

Web Mercator, introduced by Google Maps in 2005, has been widely adopted as it preserves angles and is now the standard application for web maps. A slightly modified version of Mercator's work, it ignores Earth's true ellipsoidal shape and treats it as a perfect sphere, but preserves his formulas for projecting coordinates onto a flat surface, so distortions near the poles remain. As distortions of some kind remain inevitable, map projections are categorized by the features they preserve and are chosen according to a map's intended function. ■

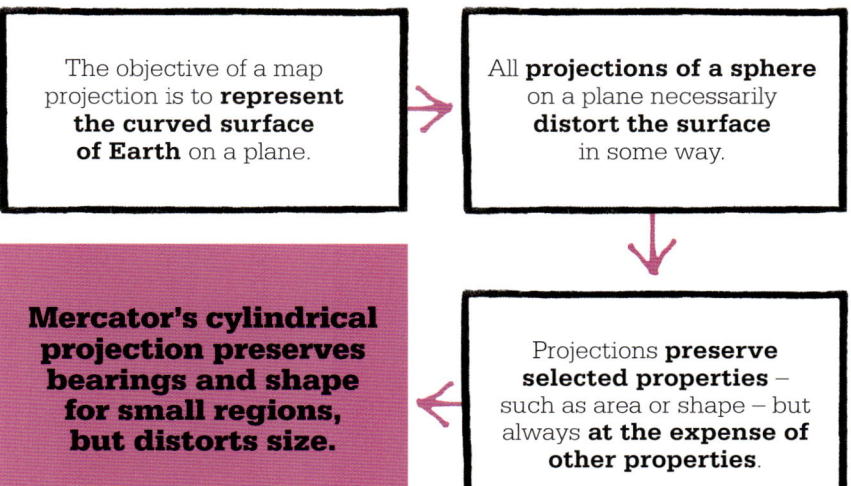

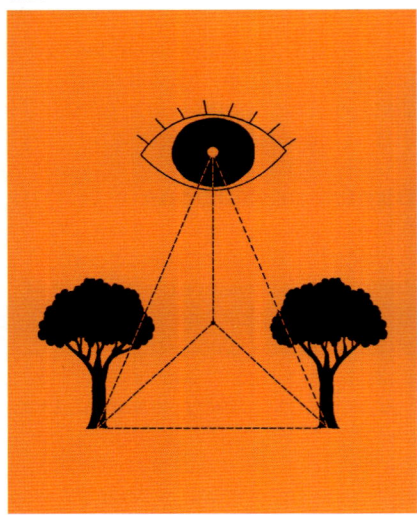

BY GEOMETRICAL CALCULATION, WE DETERMINE THE DISTANCES OF PLACES
TRIANGULATION

IN CONTEXT

KEY FIGURE
Gemma Frisius (1508–55)

BEFORE
3rd century BCE Greek mathematician Eratosthenes estimates the circumference of Earth using geometrical and astronomical calculations.

263 CE In *Haidao Suanjing* ("The Sea Island Mathematical Manual"), Chinese scholar Liu Hui describes the use of geometry to measure heights and distances.

AFTER
1829–42 The Ordnance Survey of Ireland, with a scale of 1:10,560 (6 inches to 1 mile), is the largest-scale map of a nation yet completed.

1879 An Act of Congress establishes the US Geological Survey to classify all public land and examine the geological structure and mineral resources of the entire terrain of the US.

In early 16th-century Europe, the problem of measuring long distances was a key obstacle to producing accurate maps of large regions. Surveys, especially over difficult ground or bodies of water, were usually approximate, leading to cartographic inaccuracies.
In 1533, Dutch mathematician Gemma Frisius outlined a solution. His work *Libellus de locorum describendorum ratione* ("Booklet on a way of describing places") laid out a new method of calculating distances from a known baseline – later called triangulation.

In triangulation, compass readings are taken from points A and B to a viewpoint C to establish the angles of a notional triangle in order to calculate the length of its sides – A to C, and B to C.

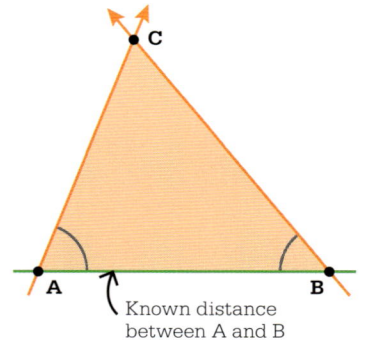

Known distance between A and B

Frisius's theoretical baseline was a recognized distance between two city observation points – a tower (point A) in Brussels and another tower (point B) in Antwerp. The location of a visible third point (point C), Frisius explained, could then be determined by taking compass bearings to it from each end of the baseline, noting the angles, then plotting on a map or chart where the two sightlines cross to form a triangle with the baseline. From the angles and the baseline length, a trigonometric calculation reveals the length of the triangle's two sides – the distances to the third point from point A and point B.

Wider adoption
Although Flemish geographer Gerardus Mercator, a pupil of Frisius, employed triangulation to create a map of Flanders in 1540, the method was not widely known until the 17th century. In 1615, Dutch mathematician Willebrord Snell used linked triangulation networks to map a large area of the Netherlands, and two years later, in the book *Eratosthenes Batavus* ("The Dutch Eratosthenes"), he explained how triangulation

MAPS AND MAPPING 33

See also: Geophysical mapping 38–39 ▪ Aerial photography and remote sensing 40–45 ▪ Mapping the ocean floor 46–47 ▪ The global positioning system 48–51

This 1734 map, based on the earlier work of Jean Picard, shows a network of triangulations to the east of Paris calculated and drawn from a series of baselines and points.

allowed him to measure distances over the surface of Earth and calculate its circumference.

Using triangulation, in tandem with increasingly precise surveying tools, gave mapmakers a framework for calculating points on Earth's surface. The technique, enabling them to measure both distances and elevation changes with hitherto unparalleled accuracy, spread across northern Europe.

Mapping a nation

French astronomer and priest Jean Picard used triangulation to improve on Snell's calculations of Earth's dimensions, and, in 1669, he joined Italian astronomer Giovanni Domenico Cassini to begin mapping France, a project initiated by the newly formed Académie des sciences. The first part, Picard's map of the Paris region, with a scale of 1:86,400, was published in 1678; a coastal survey followed in 1684.

After a delay, triangulation of the whole of France was completed by Jacques Cassini (Giovanni's son) in 1744. French monarch Louis XV then demanded more detailed maps, which were overseen initially by Jacques' son César-François Cassini de Thury (Cassini III) and later Jean-Dominique Cassini (Cassini IV), son of César-François. When finally achieved in 1815, the "Carte de France", or "Carte de Cassini", comprised 182 map sheets. This inspired a wealth of other surveying projects as European powers recognized the strategic importance of mapping their own countries and the regions they had conquered. ▪

Gemma Frisius

Jemme Reinerszoon, born in 1508 in Dokkum, Friesland, northern Netherlands, latinized his name to Gemma Frisius, as was the custom, when he attended the University of Leuven. There he gained a degree in medicine and stayed on to study mathematics and astronomy, before teaching at the university. In 1530, he published a book on the principles of astronomy and cosmography. It included descriptions of regions Europeans had begun to colonize and advice on the use of globes to locate places and determine the distances between them.

As well as being the first scholar to describe the method of accurately locating places by triangulation, Frisius created detailed terrestrial and celestial globes. He also made significant improvements to mathematical and surveying instruments, including the astrolabe and astronomical rings (also known as "Gemma's rings"). Frisius died in Leuven aged 46.

Key work

1533 *A Booklet on the Method of Describing Places*

THE CONNECTION BETWEEN THE BROAD STREET PUMP AND THE OUTBREAK OF CHOLERA
THEMATIC MAPPING

IN CONTEXT

KEY FIGURES
Charles Minard (1781–1870),
John Snow (1813–1858)

BEFORE
1607 Flemish cartographer Jodocus Hondius's world map *Designatio orbis christiani* uses symbols to show the distribution of religions.

1729 German-born cartographer Herman Moll publishes his world map, which indicates trade winds and variable climates.

AFTER
1950s Transparent overlays on maps – a technique introduced by British urban planner Jacqueline Tyrwhitt – convey specific types of information.

1990s With the development of computerized geographic information systems (GIS), geographic data can be stored, analysed, and visualized.

As cartographic and printing techniques evolved, maps could convey increasingly detailed information. As well as showing location, direction, and physical features, from the late 18th and early 19th century, thematic maps began to represent other data in a variety of different graphic styles.

One important example of thematic mapping is the dot distribution map that British physician John Snow devised in 1854 to pinpoint the source of a cholera outbreak in London. In France, civil engineer Charles Joseph Minard used a range of novel graphic forms to illustrate and

MAPS AND MAPPING 35

See also: Geographic information systems 52–59 ▪ Digital twins of Earth 60–61 ▪ Renewable energy geography 290–91 ▪ Tracking disease 310–13

explain statistics, business data, and events such as military campaigns. One of his powerful thematic maps depicted French emperor Napoleon's invasion and retreat from Russia in 1812.

On early thematic maps, simple symbols highlighted details of interest. From around the 1740s, chorochromatic maps used colours to pick out features, while later choropleth maps used shading. With these and other techniques, such as dot distribution and dot density, mapmakers could highlight information with greater precision.

Mapping the source

In the 1850s, when Snow created his cholera map, the disease was a scourge in London, with frequent outbreaks recurring in certain areas. He was keen to prove what he believed was the disease's mode of transmission, as "miasma" (noxious vapours) was popularly blamed at the time. Using official mortality reports, Snow created a map that plotted a short bar against the addresses of each of the deceased in an area of Soho. The common factor linking all of them was a pump from which they had drawn their water. Snow's map helped to confirm his theory that cholera was a waterborne disease. In this case, sewage from a nearby cesspit had leaked into and contaminated the pump water. »

John Snow's investigation into the source of the outbreak convinced the local council to disable the Broad Street pump, thereby saving more lives, and his map illustrated his conclusions.

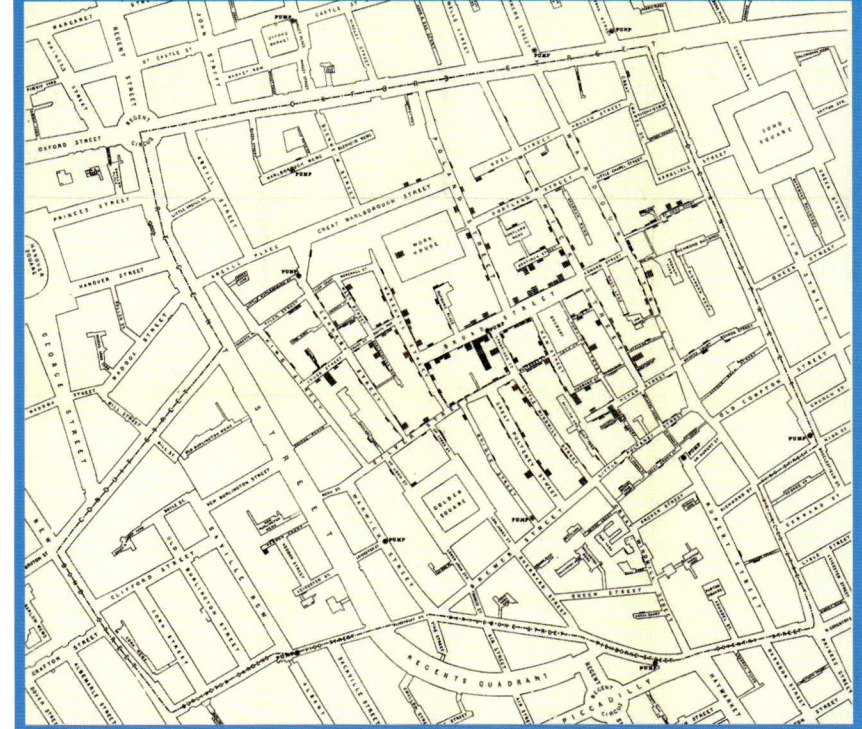

John Snow

Growing up in one of the poorest parts of York, England, John Snow experienced first-hand the unsanitary living conditions endured by the working class of his time. He was a bright child, and after a medical apprenticeship in the north of England, he moved to London to study medicine, graduating in 1844. While working as a surgeon and general practitioner, he became interested in cholera, and helped to found the Epidemiological Society of London. Snow was also a pioneer in the use of ether and chloroform in anaesthesia, especially in obstetrics, and administered chloroform to Queen Victoria during the birth of two of her children.

Snow was a lifelong bachelor and somewhat obsessively ascetic, only breaking with his teetotal, vegetarian (and sometimes vegan) lifestyle when in later life it adversely affected his health. He died in London as the result of a stroke in 1858, aged only 45.

Key work

1849 *"On the mode of communication of cholera"*

THEMATIC MAPPING

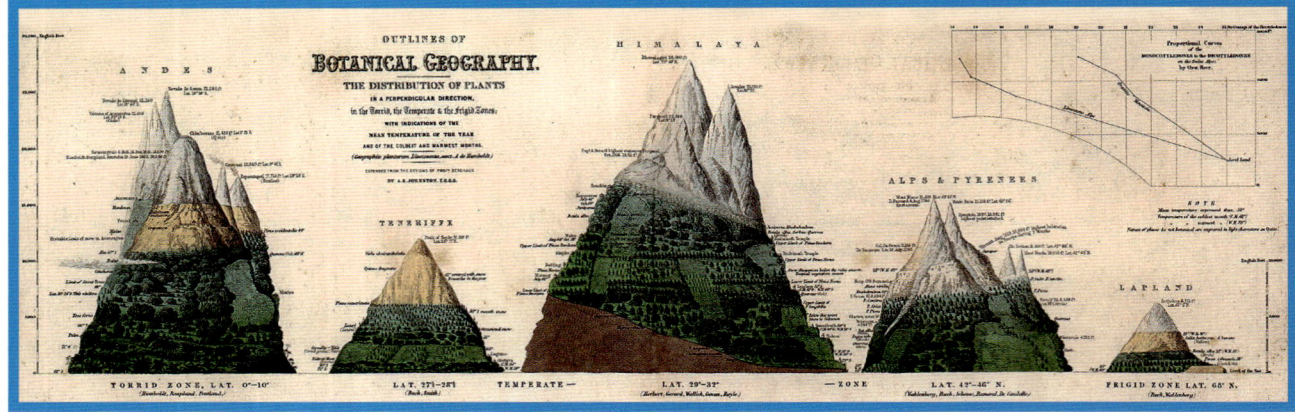

Snow's map was not the first to track disease in this way. In 1798, American physician Valentine Seaman included a dot distribution map in his account of yellow fever in New York City. Dot distribution maps are still frequently used in epidemiology to note the exact location of cases.

Presenting scientific data
Thematic maps have also proved useful for visualizing scientific data relating to phenomena such as climate, resources, geology, and the distribution of flora and fauna.

In 1807, German geographer and polymath Alexander von Humboldt pioneered a new kind of thematic map with his "Tableau Physique", which depicted vegetation at different altitudes on the South American volcanoes Chimborazo and Cotopaxi. Ten years later, he was also the first to use isothermal lines to connect areas of equal temperature on a world map. While isolines had been used on earlier navigational and topographic maps, Humboldt was the first to recognize their potential for depicting climatic conditions. Such thematic maps facilitated comparison of complex data that would otherwise have required painstaking interpretation and presented it in its geographical context, using early infographics to highlight underlying patterns.

French innovations
In 1830, Franciscan friar and schoolteacher Armand Joseph Frère de Montizon invented the dot density technique in a map that depicted the population of France's administrative departments with 1 dot for every 10,000 people. In dot density maps, now often used in demography (population studies) and resource mapping, a dot represents a specific quantity of something. By contrast, in dot distribution maps, a dot marks a single feature at a precise location. From the 1840s, Charles Minard

Vegetation across climates from the South American Andes to Lapland was the subject of Alexander von Humboldt's pioneering thematic map "The Distribution of Plants" (1850).

used a wide range of techniques to present import, export, transport, and migration statistics in an easily digestible form. By combining statistical graphs or flow diagrams with thematic mapping, he could display movement and developments over time, rather than presenting a static snapshot of a situation. Examples include a plan depicting the transport of mineral fuels in France over a 15-year period, while an 1857 map of Europe's ports is notable for its precise, proportional circles that reflect the volume of tonnage shipped through each.

A related thematic mapping technique, the cartogram, was pioneered by French geographer and economist Pierre Émile Levasseur in the 1870s. It uses the idea of size proportional to the value of a variable. However, rather than superimposing proportional symbols onto a map, the cartogram rescales the features of the map itself – the countries or regions – in proportion to the value to be illustrated, not their actual physical area. To illustrate the population of

The aim of… my *cartes figuratives* is to convey promptly to the eye the relation not given quickly by numbers
Charles Minard
"Des tableaux graphiques et des cartes figuratives", 1862

different countries, for instance, a cartogram would show India, whose population numbers more than 1.4 billion, as considerably larger than Australia, which has a population of around 28 million. Levasseur created a series of maps in which European nations appeared as squares, scaled to represent aspects of their demographics, without consideration of each nation's physical shape or size.

Colour variations

Choropleth thematic maps use colours or shades to represent data. One of the first was an 1826 literacy map created by French engineer, mathematician, and economist Charles Dupin. It divides France into its administrative departments, each shown in a shade indicating its level of literacy. Choropleth maps, like related chorochromatic maps, use colours or shades to represent data but, while a chorochromatic map of the world's religions, for example, would ignore nation boundaries and colour code regions by their major religion, choropleth maps use predefined geographical boundaries and shade levels of data within them. Choropleth maps are often used for statistical data, so that spatial patterns and regional differences can be seen at a glance.

Spurring social reform

From the 19th century onwards, thematic mapping was used to present many types of demographic information, from population size to social and economic factors such as income levels, education, and employment. In some instances, by highlighting disturbing statistics, such maps had a dramatic impact.

During the Crimean War, British nurse Florence Nightingale mapped the causes of British soldiers' deaths from 1854 to 1856, revealing that more men died from preventable diseases than from wounds. British social reformer Charles Booth's chorochromatic poverty maps of London – part of his larger 17-volume *Inquiry into Life and Labour in London* (1886–1903) – had a profound effect on British government policy-makers, even influencing the later creation of the welfare state. In a similar way, *Hull House Maps and Papers* (1895) describes a slum area of Chicago, US. Based on statistical information, census forms, and door-to-door interviews carried out by Florence Kelley and others, the maps include some homes colour-coded to show which of nine nationalities their residents belonged to, and others annotated to reveal their residents' wage levels. The project was part of a wider national study of slum areas in Baltimore, Chicago, Philadelphia, and New York. As in Britain, the stark evidence of such thematic maps helped raise public awareness and fuel social reforms.

Thematic mapmakers of the 1800s paved the way for a vast expansion of this multi-disciplined area of cartography in the 20th and 21st centuries. They supplied most of the techniques, such as colour-coding, proportional circles, grids, squares, and flow charts, that have created a rich universal language for sharing and digesting complex statistics, and they established a framework for visualizing data across geographical areas. While GIS, remote sensing, and computer software programs have since influenced the type and amount of data collected and processed, the way it is visualized and presented still owes a great deal to thematic mapmaking pioneers. ∎

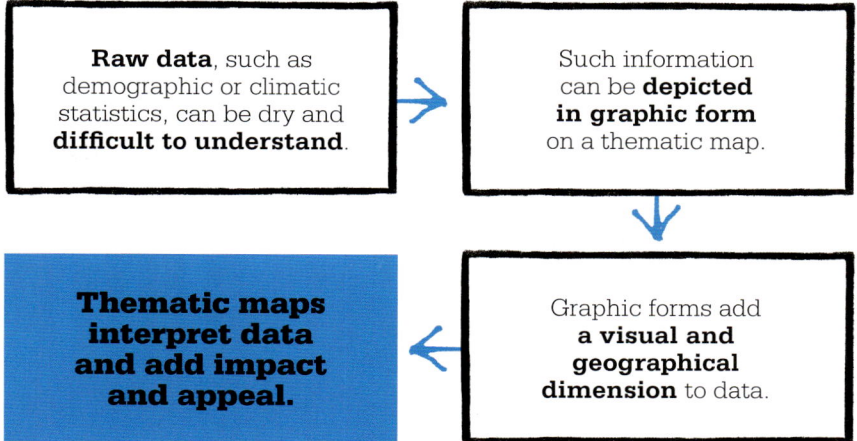

Raw data, such as demographic or climatic statistics, can be dry and **difficult to understand**.

Such information can be **depicted in graphic form** on a thematic map.

Graphic forms add **a visual and geographical dimension** to data.

Thematic maps interpret data and add impact and appeal.

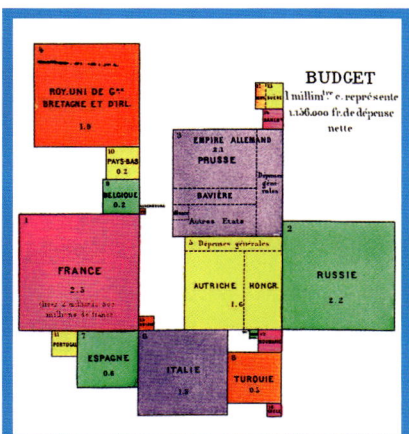

Pierre Émile Levasseur published his series of cartograms in 1876. This one scales the countries of Europe according to their national budgets. France is largest, with Russia in second place.

THE MAGNETIC PHENOMENA OF EARTH
GEOPHYSICAL MAPPING

IN CONTEXT

KEY FIGURE
Carl Friedrich Gauss
(1777–1855)

BEFORE
1600 In *De Magnete*, English scientist William Gilbert describes his studies of compass needles and spherical magnets, and concludes that Earth is a giant magnet.

1701 English astronomer Edmond Halley's "A New and Correct Chart Shewing the Variations of the Compass" is the first chart to show lines of equal magnetic variation.

AFTER
1901 German geophysicist Johann Wiechert begins to build effective seismographs at University of Göttingen.

1950s The US military develops ground-penetrating radar, later used to survey Earth's subsurface by means of high-frequency, pulsed, electromagnetic waves.

Between the 15th and 17th centuries, scientific advances greatly improved knowledge of Earth's surface. Little was known about its internal composition, however, until the 19th century. A new science – geophysics – emerged to study Earth's forces, such as magnetism, gravity, and seismic waves. Among its earliest pioneers was German mathematician, physicist, and astronomer Carl Friedrich Gauss.

Earth's magnetic poles
Around 1830, Gauss became fascinated by electrical forces, magnetism, and the study of

> **Experiments with magnets**, used mainly in navigational compasses, confirm that **Earth has a strong magnetic field**.

> The invention of a **magnetometer enables measurement** of the field and its properties.

> By detecting **variations in Earth's magnetic field** due to differing magnetic properties in subsurface materials, a **magnetometer maps underground features**.

> **Other tools that measure gravity, electrical conductivity, and seismic responses of subsurface materials expand the scope of geophysical mapping.**

MAPS AND MAPPING 39

See also: Mapping the ocean floor 46–47 ▪ The structure of Earth 112–13 ▪ Plate tectonics 114–21 ▪ Volcanic activity and hotspots 122–25 ▪ Paleoclimatology 126–27 ▪ Seismology and earthquake magnitude 132–35

Earth's magnetic field. German geographer and polymath Alexander von Humboldt introduced him to physicist Wilhelm Weber, who became Gauss's colleague at the University of Göttingen in 1831.

Gauss and Weber realized that accurate measurements and instruments were needed to assess the strength and direction of magnetic fields. They developed an innovative system of magnetic units based on length, mass, and time. In 1833 Gauss described his new invention, the magnetometer – now a vital tool in geophysical surveying.

To better understand Earth's magnetic field, Gauss and Weber decided to collect and collate data from locations worldwide. They founded the *Magnetischer Verein* (Magnetic Association), inviting observatories around the world to collaborate by recording their local geomagnetic variations. Between 1836 and 1841, this "Magnetical Crusade" collected data from observations in 61 locations, taken on precisely predetermined dates. Working from early findings and experiments with magnets, in 1839, Gauss wrote his "General Theory of Terrestrial Magnetism", correctly proposing that Earth, like a bar magnet, has two magnetic poles.

Mineral properties

In 1831, British physicist Michael Faraday had discovered that a changing magnetic field induces an electric current. When Gauss and Weber created their geomagnetic atlas nine years later, they included two charts displaying equipotential lines – lines along which the electric potential is constant.

In the 1860s, French geologist Alexandre-Émile Béguyer de Chancourtois and Russian chemist Dmitri Mendeleev listed elements by their atomic weight, enabling the rough calculation of mineral densities and compositions – crucial to later gravity surveys. Later, other geologists began to discover more about Earth's layers via seismic tracking – measuring the speed or path of seismic waves travelling through different materials. The forces of gravity, electricity, seismic waves, and heat flow became better understood, and instruments were designed to detect variations.

As seismology evolved in the 20th century, scientists began to use geophysical maps and 3-D models to visualize the properties of Earth's interior and study the forces that shaped the planet. Seismology also became a tool for prospectors of mineral deposits, oil, and gas. ▪

It is not knowledge, but the act of learning, not possession but the act of getting there, which grants the greatest enjoyment.
Carl Friedrich Gauss
Letter to Hungarian mathematician János Bolyai, 1801

Carl Friedrich Gauss

Born into a poor family in Brunswick, Germany in 1777, Carl Friedrich Gauss showed a remarkable ability in mathematics at an early age. Encouraged by his teachers, he came to the attention of the Duke of Brunswick, who provided funds for him to study at the University of Göttingen (1795–98). Gauss's major work on algebraic number theory, *Disquisitiones arithmeticae*, was published in 1801. By this time, he had also turned his attention to astronomy and successfully calculated the orbit of the asteroid Ceres, which had disappeared from sight after its discovery in 1800. In 1807, Gauss was appointed professor of maths at the University of Göttingen and director of its observatory.

Gauss conducted a geodetic survey of Hanover (1818–32) and, in 1833, collaborated with Wilhelm Weber to develop one of the first electromagnetic telegraphs. Gauss died of a heart attack in 1855.

Key work

1839 *"General Theory of Terrestrial Magnetism"*

THE EARTH UNFOLDS INTO AN IMMENSE CARPET WITHOUT BORDERS

AERIAL PHOTOGRAPHY AND REMOTE SENSING

AERIAL PHOTOGRAPHY AND REMOTE SENSING

IN CONTEXT

KEY FIGURES AND ORGANIZATIONS
Gaspard-Félix "Nadar" Tournachon (1820–1910), **Virginia Norwood** (1927–2023), **NASA** (1958), **European Space Agency** (1975)

BEFORE
1500 Italian artist Jacopo de' Barbari surveys Venice from various towers and high buildings to create a bird's-eye-view map.

AFTER
2024 Remote sensing technology reconstructs finer details from low resolution input for high resolution output.

2027 NASA's Libera satellite plans to monitor solar radiation entering Earth's atmosphere and the amount absorbed, reflected, and emitted.

2030 NASA's Landsat Next satellites will study land cover and use changes over decades.

In 1858, French photographer Gaspard-Felix Tournachon pioneered the first photographic aerial images when he captured the French village of Petit-Bicêtre (Petit-Clamart) from a tethered balloon at 80 m (262 ft) above the ground.

In the 20th century, aerial photography progressed rapidly and, from the 1960s onwards, instruments mounted on spacecraft began to take detailed readings of Earth's ever-changing surface. This revolutionized mapmaking by providing exceptionally detailed visual information about a terrain, which made it easy to identify and map natural and human-made features. Inaccessible or dangerous areas that were difficult or impossible to survey on the ground

San Francisco earthquake damage was captured in 1906 by American photographer George Lawrence, who used kites to hoist his wooden-framed cameras up to 610 m (2,000 ft) high.

now became visible from the air, plus the rapid capture of images over large areas both sped up the mapmaking process and, using the art of photogrammetry, enabled calculations to be made with enhanced precision.

A new technology

Before the 19th century, a handful of artists worked to create images of Earth from a bird's-eye view, using observation from high vantage points and trigonometric calculations. However, the invention

Gaspard-Félix Tournachon (Nadar)

Born in Paris in 1820, Gaspard-Félix Tournachon, later known as Nadar, studied medicine but rose to prominence as a caricaturist and journalist in the city's bohemian circles. In the 1850s, when photography was new and unfamiliar, he mastered the technology and opened a photographic studio. It achieved wide renown and Nadar created portraits of figures such as his friend Charles Baudelaire, Victor Hugo, Sarah Bernhardt, and science-fiction writer Jules Verne. In 1855, Nadar patented the idea of using aerial photographs in mapmaking and surveying and, three years later, produced the first successful aerial photograph. In the 1860s, Nadar experimented with techniques for underground photography and is credited with pioneering the use of artificial light to illuminate the dark spaces of the Paris sewers and catacombs. Nadar and his passion for hot air ballooning inspired Jules Verne's novel *Five Weeks in a Balloon* (1863) and Nadar's studio hosted the first exhibition of the Impressionist artists in 1874.

MAPS AND MAPPING 43

See also: The global positioning system 48–51 ▪ Geographic information systems 52–59 ▪ Natural resource management and conservation 222–27 ▪ Desertification 228–29

of air balloons in 1783, then photography in the 1820s, created the potential for aerial images of hitherto unrivalled precision.

As photographic technology advanced – lighter cameras, celluloid roll film, faster shutter speeds – pioneers experimented with affixing cameras to unmanned flying devices. Kites were used as well as hot air balloons, and even homing pigeons harnessed with a miniature camera.

Tournachon, or "Nadar" as he was known, carried a portable darkroom in the balloon's basket, and while his earliest photographs no longer survive, he later captured a famous image of Paris from a much higher altitude using his large balloon, *Le Géant*.

Widespread usage

During World War I, aerial images became essential for military reconnaissance and surveillance and, by early 1915, the use of aerial photography became widespread. Cameras attached to aircraft becoming the norm and the need

It is interesting to note that, during the past ten and a half months, photographic work has been in greater demand than ever before.
Sgt Major Frederick Laws
Experimental Photography Section, Royal Flying Corps, 1918

Photogrammetry

The process of obtaining 3-D information from 2-D photographs is termed photogrammetry. The process was developed independently by two pioneers in the field: French army engineer Aimé Laussedat in 1849, and German surveyor Albrecht Meydenbauer, who coined the term in his 1867 article "Photogrammetry". They discovered that, by utilizing multiple photographs, they could identify corresponding points in overlapping images and calculate spatial directions and object dimensions. In 1919, German physicist Carl Pulfrich created the stereo comparator. This allowed overlapping images to be viewed through a stereoscope, with an eyepiece for each eye, to create a 3-D effect and made it possible to achieve even greater accuracy. Photogrammetry advanced rapidly with the arrival of digital plotting in the 1980s. It is now largely performed by drone and satellite instruments feeding images and data to computers.

for continuous coverage of enemy lines led to the development of instruments capable of taking overlapping vertical photographs. In 1936, the founder of Fairchild Camera, Sherman Fairchild, started Fairchild Aviation, dedicated to building aircraft capable of steady, high-altitude flight for accurate aerial mapping.

Viewing terrain from above offered a new way to survey territory, gathering information about land use, vegetation type, the extent of flooding (or drought), and even archaeology. In 1935, Fairchild cameras, which could photograph 590 sq km (225 sq miles) from 7,000 m (23,000 ft) in the air, were used to evaluate the extent of soil erosion in the Rio Grande Valley of New Mexico. While working at such high altitudes, camera operators breathed oxygen through a pipe.

Satellite images

In 1935, the US Army set an altitude record of 22 km (13 miles) with the Explorer II helium balloon, and its two-person crew captured the first photograph of the curvature of Earth's horizon, in what became known as the "first photo from space." (Today's definitions put the edge of space almost five times higher.) The first images of Earth from space were taken during suborbital test flights of V2 rockets, captured from Nazi Germany at the end of World War II. The highest reached an altitude of more than 100 km (60 miles) and took a grainy black-and-white image of the New Mexico desert. The rocket crashed into the desert and the camera was destroyed, but the film was preserved in a steel box.

The Soviet Sputnik 1, the first artificial satellite, did not carry a camera but created the potential for photography in space. In 1959, *Explorer 6*, the first US spacecraft to reach orbit, travelled with a photocell optical scanner that built a photo of Earth one line at a time as the spacecraft spun around. The scanner only worked during those parts of the orbit when the satellite slowed enough to capture a single scene and the resultant crude » image was a grainy, black and white picture of Earth's surface and cloud cover. However, it demonstrated the potential of using satellites for Earth observation and, in the 1950s, when Evelyn Pruitt and Walter Bailey at the US Office of Naval Research recognized that the existing term "aerial photography" was not sufficient to describe the new technologies and data streams coming from satellite imagery, they introduced the term "remote sensing".

Multispectral scanners

In the mid 1960s, the NASA Mercury and Gemini programmes provided the first opportunities for astronauts to photograph Earth from orbit. The resulting images, particularly those showing pollution and urban growth, prompted the US Geological Survey to propose a satellite remote-sensing programme to gather data about Earth's natural resources.

Instead of ordinary cameras, the Earth Resources Technology Satellite (ERTS 1), as it was originally called, used a Multispectral Scanner System (MSS), devised by American physicist and engineer Virginia Norwood. A traditional photograph captures only visible light in the electromagnetic spectrum (red, green, blue) but Norwood's scanner could capture data in multiple bands, therefore going beyond what the human eye can see. The design consisted of an oscillating mirror that repeatedly scanned a swathe of land 185 km (115 miles) wide and

Multispectral scanner technology aboard Landsat 1 took these early images in 1972, revealing the geology of the arid Great Namaland, Namibia.

captured data from electromagnetic radiation reflected or emitted from Earth's surface in four spectral bands: green, red, and two near-infrared bands. This data was then digitally encoded and transmitted to a ground station.

The Landsat launch

The ERTS 1, subsequently renamed Landsat 1, was launched on July 23, 1972 into a Sun-synchronous orbit, meaning the satellite maintains the same relative position to the Sun and flies over the entire surface of the planet in the course of one year. MSS images produced from the green spectral band showed shallow water and water with high sediment loads, the red band showed cultural features such as cities, and two near-infrared bands showed vegetation patterns, landforms, and boundaries between land and water.

With its ability to capture both visible and invisible light, and by providing the first systematic, global, and repetitive digital imagery of Earth's land surface, Landsat 1 revolutionized Earth observation and mapping. Its imagery revealed numerous errors in existing nautical charts, led to the redrawing of coastlines and, in

---

From a **high vantage point**, more **land can be surveyed**.

The invention **of cameras and aviation** enables **precise aerial photography**.

Detailed aerial photography enables a new way of **surveying land and analysing land use**.

Using **photogrammetry**, aerial surveys of **geographical features** can include **spatial directions and object dimensions**.

Satellites equipped with remote sensors capture images of the planet in the whole electromagnetic spectrum.

1973, even revealed a previously uncharted island, now known as Landsat Island, off the coast of Labrador, Canada. The technology created baseline maps of glaciers and river systems and aided in identifying geological features and updating land-use maps.

Since 1972, a total of nine Landsat missions have provided a continuous series of images that have enabled scientists to track changes in land use, monitor natural resources, and assess the impact of human activities on the environment

Surveying the seas

Following the success of Landsat, the first remote sensing satellite to survey Earth's oceans was launched in 1978: Seasat. Although a short-circuit of its electrical system ended its mission after 105 days, the data that its novel technologies collected is still used today.

While Landsat used passive sensors to gather information from reflected light and heat, Seasat needed more active sensors, sending out beams to detect effects such as reflections, as do many current remote sensing satellites. It used radar, for example, to measure sea surface topography and infer the shape of the seafloor, and used a scatterometer to send out pulses of microwaves to detect wind speeds over the ocean.

Super-sensors

In recent years, remote sensing satellites have further contributed to our understanding of the Earth. The European Space Agency's Aeolus mission (2018–23) used ultraviolet LiDAR (light detection and ranging) to map wind speeds across the globe. Europe's Copernicus programme includes radiometers to detect thermal radiation and provide crucial data on melting sea ice, particularly in the Arctic. In 2002, the American-German GRACE (Gravity Recovery and Climate Experiment) project created a map of Earth's gravitational anomalies, which can be used to investigate the precise shape and varied composition of the planet.

As remote sensing technology evolves, and as scientists analyse ever-increasing data sets, our knowledge of what is happening to Earth will continue to expand. ∎

A drone records radar data, enabling scientists to create a 3-D map of Austria's Schmiedingerkees glacier and monitor its thawing.

Virginia Norwood

Born in New York City, US, in 1927, Virginia Norwood won a place at Massachusetts Institute of Technology in 1944, graduating in mathematical physics. She worked on weather radar at the US Army Signal Corps Laboratories, then took up a post at the aerospace and defense contractor Hughes Aircraft Co. in Los Angeles. By the early 1960s, Norwood was working in its Space and Communications Division and, aware that NASA was interested in multispectral images of Earth from space, she set about designing an instrument that satisfied most of their requirements.

In 1972, Landsat 1 launched with Norwood's Multispectral Scanner System (MSS) aboard, astonishing everyone with the quality of images transmitted. Norwood continued to work as a senior scientist and laboratory engineer at Hughes Aircraft Co. until her retirement in 1989. She filed and held three patents, two for a radar reflector and one for a folding tracking antenna. After her death in 2023, aged 96, she was inducted into the National Inventors Hall of Fame.

A REVOLUTION IN GEOLOGICAL THINKING
MAPPING THE OCEAN FLOOR

IN CONTEXT

KEY FIGURES
Marie Tharp (1920–2006),
Bruce Heezen (1924–77)

BEFORE
1875 HMS *Challenger* uses lead-weight sounding to locate the world's deepest point in the west Pacific Ocean. It is named Challenger Deep.

1923 Herbert Groves Dorsey, principal engineer of the US Coast and Geodetic Survey, invents the fathometer, the first practical echosounder.

AFTER
1970s America's National Oceanic and Atmospheric Administration (NOAA) starts to survey the seabed using multibeam sonar to create more detailed maps.

2016 The Seabed 2030 Project is launched to create a new GEBCO (General Bathymetric Chart of the Oceans) to a resolution of 1 km (0.6 mile).

Until the 1920s, depth had been measured by carrying out soundings (dropping a weight on a line and lowering it until the weight touched the sea floor). In the late 1940s, Maurice Ewing and Bruce Heezen, oceanographers at Columbia University, New York, US, began an echo-sounding survey of the North Atlantic ocean floor. Geologist and cartographer Marie Tharp mapped their findings, which included the Mid-Atlantic Ridge. She posited – controversially at the time – that this was evidence for continental drift and seafloor spreading. Echo-sounding – sending a pulse of sound towards the sea floor, then calculating depth by measuring the time the pulse takes to return – was vital to the survey, greatly improving the accuracy of depth readings.

Austrian artist Heinrich Berann celebrated Heezen's and Tharp's discoveries in 1977 with a hand-painted topographic map, "World Ocean Floor".

The mid-ocean ridge
Mariners had first begun to report geological features on the sea floor in the 19th century. Britain's HMS *Challenger* research vessel, for

See also: Aerial photography and remote sensing 40–45 ▪ Continental drift 110–11 ▪ Plate tectonics 114–21 ▪ Seismology and earthquake magnitude 132–35

Marie Tharp

Born in Ypsilanti, Michigan, US, in 1920, Tharp first learned about maps from her father, a soil surveyor. Although her bachelor's degree was in English and music, in 1944 she earned a masters in geology from the University of Michigan. Tharp then worked in the oil industry and gained a further masters in mathematics.

In 1948, Tharp took up a post as research assistant at the Lamont Geological Observatory at Columbia University. There she met Bruce Heezen, with whom she worked for 30 years, mapping the data Heezen collected. In 1957, they completed the first map of the North Atlantic ocean floor, revealing hitherto unknown valleys, mountains, and canyons. Twenty years later, they created the first-ever map of the entire ocean floor. In 1997, America's Library of Congress named Tharp one of the 20th century's top four cartographers. She died in Nyack, New York, in 2006.

Key work

1965 "*Tectonic fabric of the Atlantic and Indian oceans and continental drift*" (with Heezen)

example, which sailed the globe collecting depth measurements in the 1870s, confirmed earlier reports of a ridge running down the middle of the Atlantic's ocean bed.

Ewing, director of Columbia University's Lamont Geological Observatory, initiated the echo-sounding survey of the North Atlantic in 1947, and hired Heezen, who specialized in marine geology. Tharp joined Heezen and over several years turned data from the survey into hand-plotted maps of the Atlantic and then the whole ocean basin. By 1953, her maps were showing that the Mid-Atlantic Ridge was part of a much longer underwater mountain range, now known to extend across the whole world, stretching some 65,000 km (40,400 miles).

Continental drift

Tharp and Heezen noticed how the subsea ridges and valleys were similar to East Africa's Rift Valley. Tharp saw this as evidence of continental drift, with the crust pulling apart at the mid-ocean ridge, creating new regions of sea floor (sea-floor spreading). Heezen initially disagreed, but later changed his mind. Together the pair refined their maps, adding data from many sources including historical surveys. By the 1970s, the extensive detail of their ocean floor maps had convinced sceptical scientists of the legitimacy of continental drift and its newer, more refined theory of plate tectonics.

Heezen's and Tharp's work also paved the way for modern sea-floor mapping, now carried out largely via multibeam sonar. A global ocean floor map to an unprecedented resolution is expected by 2030. ∎

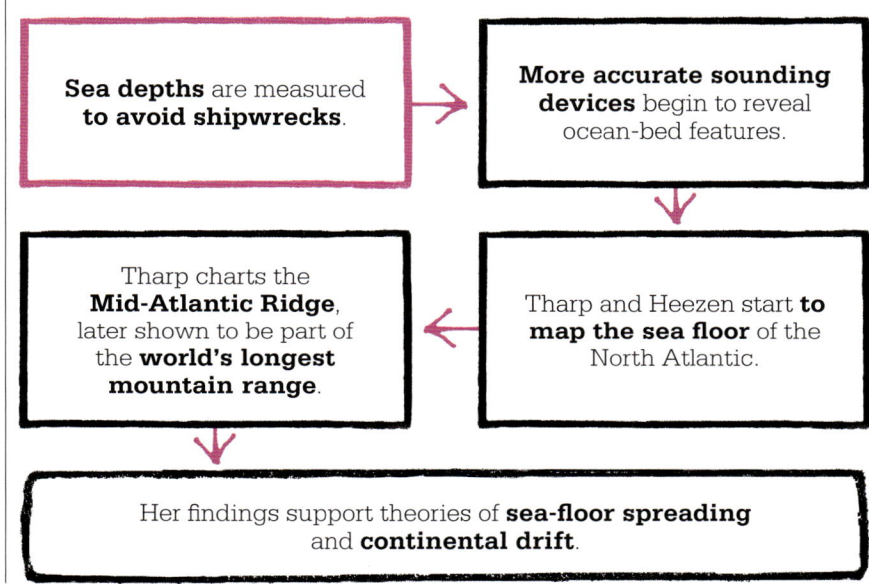

ACCURACY, INTEGRITY, AND AVAILABILITY
THE GLOBAL POSITIONING SYSTEM

IN CONTEXT

KEY FIGURES
Roger Easton (1921–2014),
Gladys West (1930–),
Bradford Parkinson (1935–)

BEFORE
1714 The British government passes the Longitude Act to encourage better navigation technology.

1920s Pilots begin to use signals from radio beacons as navigational guides.

1960 The US Navy launches the first of 36 Transit satellites for navigation purposes.

AFTER
1983 US president Ronald Reagan allows civilian airlines to use GPS.

2024 Britain tests "unjammable" quantum navigation, using quantum sensors and AI to provide positioning.

Until the 1940s, navigators, especially those at sea, generally relied on pre-industrial technologies – such as the sextant and chronometer – to determine their position. However, during World War II, combatant navies adopted more effective radio-based navigation systems.

The concept of space-based navigation emerged when the USSR launched its Sputnik satellite in 1957. Tracking its radio signal, American scientists observed a shift in frequency as the satellite moved closer and further away (known as the Doppler Effect). By carefully measuring the frequency shifts and knowing their own

MAPS AND MAPPING   49

See also: A geographical coordinates system 22–23 ▪ Aerial photography and remote sensing 40–45 ▪ Geographic information systems 52–59

---

ground location, scientists discovered they could calculate the satellite's position in orbit. This demonstrated the potential of using satellites for navigation and, throughout the 1960s and '70s, the US developed and refined the technology. On February 22, 1978, the first Navstar GPS satellite, Navstar 1, was launched, marking the beginning of the global positioning system (GPS).

The path to GPS
During World War II, radio-based navigation, such as the American Loran (Long Range Navigation) system, used multiple ground-based transmitters, working in pairs (master and secondary stations) to send out pulsed radio signals. By measuring the time delay between signals, navigators could plot a hyperbolic line on a map. The signals between other pairs of towers provided further hyperbolic lines and the point at which they crossed indicated the location. Loran had a range of some 2,414 km (1,500 miles).

After the successful launch of Russian and US space satellites in the 1950s, the US Navy began to develop a satellite navigation »

Bradford Parkinson

Born in 1935 in Madison, Wisconsin, US, Bradford Parkinson graduated from the United States Naval Academy in 1957 with a degree in engineering. While pursuing a career in the US Air Force, he also gained an MA in aeronautics from MIT (1961) and a PhD from Stanford University (1966) in guidance control navigation. Early in his career, he was a key developer of the modernized AC-130 Spectre Gunship and flew combat missions in the Vietnam War to assess it.

In 1972, Parkinson was asked to create a global navigation system (GPS) with satellites. He became the first Director of the Navstar GPS Joint Program Office in 1973, which led to the development of the GPS spacecraft, Master Control Station, and eight types of user equipment.

Parkinson retired from the Air Force in 1978 and, moving into the aerospace industry, worked on the development of software used by NASA's Space Shuttle. In 1984, he joined Stanford University as a research professor, retiring in 2001.

By tracking the **signals of the USSR's Sputnik satellite** and observing the Doppler shift in frequency, American scientists **locate its position**.

Using the same method to **plot the position of a receiver on Earth**, the scientists **discover the potential for satellite navigation**.

Using satellites, the Doppler effect, and ground receivers, the **US Navy creates the Transit system**.

The US creates **the Timation (time navigation) system**, leading to more **precise timekeeping in satellites**.

The Transit and Timation projects enable the creation of a worldwide navigational and surveying system: GPS.

50 THE GLOBAL POSITIONING SYSTEM

programme that tracked signals between satellites and ground station receivers and calculated the Doppler shift. The first programme, known as Transit, was designed to aid the navigation of ballistic missile submarines. However, a series of technical failures and design flaws hindered Transit's effectiveness and reliability. It was also too slow a system to be used by aircraft – or missiles.

Proof of concept

Physicist Roger Easton invented the GPS concept. In the 1960s, he led scientists at the US Naval Research Laboratory to develop a time-based navigation concept – the Timation (Time-navigation) programme. Satellites were fitted with an atomic clock and used the time of flight (ToF) of radio signals – the time signals took to reach the receiver – to determine distance, instead of using the Doppler shift. By knowing the speed of light and measuring time delay, the distance to the satellite could be calculated with extreme speed and accuracy.

The chief architect of GPS was US Air Force engineer Bradford Parkinson. In the early 1970s, the US military began planning for a comprehensive, worldwide navigational system that became known as the global positioning system (GPS).

The Navstar/GPS launch

Parkinson integrated the best elements of the Transit and Timation systems into a design that utilized a constellation of 24 satellites, circling the globe in precisely defined orbits. American computer scientist Gladys West created a geoid model of Earth's true irregular shape and enabled highly accurate calculations of satellite orbits, which became fundamental to GPS functionality.

Between 1978 and 1985, the first Navstar/GPS satellites were launched and tested and, by 1993,

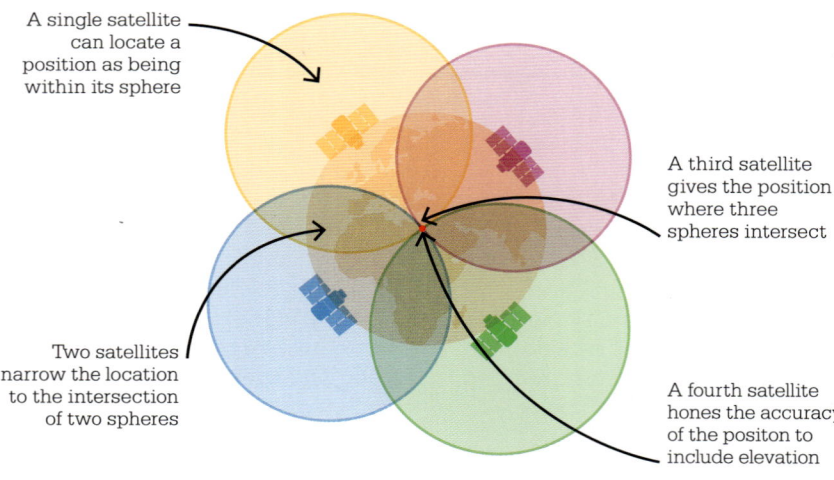

A single satellite can locate a position as being within its sphere

A third satellite gives the position where three spheres intersect

Two satellites narrow the location to the intersection of two spheres

A fourth satellite hones the accuracy of the positon to include elevation

In a process called trilateration, a GPS receiver uses the signals of at least three satellites to determine a two-dimensional position (latitude and longitude) on Earth. To calculate altitude (three-dimensional position), a fourth satellite is needed.

Gladys West

Born in Sutherland, Virginia, US, in 1930, Gladys West grew up on a farm, then won a full scholarship to Virginia State College (now Virginia State University), where she earned bachelor and master degrees in mathematics. In 1956, she joined the US Naval Weapons Laboratory, where she worked on the Naval Ordinance Research Calculator (NORC), tracking the movements of Pluto in relation to Neptune.

In 1978, West was a project manager on the Seasat programme, which was the first project to demonstrate that satellites could be used to observe oceanographic data. In the 1980s, she led the team that programmed the Geosat satellite to create computer models of the Earth's surface, taking into account gravitational, tidal, and other forces. This resulted in a highly accurate geodetic model of Earth that was used to develop GPS satellite orbits.

West retired in 1998 and completed a PhD in 2000. She was inducted into the US Air Force Space and Missile Pioneers Hall of Fame in 2018.

MAPS AND MAPPING 51

Europe's Galileo system is a network of 28 satellites at an altitude of 23,222 km (14,429.5 miles) in three orbital planes and provides highly accurate global positioning.

the full 24-satellite constellation became operational. The network was arranged in six orbital planes, with four satellites in each plane and each satellite completing an orbit around Earth in approximately 12 hours. This ensured that, at any time, at least five satellites were visible from any point on Earth and resulted in a highly accurate positioning system. Navstar was able to determine location within metres and time to within nanoseconds, which was unprecedented at the time, and it was quickly adopted by the US military and selected US allies.

Public benefit
In 1983, Korean Air Lines flight 007 accidentally strayed into Soviet airspace and was shot down by Soviet air-to-air missiles. If the airliner had been equipped with a GPS receiver, it would never have gone off course and, shortly after, US President Ronald Reagan authorized the use of GPS for civilian use. Initially, the civilian version was programmed with Selective Availability (SA) with reduced accuracy, but in 2000 the full service became accessible.

GPS is the best known of the GNSS (Global Navigation Satellite Systems) technologies that enable anyone with a receiver to pinpoint exact locations on Earth's surface, or in the air above it. The three others are Russia's GLONASS, China's BeiDou, and Galileo, run by the European Union (EU).

GPS mapping
The accuracy of GPS (with latitude, longitude, and altitude coordinates), has led to enhanced precision in mapping and surveying. For example, with global positioning, each tree in a forest or building in a city has its own point location, and each point can be mapped. New maps can also be created from multiple-source GPS data, such as hiking trails or remote regions, and these aggregated maps are often more accurate than those created from a single GPS track.

GPS has introduced a form of dynamic mapping, meaning maps can be updated in real time to reflect changes, such as in building construction or traffic flow, and this data can be collected and analysed in a geographic information system (GIS) and used to improve logistics planning and emergency response.

Scientific application
GPS technology plays a crucial role in enabling scientists to map and monitor changes in Earth's surface. By precisely tracking ground movement, GPS data helps scientists understand and assess various geological phenomena, including earthquakes, volcanic activity, and glacial movements. For example, in 1989, GPS data provided insights into the San Andreas fault rupture that caused the 6.9-magnitude Loma Prieta earthquake in San Francisco. In other earthquake-prone areas, GPS receivers are now estimating magnitudes faster and more accurately than seismometers and have significantly improved tsunami early-warning systems.

GPS satellites are also being used in scientific research. Satellite laser-ranging (SLR) stations on Earth reflect lasers off the satellites to check how precisely they are following their prescribed orbits and to monitor any variations in Earth's rotation and gravity ∎

GPS-enabled devices in cars, phones, and other devices provide turn-by-turn directions and real-time location tracking.

THE POWER OF GEOGRAPHY

GEOGRAPHIC INFORMATION SYSTEMS

54 GEOGRAPHIC INFORMATION SYSTEMS

IN CONTEXT

KEY FIGURES
Roger Tomlinson (1933–2014), **Jack Dangermond** (1945–)

BEFORE
1950 British urban planner Jacqueline Tyrwhitt combines four thematic maps into one map with transparent overlays.

1959 American geographer Waldo Tobler develops MIMO (map in–map out), a system that converts maps into a computer-usable form.

AFTER
1986 Mapping Display and Analysis System (MIDAS), the first desktop GIS, is released.

2007 Apple launches the iPhone, adding GIS potential to hand-held computer power.

2020 Johns Hopkins University in the US creates an ArcGIS-powered COVID-19 tracking dashboard that helps to monitor the global pandemic.

Since the 1960s, advances in computer technology have revolutionized the way data can be processed. In the field of geography, this has been applied to the mapping of land and sea to gain deeper knowledge of Earth's physical and human environments, helping to visualize how they relate to each other in space and time.

This rapid progress has been led by the development of geographic information systems (GIS), a computer-based way of capturing, manipulating, analysing, and presenting data linked to a place and time. It emerged in the 1960s, when British-Canadian geographer Roger Tomlinson had the idea of combining federal data in Canada with survey maps. Later, American environmental scientist Jack Dangermond advanced the field of commercially available GIS systems.

Using a variety of map types and layering techniques, GIS can represent data visually, allowing a greater understanding of "what" is happening "where". This allows governments, businesses, and individuals to make informed decisions on issues ranging from the placement of new housing or the planning of crop planting, to disaster prediction or access to sources of food and healthcare.

The early days of GIS were very lonely. No one knew what it meant.
Roger Tomlinson
Newspaper profile, 2007

Linking maps to data
The first examples of combining data with maps date from the 19th century, related largely to a need to track the spread of disease among urban populations. In 1834, French cartographer Charles Picquet published a map of Paris indicating deaths from cholera, and British physician John Snow mapped the disease's spread in London in 1855. Other

---

Hardware
Any device, such as a computer, server, or scanner, used to run GIS software and store data.

Software
Programs or applications used to perform GIS operations and functions.

Data
Spatial and temporal information stored digitally on computers or servers

Methods
Techniques, including data collection, data processing, and analyses, that turn data into effective visualizations.

The five elements of GIS

People
GIS analysts, database managers, cartographers, and end-users.

MAPS AND MAPPING 55

See also: Thematic mapping 34–37 ▪ Aerial photography and remote sensing 40–45 ▪ Global positioning system 48–51 ▪ Time-geography 266–67 ▪ Monitoring biodiversity 306–07 ▪ Tracking disease 310–13

Landscape and physical features can be isolated and studied using GIS, such as these valley contours (in pink) and Roman stadium remains (in red) in Italy.

cities followed, as they struggled to manage industrial-age health problems.

In the early 20th century, new photographic techniques using glass plates allowed thematic layers to be added to a map. The process was made more flexible and less cumbersome by the advent of transparent plastic layers, each of which could show a different feature, such as rivers, farmland, and buildings. These layers could be used alone or variously grouped.

Data meets the computer
In the 1950s, the Cold War that had broken out between Communist Eastern Europe and the US and other Western nations saw a surge in the military demand for detailed mapping in strategic planning, such as early warning systems and the targeting and control of missiles. This encouraged the development of mainframe computers that could process huge amounts of data, including geographic information, and by the early 1960s these bulky machines had entered commercial use. Technological advances in data storage, measurement, analysis, and processing speed were joined by new computer graphics capabilities, which could visualize the outcomes of geographic data entered into a mainframe and present them in a readable and printable form.

At this time, Roger Tomlinson, who was working in Ottawa, Canada, for an aerial survey company, began to see the potential of harnessing the mapping of land use with emerging computer technologies. In 1962, in tandem with the American computer company IBM, he began applying data from the Canadian Federal Department of Forestry and Rural Development to land maps. Plotting and layering information on soils, agriculture, wildlife, forestry, and recreation, he was able to provide highly detailed levels of analysis.

Tomlinson continued to develop his system over the next six years, and applied the term "geographic information system" to it for the first time in a 1968 symposium paper, later published in 1969.

The resulting Canada Geographic Information System (CGIS) became the first true GIS, run on a mainframe computer based in Ottawa. It was used to store, analyse, and manipulate data collected for the Canada Land Inventory and land planners across Canada could access its resources. Similar government GIS systems were soon set up in other countries, including the US, UK, Japan, and Sweden, harvesting extensive pre-existing data sets collected by geological surveys, national cartographers, and government departments.

The reach of GIS widens
While Tomlinson developed the CGIS in Canada, other GIS foundations were being laid. »

Through GIS, geography is actually becoming an organizing tool.
Roger Tomlinson
Thinking About GIS, 2003

GEOGRAPHIC INFORMATION SYSTEMS

Roger Tomlinson

Widely regarded as the "father of GIS", Roger Tomlinson was born in Cambridge, England, in 1933. After service in the Royal Air Force in the early 1950s, he earned a degree in geography from the University of Nottingham, UK, followed by a degree in geology from Acadia University, Canada, and a Master's degree in geography from McGill University, Montreal.

Adopting Canadian citizenship in 1957, Tomlinson initiated and led the world's first GIS development programme for the Canadian government throughout the 1960s. In 1974, Tomlinson completed his doctoral thesis at University College London titled "The application of electronic computing methods and techniques to the storage, compilation, and assessment of mapped data". His unique methods inform all GIS systems today.

In 1977, Tomlinson founded Tomlinson Associates Ltd, in Ottawa, through which, for the next three decades, he worked as "consulting geographer" on the use and development of GIS software. His clients included many international and national bodies, including the United Nations, the World Bank, and US Census Bureau. In 2001, the Canadian government granted him the Order of Canada, its highest civilian honour. Tomlinson continued to advise on GIS until his death in 2014.

Key work

2003 *Thinking About GIS*

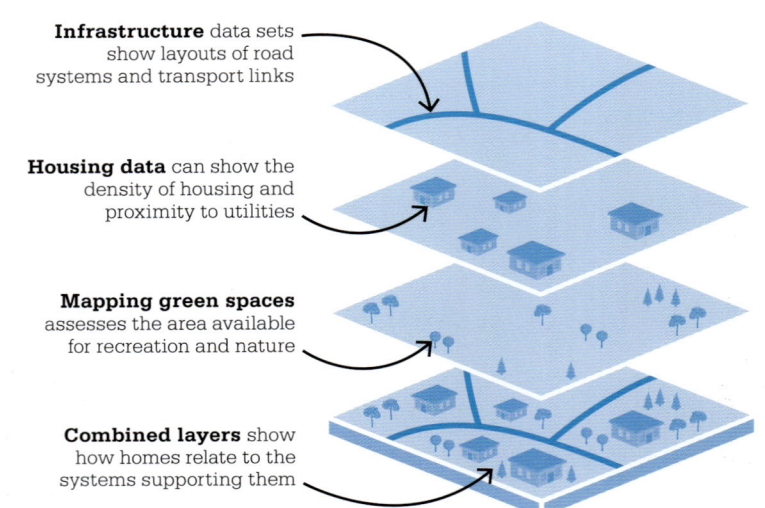

Infrastructure data sets show layouts of road systems and transport links

Housing data can show the density of housing and proximity to utilities

Mapping green spaces assesses the area available for recreation and nature

Combined layers show how homes relate to the systems supporting them

Layers of information, in the form of data sets relating to different geographical aspects, can be identified by GIS mapping and data analysis software, then integrated into a map or other visualized form.

American architect and city planner Howard Fisher worked on SYMAP, an early computer mapping programme, in 1964, and a year later founded the Harvard Laboratory for Computer Graphics and Spatial Analysis, which set about developing computer mapping software. In his book *Design with Nature* (1969), Scottish landscape architect Ian McHarg introduced the "layer cake" method of stacking and integrating geographic information. In order to analyse landscapes, he overlaid transparent maps of natural features – such as vegetation and topography – and this became the template for GIS mapping.

A number of data-gathering breakthroughs soon followed, including, in 1970, a move by the US Census Bureau from manual to computerized address coding and the use of computer tape files for data storage. On a global level, in 1972 NASA launched the first Landsat satellite – initially known as Earth Resources Technology Satellite (ERTS) – observing Earth from space and gathering a constant supply of new information about the whole planet in a process known as remote sensing.

Further progress in computing power and computer graphics led to the launch of commercially available GIS systems in the 1980s. The most successful was ArcGIS, created in 1982 by the Environmental Systems Research Institute, Inc. (Esri), a company that had been set up in 1969 by Jack Dangermond and his wife, Laura, who met while working at the Harvard Laboratory. ArcGIS built on early GIS tools to create a standardized GIS that could be used across government, business, and education.

The growth of the internet and World Wide Web in the 1990s had a profound impact on GIS, taking it from a desktop-based system to one that could be accessed from anywhere online, allowing any number of regional actors to access the technology. Further development of the internet in the early 2000s prompted much greater sharing of

> GIS is waking up the world to the power of geography… and creating a better future.
> **Jack Dangermond**
> Esri User Conference, 2015

data between users, leading to more public participation in GIS. For instance, Google Earth has, since it was founded in 2005, provided high-definition satellite maps of most of the planet, which can form base layers for GIS. Users create a data set (a body of information treated as a single unit by a computer), such as weather conditions, points of interest, or pedestrian or vehicle routes, and share them publicly.

Analysing change

The majority of information that can be collected, such as crime statistics, rainfall levels, or voting intentions, has a geographical element. GIS can take all this data and plot it on a map, making it easy to observe and analyse the relationships a data set has to the geography of an area. Data collection also captures a time component. Crime rates or rainfall levels, for instance, rise and fall across hours, days, and months. GIS can move backwards or forwards in time to show potential links in data and detect patterns in those relationships to predict how they may change in the future.

GIS can model any data with a spatial (place) or temporal (time) component. Spatial data (also known as geospatial data) might include latitude and longitude coordinates, addresses, or natural features, while temporal data might include dates, time, and time ranges. Combining the two components – a process known as geotemporal analysis – creates an understanding of how features or events may change over space and time.

Raster and vector

GIS generally uses two types of data: raster (from the Latin *rastrum*, meaning "rake") and vector. Raster data is based on a grid, with the base map divided into equal sections, or cells – each given a particular value from that data set, such as temperature or elevation. Raster data sets can then be integrated into a multidimensional mosaic. For example, the remote sensing data received from Landsat satellites creates a mosaic of data-filled cells that can be analysed to ascertain the type of habitat, such as forest or desert, being picked up by the cameras orbiting Earth. Vector data is presented on the base map as geometric elements, including points and elements constructed from points, such as lines and polygons. The points in vector data may be longitude and latitude, or another coordinate system. A point might represent a mountain summit or something as small as a fire hydrant. The line data represents roads, paths, and watercourses, while the polygons cover more extensive features, which, depending on the scale used, could be a building or a national park.

Both raster and vector are essential parts of GIS but offer different qualities. Raster data is useful for covering continuous or contiguous features, such as changes in terrain or climate, allowing for complex analysis of land forms and habitats. Vector data, which requires much less computer power, is more suited to zooming in on discrete detail, such as roads, boundaries, and networks, which works well for more compact areas, where the precise shapes and relationships of streets, utilities, and other features are particularly important. »

A raster layer represents geographic data as cells – like photograph pixels. A vector layer locates data in space, using coordinates linked to geometric objects (points, lines, and polygons), such as rivers or buildings.

Key:
- Urban area
- River
- Parkland

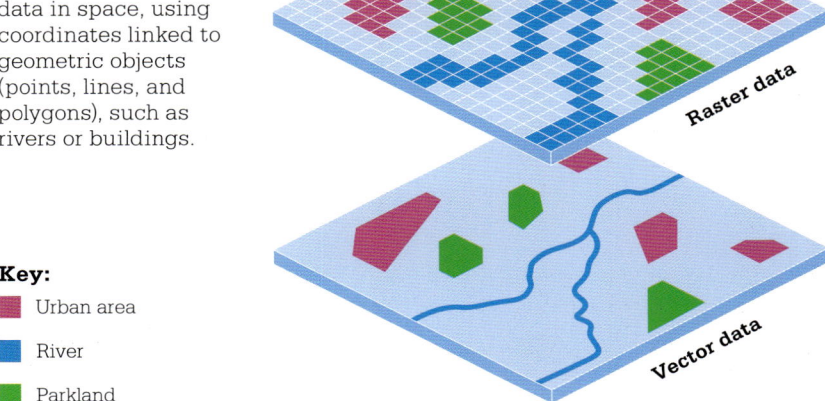

GEOGRAPHIC INFORMATION SYSTEMS

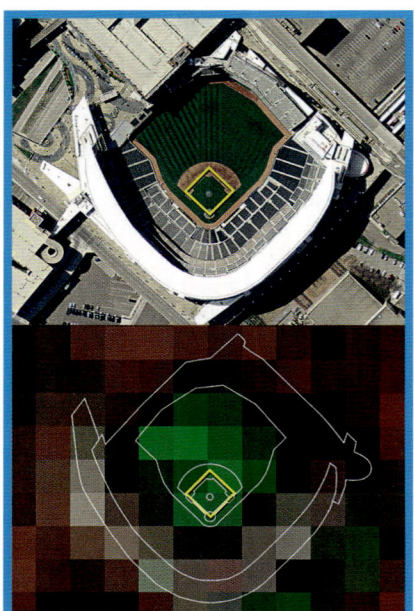

A baseball ground (top) is captured in a GIS satellite image (bottom), divided into 30 x 30 m (98 x 98 ft) raster data cells that gather information on features such as land cover or vegetation.

The changing elevations and terrain across a mountain range, for example, are best understood on a raster map, whereas a road's key features, such as direction and interaction with its surroundings, can be most accurately traced on a vector map. Unlike raster, vector data can also add layers of non-geographical information, such as the name, speed limit, and surface material of a road.

Geocoding

GIS relies on accurate, reliable data, which may come from a variety of sources – international, national, regional, local, and even individual. Many countries already have detailed maps of their territories, including high levels of physical detail. Large data sets covering Earth's surface can come from remote sensing by satellite, aerial photography, and surveillance. The most local data of all, such as locations of particular buildings and facilities, is collected on foot or in vehicles. Local data collectors on the ground will use global positioning system (GPS) devices or applications to locate features as they tour an area, and can take measurements from handheld light detection and ranging (LidAR) devices and laser rangefinders.

Location details recorded as text, such as an address, postal code, or place name, are converted into geographic coordinates that can be shown on a map – a process known as geocoding. This potentially allows all traditional location data to be turned into digital spatial data, providing many planning and decision-making benefits. Businesses, for example, can see exactly where their customers live and use that information to manage delivery routes or marketing and advertising strategies, while individuals can work out optimum travel routes or find the most favourably reviewed restaurants or shops nearby.

An opposite process – known as reverse geocoding – can also be applied to GIS spatial data to convert it back into a specific address or point. When, for example, a call is made to an emergency service from a mobile phone, the service can use the spatial data within the device to locate the relevant address.

Not all data, however, is accurate – it can be imprecise or may not exist at all. In these instances, GIS technicians use two techniques – interpolation and

GIS works best when the computer and the brain combine forces… in ways that reveal things that would otherwise be invisible.
Dr Michael F. Goodchild
British-American geographer, 1999

Local data collectors drive around in vehicles equipped with LidAR, GPS, and panoramic cameras to record street-level detail for GIS mapping.

Shared national data

Many countries, including Canada, Australia, Kenya, and Peru, offer access to a range of GIS data through their national mapping agencies. In 2009, the United States Geological Survey (USGS) released an online National Map. This GIS-based service allows anyone to explore both the natural and built environments of the US in great detail. The map has eight layers of information: elevation, land cover, hydrography (bodies of water), geographic names, boundaries, transportation, structures, and orthoimagery (a type of aerial or satellite imagery). Users can search for and identify features, measure distances and areas, and download high-definition maps. In Britain, the Ordnance Survey (OS) – the world's first national mapping organization, founded in 1791 – provides GIS data sets, including road, land, and topographic features, with similar data offered Europe-wide by Open Maps for Europe, a project launched in 2021 and co-funded by the European Union.

geostatistics – to enhance, predict, or estimate values. Interpolation uses interpretational tools and algorithms built into software to create a continuous surface from a set of discrete points. This is particularly useful for modelling features such as elevation or temperature, where it is impossible to measure every point.

Geostatistics uses statistical models to analyse spatial, and sometimes temporal, data to predict values that can be applied to locations that are lacking in data. For example, if a road junction is known to be particularly polluted, geostatistic techniques can be used to estimate the pollution in the surrounding areas.

Analytical power

GIS systems have turned maps from static geographical representations of elements of Earth's surface into interactive visualizations of data that allow close analysis of most aspects of the natural and human worlds. GIS also widens mapping analysis by bringing together different data types and placing them in the same time and space to see how they affect each other. Computer modelling techniques can then be used to understand how one factor might impact another in different scenarios.

The power of GIS analysis has a wide impact on all levels of decision-making where at least some geographical element is involved. Conservationists might use GIS data and mapping to follow the migration patterns of birds or butterflies to evaluate habitat changes and environmental threats. A local authority might add a tree density layer to a GIS map to help its residents find the shadiest spots during a heatwave. An ice cream company, knowing its sales depend on the weather, could work out where best to place its vendors by using GIS to show how pedestrian footfall changes as temperatures fluctuate. A homeowner thinking of installing a solar panel might might base their decision on GIS data, such as local energy costs, tree cover, and average sunlight hours.

Democratizing mapping

One of the most remarkable attributes of GIS is the enormous scope of its potential usage and relevance across all sectors of society – from governments, large organizations, and multi-national companies down to any individual with access to a smartphone. GIS is becoming increasingly accessible, both through map making or data-analysing software, such as QGIS or GRASS GIS, and data sources, such as OpenStreetMap and Natural Earth. Anyone can use these "open sources" and also contribute data to them.

Rapid increases in computer resources and technological developments mean that GIS can be made ever more complex and comprehensive without the risk of it becoming sluggish or beyond the reach of most computer users. Also, GIS has become even more accessible and easy to interpret through further developments in the fields of visualization, such as interactive animation, virtual reality (VR), and three-dimensional (3-D) systems. Perhaps most significantly, the development of artificial intelligence (AI) may further enhance the power of GIS, through faster and more accurate identification and processing of data, patterns, and predictions. ∎

Much of the current action in GIS uses citizen sensing of everything, from the effects of disasters to the quality of restaurants.
Dr Paul Longley
British geographer, 2011

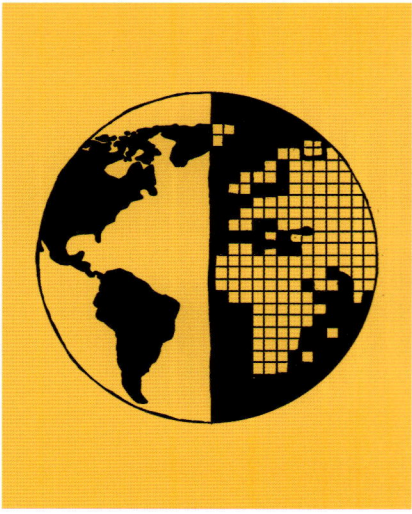

A PARADIGM SHIFT IN EARTH SYSTEM MODELLING
DIGITAL TWINS OF EARTH

IN CONTEXT

KEY BODY
European Commission
(1957–)

BEFORE
1956 US meteorologist Norman Phillips develops a mathematical model of the global climate that can predict monthly weather patterns.

1972 NASA launches the Landsat 1 satellite to gather information about the changing nature of Earth's surface.

1998 US vice president Al Gore proposes a Digital Earth with georeferenced scientific and cultural information.

AFTER
2030 Destination Earth (DestinE), a European Union-funded initiative, aims to build a comprehensive digital replica of Earth.

In 2024, the European Commission, working with many partners including the European Space Agency, marked the beginning of its ambitious Destination Earth (DestinE) project with the launch of its first two highly accurate and interactive digital simulations of Earth. By simulating current weather and climate patterns, these "digital twins" (DT) are able to monitor the effects of natural and human activity on the planet in order to help predict and manage the impacts of climate change.

In 1904, Norwegian scientist Vilhelm Bjerknes calculated that atmospheric fluctuations could be expressed mathematically in the form of equations. These described the balance of energy and matter in small parcels of air. Half a century later, early computers made it possible to handle the vast number of equations that must be solved

Italy's Po River valley has a complex hydrological pattern and is a focus for DestinE hydrology modelling, which aims to improve management strategies.

simultaneously to produce a weather prediction. From the 1950s, as digital technology advanced, meteorologists began modelling global weather patterns with increasing accuracy.

Climate-computing at work

The European Commission's DestinE and other less complete global climate models, such as NASA's Integrated Digital Earth Analysis System (IDEAS), rely on artificial intelligence and a continuous flow of meteorological and climate data from around the world. DestinE has access to Europe's largest supercomputers so that its digital twins can be refined and maintained.

Weather and climate models work by dividing the atmosphere into grid cells – cubes of a specific volume. All the activity inside a cell, such as temperature, humidity, cloud formation, and wind speed, is represented by a single series of data points. The size of the cell is the model's "resolution"; the smaller the cube, the greater the accuracy.

> Our understanding of the present, and simulation and evaluation of potential futures, will be changed beyond recognition.
> **Peter Bauer**
> **Director of DestinE, 2022**

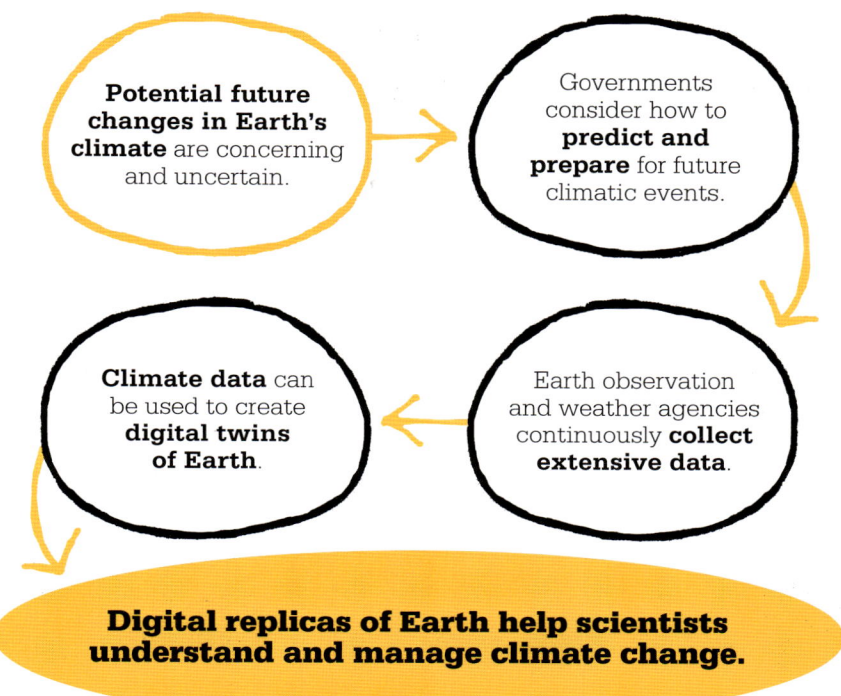

Potential future changes in Earth's **climate** are concerning and uncertain.

Governments consider how to **predict and prepare** for future climatic events.

Earth observation and weather agencies continuously **collect extensive data**.

Climate data can be used to create **digital twins of Earth**.

Digital replicas of Earth help scientists understand and manage climate change.

Today's regional weather forecasts have a resolution of 9 cu km and are accurate up to around 15 days into the future. Most other climate models are around 25 cu km (but most are closer to 100 cu km). However, the model underpinning DestinE's twins currently has a resolution of 2.8 cu km, with the aim to reduce that to 1 cu km. According to Peter Bauer, who spearheaded the development of DestinE, this level of data represents "a paradigm shift" as it can "create a digital replica of Earth that evolves as our planet does".

The first digital twins

The project's initial simulations went live in 2024. The Weather-Induced Extremes DT provides close to real-time predictions of extreme weather events on a timescale of 2–4 days in advance. It also simulates their potential impact, which allows governments in affected areas to take preventive action and to prepare and prioritize resources. The Climate Change Adaptation DT looks further into the future, providing information on how individual regions will be affected in decades to come. This helps governments and corporations to integrate resilience into all areas of planning.

In time, the goal of DestinE is to integrate additional digital twins, covering areas like oceans and biodiversity, to produce a comprehensive digital twin of Earth that incorporates all aspects of climate activity. ∎

PHYSICAL GEOGRAPHY

PHY

Bernard Palissy **describes the water cycle**, in which water vapour falls as rain or snow, the source of all Earth's freshwater.

1580

George Hadley's **theory of atmospheric circulation** provides an explanation of the trade winds.

1735

Gaspard-Gustave de Coriolis discovers the **the apparent curved path of objects** as they move across Earth's surface: the Coriolis effect.

1835

James D. Dana **classifies minerals** primarily based on their chemical composition and atomic structure.

1837

1669

Nicolas Steno's **principle of superposition** explains the layering of sediment in the formation of rock.

1824

Joseph Fourier observes that certain gases trap heat in Earth's atmosphere, causing a **greenhouse effect**.

1837

Louis Agassiz demonstrates how **glaciers shaped regions** of the world during ancient ice ages.

1855

George B. Airy describes **isostasy**, the way Earth's lithosphere or crust "floats" on the asthenosphere, the upper layer of the mantle.

When geography first emerged as a field of study, Ancient Greek scholars such as Aristotle and Plato proposed abstract philosophical explanations of Earth's physical characteristics. Rivers, for example, were believed to be replenished by water forced up from a subterranean source called the "Tartarus", with the water filtered through rocks to remove salt. However, from the mid-15th century, the "Scientific Revolution" in the West prompted a quest for rational scholarship to explain the natural processes that shape Earth's physical geography.

Early theories
In the 16th century, French potter and craftsman Bernard Palissy observed how freshwater wells could exist on islands surrounded by the sea and deduced that these must be replenished from rainfall. In 1580, he published the first clear description of the continuous water cycle, through stages of evaporation, condensation, and precipitation, and established the science of hydrology. A century later, Danish scientist Nicolas Steno analysed how rock strata formed in time-sequenced horizontal layers and established foundational principles for the study of modern geology.

Land formation
In the 19th century, the study of geomorphology developed out of the field of geology to describe the creation of Earth's landforms. French naturalist Georges Cuvier popularized the concept of catastrophism, which hypothesized that sudden, catastrophic events had shaped Earth's landscapes and life forms. However, this was largely superseded by the theory of uniformitarianism, developed by Scottish geologists James Hutton and Charles Lyell, which argued that consistent, gradual changes were the primary drivers of geological evolution over vast time spans.

In Lyell's influential book *Principles of Geology* (1830–33), he showed how geological history could be interpreted by observing current processes. From studying the landscape features left by glaciers in the Swiss Alps, for example, French scientist Louis Agassiz postulated that vast ice sheets once covered much of the northern hemisphere and shaped its landscape. In the same way, American geologist William M.

PHYSICAL GEOGRAPHY

 Wladimir Köppen proposes a **climate classification system** based on temperature, precipitation, and seasonal patterns.

 Alfred Wegener's description of **continental drift** lays the foundations for the theory of plate tectonics.

 Charles F. Richter establishes a **logarithmic scale** to measure the magnitude of earthquakes.

 Harry Hess and John Tuzo-Wilson develop the **theory of plate tectonics**, explaining events such as mountain formation and earthquakes.

1884 — **1912** — **1935** — **1960s**

1899 — **1920s** — **1941** — **1987**

 William M. Davis explains how the **cycle of erosion and the stages of river evolution** shape the landscape.

 Vladimir Vernadsky shows how both **environmental and biological factors** shape ecosystems.

 Hans Jenny identifies climate, organisms, parent material, topography, and time as **factors in soil formation**.

 Wallace S. Broecker describes how Earth's **deep ocean currents** transport heat and salt, regulating climate.

Davis described how a landscape evolved through a "geographical cycle" of erosion, divided into stages of youth, maturity, and old age.

The concept of "deep time" continued to influence scientists in the 20th century. In 1912, German scientist Alfred Wegener proposed the idea that Earth's continents were once joined as a single landmass and moved apart. In the 1960s, this led to the development of a comprehensive theory of plate tectonics, which provided a foundational explanation for many of Earth's physical features, such as mountain ranges and ocean basins.

Weather and climate

The application of scientific observation and experimentation led to discoveries in the other major fields of physical geography, the study of weather, climate, and ecosystems. In 1735, English physicist and meteorologist George Hadley created a model of Earth's atmospheric circulation, which consisted of a single wind system in each hemisphere. A century later, Gaspard-Gustave de Coriolis observed how Earth's rotation created the curved paths and spiral patterns of ocean currents, known as gyres. Both theories laid the groundwork for understanding global weather patterns.

In the 3rd century BCE, Eratosthenes had divided Earth into five climatic zones. In 1884, German-Russian climatologist Wladimir Köppen used the scientific criteria of temperature and precipitation, and the empirical observation of native vegetation, to classify the world's climates into five major groups. The idea that identifiable zones, or ecosystems, are created by an interaction between a region's environment and its flora and fauna spurred further developments in the field of ecology throughout the 20th century, especially in soil science and the classification of biomes.

In the 1970s, French climatologist Claude Lorius pioneered the study of palaeoclimatology by drilling ice cores, which revealed the varying impacts of the greenhouse effect through time. The evidence threw the rapid climate change of post-industrial history into sharp contrast and the need for greater understanding of environmental pressures continues to shape the study of physical geography today. ∎

NEITHER FROM SPRINGS NOR FROM THE SEA: BUT FROM THE RUN-OFF OF RAINS
THE HYDROLOGICAL CYCLE

IN CONTEXT

KEY FIGURE
Bernard Palissy (1510–90)

BEFORE
4th century BCE Valmiki's epic Hindu tale *Ramayana* mentions that the Sun heats up water and sends it down as rain.

80 CE Chinese scientist Wang Chong describes the water cycle, but he is disregarded.

AFTER
1841 Heinrich Berghaus, a German geographer, produces a map showing the global distribution of rainfall.

2017 Canadian water activist Maude Barlow reveals that half of China's rivers have disappeared since 1990.

2022 The US and French space agencies launch a satellite to survey how oceans and other bodies of water change over time.

For millennia, humans pondered why seas and oceans never seemed to overflow despite the volume of water pouring into them from rivers. They also questioned whether rain was a sufficient source to replenish watercourses. In the 4th century BCE, the Greek philosopher Aristotle recognized that rainwater did contribute to the flow of rivers; however, he also believed that a river's primary sources were reservoirs within Earth, fed from the oceans via a network of underground passages.

These false assumptions persisted until Bernard Palissy, a potter and self-educated naturalist,

PHYSICAL GEOGRAPHY 67

See also: Atmospheric circulation and winds 72–73 ▪ Climatic zones 96–101 ▪ The cycle of erosion 102–07

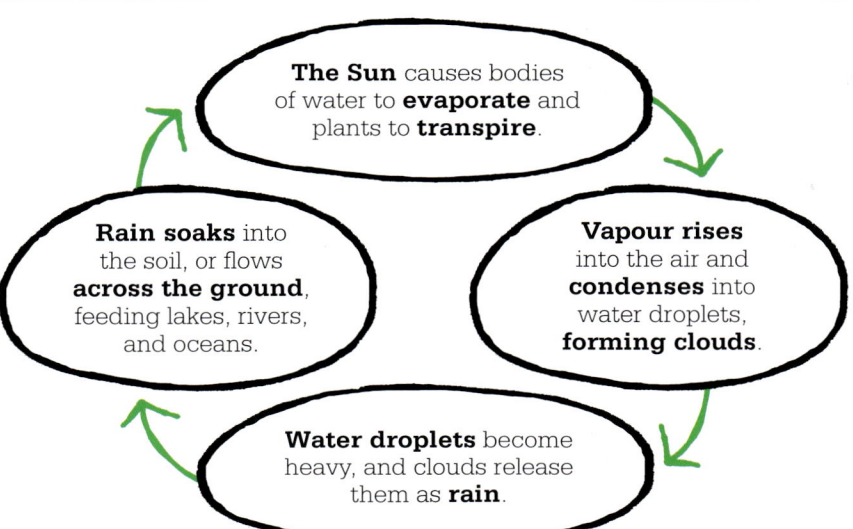

Bernard Palissy

Perhaps better known for his art pottery than his scientific contributions, Bernard Palissy was born in 1510 in the ancient Agenais region of western France (modern-day Lot-et-Garonne). He began his working life as a glass painter, later developing an interest in pottery, which led him to research ways to replicate the enamel of Italian ceramics.

Palissy was also a self-taught hydraulics engineer. In 1565, Catherine de' Medici, mother of the French king Charles IX, commissioned him to design a large garden grotto with an elaborate fountain for her Palace of the Tuileries, in Paris. Palissy moved to the French capital to undertake the project, but it was abandoned in 1572.

As a Huguenot, Palissy was a Protestant, but the practice of his religion was outlawed in France in 1585. In 1588 he was incarcerated in the Bastille, where he died in 1590, probably as a result of malnourishment and disease.

Key work

1580 *Admirable Discourses*

proposed an early model of the hydrological cycle. In 1575, he gave a series of public lectures on natural history, which he published five years later as *Admirable Discourses*. Through observation, experimentation, and critical thinking, Palissy suggested that the sole source of rivers was rainwater falling on mountains, soaking into the ground, feeding underground reservoirs, and flowing out on the surface as springs. He dismissed the idea that seawater flowed in veins to the tops of mountains, arguing that for this to happen, the sea level would have to be higher than the mountaintops.

Further proof emerges

Accurate though they later proved to be, Palissy's ideas were generally overlooked by his contemporaries. It was not until the mid-17th century that efforts were made to demonstrate scientifically whether or not rainwater was sufficient to maintain the flow of rivers.

In 1674, French scientist Pierre Perrault published *On the Origin of Springs*, in which he detailed his studies of the River Seine. He estimated that the river's annual run-off was only one sixth of the amount of water that fell as rain or snow over its drainage basin (the area of land from which water flows into a river) during the same period. This proved that rainfall alone was enough to account for »

If the fountains and rivers came from the sea, their waters would be salty.
Bernard Palissy
Admirable Discourses, 1580

THE HYDROLOGICAL CYCLE

the flow of a river. Palissy had believed that rainfall and run-off were roughly equivalent.

In 1687, English scientist Edmond Halley published his first paper on hydrology, "An Estimate of the Quantity of Vapour Raised Out of the Sea by the Warmth of the Sun". In it, he described an experiment designed to estimate the magnitude of water evaporating from Earth's oceans. Extrapolating from a thermometer in a pan of water heated "to the same degree of heat [as] that of the air in our hottest summers", Halley estimated that 5,280 million tonnes (5,820 million tons) of water vapour evaporated from the Mediterranean Sea on a summer's day.

Halley's 1691 paper "An Account of the Circulation of the Watry Vapours of the Sea, and of the Cause of Springs" attempted to explain what happens to the water that evaporates from the ocean. He postulated that water vapour is carried by the wind inland, where it precipitates on the tops of mountains. There, it would gather in "caverns of the hills" from which springs, streams, and ultimately rivers would flow.

A modern understanding

The processes of the hydrological cycle became better understood in the early 19th century, thanks to the studies of British scientist John Dalton. During his lifetime, he made and recorded more than 200,000 meteorological observations, which formed the basis of a paper published in 1802.

In the paper, Dalton detailed his efforts to calculate the hydrological balance of England and Wales. After dividing the two countries up into large catchment areas, he tried to quantify precipitation,

In the water cycle, the Sun turns water from oceans, rivers, and lakes into vapour. Vapour rises, then it cools and turns into liquid droplets, forming clouds. When the droplets become too heavy, precipitation occurs. Rain falls to the ground, infiltrating the soil and feeding rivers and oceans.

river outflow, and evaporation. Using the data available at the time, he estimated the annual rainfall for all catchment areas as 787 mm (31 in), to which he added 127 mm (5 in) for dewfall. Dalton then calculated the total river outflow of the area under study, estimating it as 330 mm (13 in) annually. From his own observations of soil-filled containers, he estimated annual evaporation at 762 mm (30 in). These calculations gave a discrepancy between input and output of 178 mm (7 in), which Dalton said could be accounted for from errors in his estimate of evaporation.

Dalton concluded that "the rain and dew of this country are equivalent to the quantity of water carried off by evaporation and by the rivers". His scientific-based approach confirmed Palissy's observations, establishing the correct understanding of the hydrological cycle.

The cycle above ground

One of the fundamental concepts of hydrology, the science that aims to understand Earth's water resources, is the hydrological, or water, cycle. This is a natural sequence of events in which water circulates continuously between the surface of Earth and the atmosphere through the processes of evaporation, transpiration, condensation, precipitation, and run-off.

The vast majority of water on Earth is held in the oceans. Only 2.5 per cent of all the water on Earth is freshwater, and nearly all of that is locked up in glaciers and groundwater. A mere 1,000th of a per cent of Earth's water is held in the atmosphere.

Evaporation takes place when a liquid changes state and turns into a gas. The energy required for this phenomenon comes from the Sun. Transpiration is the evaporation of water from plants. This happens through small pores known as stomata, which are found on the underside of leaves. Only 1 per cent of the water passing through a plant is used for growth; the rest passes out into the atmosphere.

Condensation is the reverse process to evaporation, whereby water vapour – that is, water in a gaseous state – changes into water, or a liquid state. When the temperature cools to a certain level, the air becomes saturated and cannot hold any more water vapour. At this "dew point", the water vapour then condenses into droplets of dew and forms clouds. Water vapour has more energy than liquid water, so when condensation occurs, the excess energy is released in the form of heat energy. When released, this heat energy has the ability to generate storms.

Precipitation, in the form of rain, snow, sleet, or hail, occurs when particles of condensing water in clouds become too large for air currents to support them and so fall to the ground. Precipitation is the primary mechanism by which freshwater reaches the ground.

The cycle below ground

By the process of infiltration, water penetrates the soil. If there is an excessive amount of rainfall, the ground may become saturated and unable to absorb any more water. In this instance, surface run-off

A glacier-fed river flows from the mountain of Manaslu, in the Nepalese Himalayas. The majority of Earth's minute percentage of freshwater is stored as snow and ice.

> The ordinary water sometimes will be absorbed by the earth and will descend lower or else will evaporate and go off as vapours into clouds.
> **Bernard Palissy**
> *Admirable Discourses*, 1580

occurs, which leads to the creation of rivers and lakes, along with factors such as groundwater seepage and springs. Most water in rivers and lakes eventually returns to the oceans.

Water also moves through the soil via the gravity-driven process of percolation. After seeping into the ground, water trickles through cracks and pores in the soil. Some of it might reach a subterranean aquifer, while some is drawn into plants to be returned to the atmosphere through transpiration, which, along with the evaporation of run-off, starts another turn of the hydrological cycle.

A grim outlook

Data gathered by Earth observation satellites show that the global water cycle is changing. Rising temperatures and shifting weather patterns are resulting in droughts in some areas and intense rainfall and flooding in others.

The amount of water stored in ice sheets and glaciers is decreasing year on year, contributing to rising sea levels and reducing the availability of fresh water. Rivers that rely on glacial meltwater in regions such as the Himalayas are drying up, causing problems for millions of people who rely on them for their water supply. The growing demands of agriculture and industry have also depleted water resources, throwing the natural system out of balance. Largely driven by human activities, these changes could significantly affect ecosystems and potentially reduce the habitability of some areas. ■

AT THE TIME THE LOWER STRATUM WAS BEING FORMED, NONE OF THE UPPER STRATA EXISTED
STENO'S LAW OF SUPERPOSITION

Objects in rocks, such as fossils, show that a layer of **rock formed** at a particular time in **the past**.

Rocks form in **horizontal layers** that may become **folded** or **faulted** later on.

A new layer of rock always forms **on top** of a **pre-existing layer** of rock.

The lower a layer of rock is in a sequence of strata, the older the rock.

IN CONTEXT

KEY FIGURE
Nicolas Steno (1638–86)

BEFORE
1074 To explain the presence of marine fossils in mountain rocks, Chinese scientist Shen Kuo proposes that landscapes are shaped by very slow but powerful forces that move rivers and oceans.

1510 Italian Renaissance polymath Leonardo da Vinci suggests that rocks can be formed by the deposition of sediments over time.

AFTER
1785 According to Scottish doctor James Hutton, the processes that formed rocks in the past are also taking place in the present.

1799 British geologist William Smith uses Steno's Law (briefly known as Smith's Law in his day) to produce the world's first geological map, which covers Wales and the west of England.

One of the key principles of modern geology is the law of superposition. The law states that in undisturbed horizontal layers, or strata, of sedimentary rock, the oldest layers lie deeper than younger ones. These strata may subsequently be folded, faulted, or eroded away, but the law of superposition is unbroken. As such, it can be used as a reliable aid to piecing together the history of a rock sequence and the geological processes that shaped it.

The law is attributed to Danish scientist Niels Steensen, who is better known by the Latinized name Nicolas Steno. A career in anatomy led him to Florence,

PHYSICAL GEOGRAPHY

See also: Triangulation 32–33 ▪ Uniformitarianism 75 ▪ Orogeny 76–79 ▪ Glaciation and ice ages 86–89

Fossilized shark teeth embedded in rock, such as this one, confounded early scientists, with some suggesting that these "tongue stones" had fallen from outer space.

Italy, where he became the personal physician of the Grand Duke of Tuscany, Ferdinand II de' Medici.

In 1666, while examining the teeth of a huge shark landed by a Tuscan fisherman, Steno was struck by their similarities to "tongue stones", which were sharp, tongue-shaped objects embedded in rocks. He believed – as Italian naturalist Fabio Colonna had already posited 50 years earlier – that the tongue stones were in fact fossil teeth left by sharks that had lived before the rock had formed.

Rock formation

The process of how solid objects – whether fossils, crystals, or stones – could get inside another solid, such as a layer of rock, led Steno to ponder about the formation of rocks more generally.

The layered strata of The Wave, a distinctive sandstone rock formation in the Vermilion Cliffs of northern Arizona, US, are a clear illustration of Steno's geological principles.

His 1669 treatise *Prodromus* set out his conclusions and the four fundamental principles that came to be known as Steno's laws.

The first principle is the law of superposition, which states that when a rock layer is still forming, there must be a pre-existing – and therefore older – stratum beneath it. This stratum prevents the materials of the new rock from seeping any deeper.

Other geological principles

The other three principles expand on the first law. The second is the principle of original horizontality, which reiterates that strata form in horizontal layers, parallel to the surface of Earth. Any deviation from the horizontal plane is therefore due to a subsequent event, such as tectonic activity.

The third principle, the law of lateral continuity, states that sediment spreads in all directions until it tapers out or until it encounters a pre-existing solid barrier. It follows, therefore, that rock sequences now exposed in cliffs or outcrops were not formed like that; the original strata have been eroded away.

The final law is the principle of cross-cutting relationships, which states that faults, dykes, and other discontinuities that run through strata occurred after the rocks formed. This allows for strata to be dated correctly despite having been thrust up or down out of sequence with surrounding rocks. ■

The strata of the earth, as regards the place and manner of production, agree with those strata which turbid water deposits.
Nicolas Steno
Prodromus, 1669

THE ACTION OF THE SUN IS THE ORIGINAL CAUSE
ATMOSPHERIC CIRCULATION AND WINDS

IN CONTEXT

KEY FIGURE
George Hadley (1685–1768)

BEFORE
1565 Spanish navigator Andrés de Urdaneta theorizes that the winds and currents move in a circular pattern around the Pacific Ocean, and he pioneers a route from the Philippines to Mexico.

c. 1615 According to Italian polymath Galileo Galilei, the existence of the trade winds is proof that Earth rotates.

1686 Edmond Halley publishes the first chart of the world's trade winds.

AFTER
1835 French scientist Gaspard-Gustave de Coriolis explains that the rotation of Earth is responsible for the deflection of winds.

1847–49 Matthew Maury, an officer in the US Navy, produces wind and current charts for the world's oceans.

From the earliest times, sailors have attempted to harness wind patterns to their advantage when plotting a course across the ocean. During the Age of Exploration, between the 15th and 17th centuries, European navigators made use of the trade winds – persistent, steady winds that, in the northern hemisphere, blow from the east in the tropics – to sail across the Atlantic Ocean. To return to Europe, they relied on the westerlies, which blow in the mid-latitude regions north and south of the tropics.

Early trading voyages were reliant on a knowledge of the winds. The term "trade winds", however, is not related to commerce but derives from the medieval word for path or direction.

At the time, no one could explain why the winds blew one way at one latitude and in the opposite direction at another. In 1686, English scientist Edmond Halley suggested that warm air rising at the equator and moving towards the poles was causing the trade winds. However, his explanation did not take into account Earth's rotation, so it could not fully explain the observed direction of the wind.

Hadley's theory

Barrister and amateur meteorologist George Hadley sought to provide a scientific explanation for the trade winds and the westerlies. In 1735, he published a paper called "Concerning the Cause of the General Trade Winds". In his article, Hadley built on Halley's ideas, offering an explanation for the trade winds that was based on a combination of solar heating and Earth's rotation.

Hadley suggested that intense solar heating causes air to warm and rise at the equator. This warm, rising air moves at high altitudes towards the poles, subsequently sinking at subtropical latitudes (around 30 degrees north and south) as it cools. The cooler,

PHYSICAL GEOGRAPHY

See also: The Coriolis effect 84–85 ▪ Global ocean circulation 140–41

Atmospheric circulation is the way in which air moves across Earth, creating reliable wind patterns, such as the polar easterlies, the westerlies, and the trade winds.

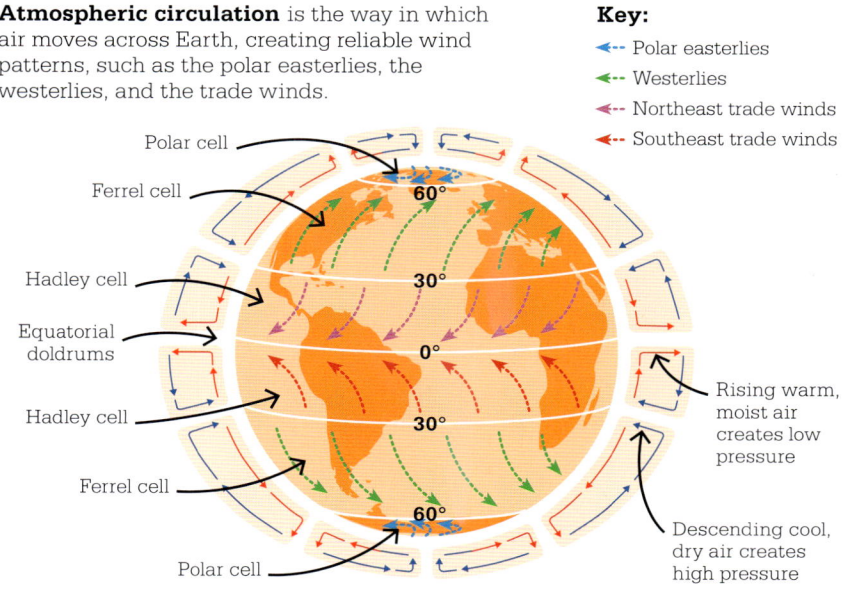

Key:
- Polar easterlies
- Westerlies
- Northeast trade winds
- Southeast trade winds

Rising warm, moist air creates low pressure

Descending cool, dry air creates high pressure

poles. Cold air in polar cells sinks and moves towards lower latitudes. At around 60 degrees latitude, the air warms, rises, and moves back towards the poles.

Between Hadley and polar cells lie Ferrel cells, first proposed by US meteorologist William Ferrel in 1856. These are not driven by temperature but are powered by the motion of the adjacent polar and Hadley cells, and by the Coriolis effect.

Jet streams

A phenomenon that is closely related to the circulation cells is that of jet streams. These four narrow bands of powerful winds form at the boundaries between cells in both the northern and the southern hemisphere. The subtropical jet stream forms between the Hadley and Ferrel cells, while the polar jet stream forms at the boundary of the Ferrel and polar cells. Flowing west to east, jet streams have an important influence on weather patterns. ∎

sinking air then flows back towards the equator, replacing the rising air and creating a loop.

Circulation cells

Hadley's model paved the way for the modern concept of atmospheric circulation cells. This concept divides each hemisphere of Earth's atmosphere into three major circulation cells. As well as Hadley cells between the equator and 30 degrees latitude north and south, there are polar cells, extending from about 60 degrees latitude to the North and South

George Hadley

The younger brother of astronomer John Hadley, George Hadley was born in London, England, in 1685. At the age of 15, he entered Pembroke College, Oxford, and in 1701, he started training for a legal career at Lincoln's Inn, where his father had bought chambers for him. Eight years later, Hadley qualified as a lawyer, but he remained more interested in mechanical and physical studies than in legal concerns.

In 1735, Hadley was elected a fellow of the Royal Society, an academy of science where he was in charge of meteorological observations for seven years.

Later in life, Hadley, who had never married, left London and went to live with one of his nephews in Flitton, Bedfordshire. He died there in 1768.

Key work

1735 "Concerning the Cause of the General Trade Winds"

Generally all Winds from any one Quarter must be compensated by a contrary Wind some where or other.

George Hadley
"Concerning the Cause of the General Trade Winds", 1735

THE TRACES OF REVOLUTIONS WHICH HAVE TAKEN PLACE
CATASTROPHISM

IN CONTEXT

KEY FIGURE
Georges Cuvier (1769–1832)

BEFORE
540 BCE The Greek philosopher Xenophanes studies the presence of marine fossils on mountains.

1696 In *A New Theory of the Earth*, English theologian William Whiston attempts to prove that the rock strata of Earth were created by a great flood.

AFTER
1923 A paper by US geologist J. Harlen Bretz suggests that the Channeled Scablands in Washington State were formed by catastrophic flooding rather than gradual geological events.

1975 US scientists William Hartmann and Donald Davis suggest that the Moon is the result of a catastrophic collision between Earth and a minor planet more than 4 billion years ago.

The earliest attempts to explain how Earth came to be the way it is were rooted in religious beliefs. In Western Europe, in particular, 17th- and 18th-century geologists tried to mould their observations to be consistent with biblical interpretations by suggesting that catastrophic events such as Noah's Flood had really taken place.

Influential catastrophists

In the 1780s, German geologist Abraham Werner promoted neptunism, a theory stating that surface rocks were formed by sediment left by a vast ocean, since receded. He cited catastrophism as proof that Earth had experienced mass floods throughout its history.

Thirty years later, French naturalist Georges Cuvier based his theory of catastrophism on his observations of fossils in the Paris Basin. He had noticed that, instead of a continuous succession of fossils, there were several gaps in the fossil record where all evidence of life disappeared before suddenly reappearing at a later time. Cuvier believed these gaps indicated mass-extinction events.

Cuvier proposed that these mass extinctions could have resulted from the flooding of lowland areas, but he never explained what might have caused such floods. This left it open for later geologists – such as Britain's William Buckland in the mid-1820s – to embrace the diluvian theory, which suggested that the most recent such event had been Noah's Flood. ∎

Life, therefore, has often been disturbed on this earth by terrible events.
Georges Cuvier
Essay on the Theory of the Earth, 1827

See also: Steno's law of superposition 70–71 ▪ Orogeny 76–79 ▪ Glaciation and ice ages 86–89 ▪ Continental drift 110–11

PHYSICAL GEOGRAPHY

A CONTINUAL SUCCESSION
UNIFORMITARIANISM

IN CONTEXT

KEY FIGURES
James Hutton (1726–97),
Charles Lyell (1797–1875)

BEFORE
c. 1750 Italian geologist Anton Moro promotes the theory of plutonism, suggesting that molten rock is formed beneath Earth's surface.

1787 Abraham Werner, a German geologist, publishes his theory of neptunism, which states that all rocks have crystallized from a primeval ocean since receded.

AFTER
1859 The publication of British naturalist Charles Darwin's *On the Origin of Species* strengthens the argument for gradual geological changes.

1980 US professors Luis and Walter Alvarez propose that the impact of a massive asteroid brought about the extinction of the dinosaurs 66 million years ago.

Uniformitarianism proposes that geological forces and processes observable today, such as erosion and deposition, are the same ones that have shaped Earth throughout its history.

In a 1788 paper, Scottish geologist James Hutton suggested that Earth goes through a constant cycle: rocks and soil are eroded and washed into the sea; there, the sediment is compacted into bedrock, pushed up to the surface by movements of Earth's crust, and worn away into sediment again.

Realizing that each cycle took place over huge gulfs of time, Hutton concluded that Earth must be much older than was allowed for by biblical interpretations. This led to the concept of deep time, the idea that Earth had existed for billions of years.

In his *Principles of Geology* (1830–33), Charles Lyell, another Scottish geologist, developed Hutton's idea that Earth was shaped by slow-moving forces over a very long period of time. The term "uniformitarianism" was coined in a review of Lyell's book by British philosopher William Whewell.

Balancing two theories

Uniformitarianism is widely accepted as the overarching way Earth evolves geologically. However, the doctrine of neocatastrophism accepts that rare catastrophic events, such as super-volcano eruptions or the impact of massive asteroids, can also play a significant role. ■

The Bunda Cliffs of South Australia, battered by the waves of the Great Australian Bight, have been subject to a slow process of erosion for millennia.

See also: Steno's law of superposition 70–71 ▪ Glaciation and ice ages 86–89 ▪ The cycle of erosion 102–07 ▪ Continental drift 110–11

RAISED FROM THE DEEP BY A SUCCESSION OF UPWARD MOVEMENTS
OROGENY

IN CONTEXT

KEY FIGURES
Charles Lyell (1797–1875),
Edward Suess (1831–1914)

BEFORE
1669 Danish scholar Nicolas Steno explains that rock strata form horizontally and may subsequently be tilted.

1788 In *Theory of the Earth*, James Hutton argues that mountains have formed slowly over very long periods of time.

c. 1800 German naturalist Alexander von Humboldt theorizes that South America and Africa were once joined.

AFTER
1912 Alfred Wegener proposes the theory of continental drift.

1960s US geologist Harry Hammond Hess and Canadian geophysicist John Tuzo Wilson find evidence that proves the theory of plate tectonics.

An orogeny is the process of mountain building through deformation of Earth's crust. Before our modern understanding of tectonic plates, geologists proposed many theories to explain it. In the late 18th century, Scottish geologist James Hutton argued that mountains were not fixed features but products of continual uplift and erosion. In the early 1830s, Charles Lyell concurred that mountains and other landforms have formed very slowly through ongoing processes of sedimentation, erosion, and Earth movements. Later, he noted that the 1855 Wairarapa earthquake in New

PHYSICAL GEOGRAPHY 77

See also: Steno's law of superposition 70–71 ▪ Uniformitarianism 75 ▪ Continental drift 110–11 ▪ Plate tectonics 114–21 ▪ Volcanic activity and hotspots 122–25

Uniformitarianism: mountain-building processes – sedimentation, uplift, folding, and erosion – have been at work throughout Earth's history.

Geosynclinal: mountains are thought to form when deep, sediment-filled troughs subside and are later compressed and uplifted.

Contractionism: a theory suggesting that, as Earth's interior cools, the surface of the planet contracts and wrinkles, forming mountain ranges.

Continental drift: mountains are believed to be the result of continents drifting and colliding along their margins.

Plate tectonics: mountains are understood to form through horizontal plate-tectonic movement.

Charles Lyell

While studying classics at Oxford University, Charles Lyell, who had been born in Kinnordy, Scotland, in 1797, developed a keen interest in geology and started to attend William Buckland's lectures.

In 1824, Lyell presented a paper to the Geological Society of London explaining that the modern lake sediments in his native Kinnordy were similar in composition to Tertiary limestones in the Paris Basin, highlighting an example of uniformitarianism. Three years later, he married fellow geologist Mary Horner; their honeymoon was a geological tour of Switzerland.

In 1833, after his *Principles of Geology* drew criticism for contradicting biblical chronology, Lyell resigned as professor of geology at King's College, London, just two years after being appointed. He later helped Charles Darwin publish his *On the Origin of Species* (1859). Lyell died in 1875.

Key work

1830–33 *Principles of Geology*

Zealand had created subsidence, uplift, and faulting, and concluded that not all geomorphological processes were gradual.

Looking for a mechanism

Although Lyell proved that mountain building had taken place, he was not able to identify the mechanisms responsible. In 1859, American geologist James Hall published the results of his studies of the Appalachian Mountains in North America, which showed that their folded strata were at least 10 times as thick as the unfolded sediments in the Mississippi Basin to the west. This indicated that enormous subsidence must have occurred within Earth's crust to accommodate such a thickness of sediment before uplift created a great mountain range. However, Hall did not propose a viable mechanism for the uplift, either.

In 1873, after examining Hall's findings, US geologist James D. Dana published "On the Origin of Mountains". In the article, he coined the term "geosynclines" to define the great linear, subsiding troughs such as that which had produced the Appalachian sediments, and he proposed what later became known as the geosynclinal theory. As for the question of how such troughs form in the first place, both Dana and Austrian geologist Eduard »

78 OROGENY

Suess believed that the cooling of Earth had caused it to contract. The subsequent wrinkling and shortening of the crust, they proposed, created parallel deep troughs and mountain ranges. This theory became known as thermal contraction. In his three-volume *The Face of the Earth* (1883–1909), Suess applied this theory to the formation of the Alpine-Himalayan mountain chains and their curved structure. However, thermal contraction failed to provide an explanation for the evidence of immense horizontal forces in mountain chains; and while it is true that Earth's interior is slowly cooling, the planet was subsequently shown not to be contracting at any significant rate.

A modern understanding

The mechanisms behind orogenesis (mountain-building) were not understood until much later. In 1912, German geologist Alfred Wegener proposed that all Earth's landmasses were once part of a super-continent, Pangaea, that later split apart to create the modern continents. Then, in 1919, British geologist Arthur Holmes suggested that convection currents in Earth's mantle could drive the movement of continents. Further discoveries confirmed that the planet's outer shell, or lithosphere, is divided into seven vast, and several smaller plates, which are constantly moving – converging, diverging, or sliding past each other.

The discovery of plate tectonics explained why, when, and where mountain-building takes place. Orogenesis usually occurs when two or more tectonic plates, driven by convection currents in the mantle, converge along a continental margin. Horizontal compression generates pressure, which folds and faults rock strata

At Lulworth Cove, on the Dorset coast of Britain, the Alpine orogeny has forced horizontal strata into vertical positions, creating a distinctive feature known as the Lulworth Crumple.

and creates mountain uplift. Pressure, heat, and volcanism change the character of the rocks, a process called metamorphosis.

When two continental plates collide, huge volumes of rock are forced upwards to form mountain belts such as the Himalayas or the Alps. This is called collisional orogeny. Non-collisional orogeny occurs where continental and oceanic plates converge, such as along the west coast of South America. Here, the denser oceanic crust subducts, or slides beneath, the less dense continental rocks, forcing them up. The oceanic crust is consumed as it descends into the hot mantle, leading to earthquakes and volcanism.

A long history of orogenies

Some geologists believe that the first mountain formation occurred during the early Precambrian (4.6 billion to 541 million years ago); others doubt this, because it is not yet known when tectonic plates first formed. There is, however, a consensus that, around 2,500 mya, the Kenoran orogeny involved continents colliding and subducting in what is now North America.

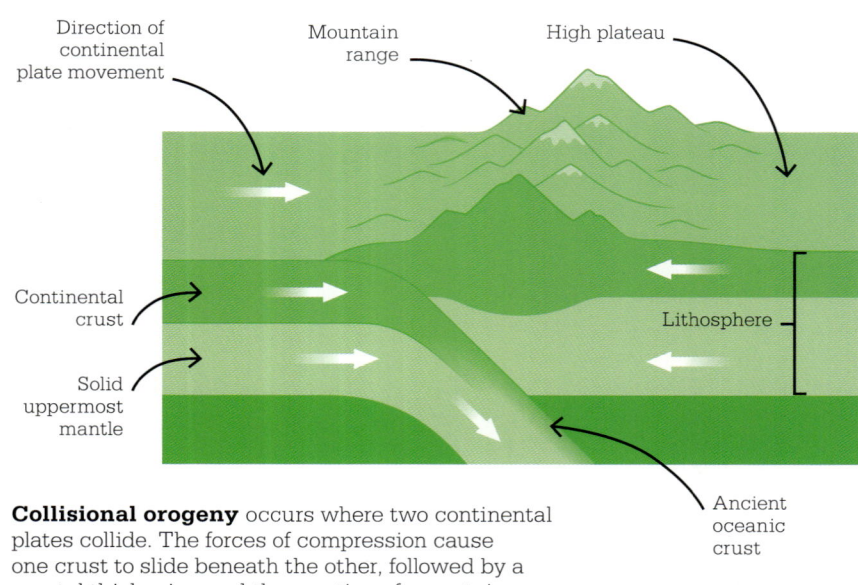

Collisional orogeny occurs where two continental plates collide. The forces of compression cause one crust to slide beneath the other, followed by a crustal thickening and the creation of mountains.

PHYSICAL GEOGRAPHY

Continents… may have originated in movements of this kind, continued throughout incalculable periods of time.
Charles Lyell
Principles of Geology, 1830

Between 542 and 416 mya, the multiphase Caledonian orogeny resulted from the collision of the ancient continents of Laurentia (present North America) and Baltica (Scandinavia) and the micro-continent of Avalonia (England, Wales, northern France, and eastern Newfoundland). This produced the Caledonian Mountains, which once stretched from Scandinavia through the British Isles to North America.

Ancient orogenic belts such as this are typically heavily eroded and highly metamorphosed. They also contain huge bodies of intrusive igneous rock, which is solidified magma or lava.

Different mechanisms

Two ongoing mountain-building events are very different in nature. Starting around 65 mya, the northward-moving African, Arabian, and Indian plates began to converge with the Eurasian Plate and the Anatolian sub-plate (present-day Türkiye). This event is called the Alpine-Himalayan (Alpide) orogeny. Along the zone of convergence, from Morocco to Southeast Asia, rock strata have been crumpled, uplifted, and faulted, and they have undergone intense metamorphism as a result of being subjected to intense heat and pressure. Oceanic crust was forced north beneath Eurasia, generating volcanic activity in France, Italy, and the UK. The Tethys Ocean was closed, and mountain ranges were created, including the Atlas, Pyrenees, Alps, Carpathians, Armenian Highlands, Caucasus, and Himalayas. Distant ripples of this orogeny produced more gentle folds, such as the Weald-Artois Anticline in southern England and northern France.

A different mechanism drives the Andean orogeny in South America. Far to the west, on the floor of the Pacific Ocean, new oceanic crust is being created at the East Pacific Rise, a divergent plate boundary. This produces seafloor spreading, the process where ocean crust forms from rising magma. This pushes the eastern boundary of the Nazca Plate eastwards at a very fast rate – in geological terms – of 3.7 cm (1.5 in) per year. The plate subducts at a shallow angle beneath the continental South American Plate, producing the Peru–Chile Ocean Trench and the Andes Mountains.

Greater insights

In recent years, improved remote sensing and seismic reflection technologies have helped geologists more accurately map surface topography and deep features, such as faults and ore deposits within orogens. This research is important because orogens help scientists reconstruct past geological events, especially the movement of Earth's tectonic plates. It also gives key insights into the location of valuable mineral and petroleum deposits, and helps seismologists predict areas at greatest risk from earthquakes. ∎

The Southern Andes, which stretch between Chile and Argentina, are the product of a mountain-forming process that has been ongoing for about 50 million years.

A TENDENCY TO ACCUMULATE HEAT AT THE SURFACE

THE GREENHOUSE EFFECT

IN CONTEXT

KEY FIGURE
Svante Arrhenius
(1859–1927)

BEFORE
1750s Scottish physicist Joseph Black discovers "fixed air" (carbon dioxide).

1824 According to Jean-Baptiste Joseph Fourier, Earth's atmosphere traps heat much like a greenhouse.

1838 Claude Pouillet discovers that water vapour and carbon dioxide absorb heat radiation.

AFTER
1985 Ice cores from Antarctica show that carbon dioxide levels and temperatures have risen and fallen simultaneously over the past 150,000 years.

2023 Naturally occurring greenhouse gases hit record highs, according to the World Meteorological Organization.

The greenhouse effect is a natural phenomenon whereby gases in Earth's atmosphere trap heat from the Sun. Nitrogen, oxygen, and argon, which do not absorb or emit radiation, make up 99.9 per cent of the gases in Earth's atmosphere. The other 0.1 per cent consists of water vapour, carbon dioxide (CO_2), methane, nitrous oxide, and ozone; these are the "greenhouse gases" with heat-trapping properties. The greenhouse effect is essential to maintaining a habitable planet. Without it, much of life would struggle to survive the icy conditions. This was the case some 650 million years ago, when photosynthesizing cyanobacteria

PHYSICAL GEOGRAPHY   81

See also: Glaciation and ice ages 86–89 ▪ Palaeoclimatology 126–27 ▪ Extreme weather 142–43 ▪ Climate change 232–39 ▪ Climate migration 254–55 ▪ The Anthropocene 256–59 ▪ Monitoring environmental change 284–89

Garden greenhouses – even those on a more modest scale than the Victorian Palm House at Kew Gardens, UK – convert sunlight to heat in the same way that the planet does.

caused a reduction in greenhouse gases, triggering millions of years of freezing global temperatures.

In 1896, Svante Arrhenius was the first to quantify the contribution of carbon dioxide to the greenhouse effect and to speculate on its role in long-term climate changes.

The greenhouse process
Energy for the greenhouse effect comes from the Sun as sunlight. While some of that is reflected back into space by clouds or ice, most is absorbed by Earth's surface and so warms up the planet. Earth radiates some heat back into space as infrared radiation, but not all of it. Greenhouse gases absorb some of that heat and re-emit the absorbed energy in all directions, including back down towards Earth's surface. This process warms the lower atmosphere and effectively traps the heat, like the glass in a greenhouse. Put simply, the more carbon dioxide there is in the atmosphere, the hotter it gets.

Greenhouse pioneers
A full understanding of the science behind the greenhouse effect did not come until the 19th century. In the 1820s, French mathematician Jean-Baptiste Joseph Fourier calculated that a planet the size of Earth, at its distance from the Sun, should be considerably colder than it was if it were warmed solely by »

[Earth's] temperature can be augmented by the interposition of the atmosphere.
Joseph Fourier
"On the Temperature of the Terrestrial Globe...", 1827

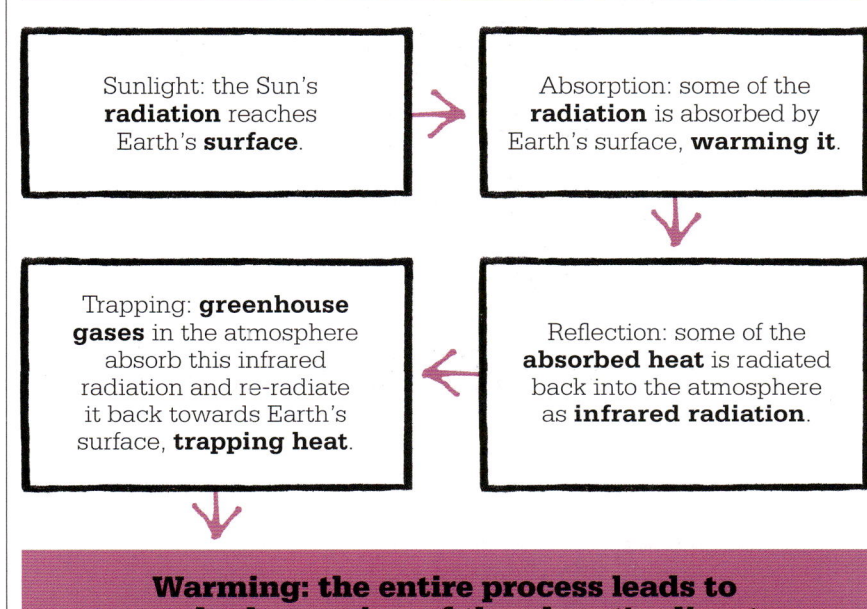

Sunlight: the Sun's **radiation** reaches Earth's **surface**.

Absorption: some of the **radiation** is absorbed by Earth's surface, **warming it**.

Reflection: some of the **absorbed heat** is radiated back into the atmosphere as **infrared radiation**.

Trapping: **greenhouse gases** in the atmosphere absorb this infrared radiation and re-radiate it back towards Earth's surface, **trapping heat**.

Warming: the entire process leads to a gradual warming of the planet's climate.

direct solar radiation. In what is perhaps the first theory about the greenhouse effect, Fourier suggested that Earth's atmosphere might act as some kind of insulator.

Fourier himself never used the expression "greenhouse effect", although he did observe a greenhouse when studying the experimental work of Horace Bénédict de Saussure, a Swiss physicist. In 1767, de Saussure had lined the inside of a vase with blackened cork before inserting small panes of clear glass at regular intervals, with a thermometer in each compartment. The top of the vase was then exposed to midday sunlight. De Saussure discovered that the temperature of the air in the compartments lower down the vase was higher than in those at the top. Fourier realized that if gases in the atmosphere could form a barrier like de Saussure's glass panes had, they would have a similar effect on the temperatures of the planet.

In 1838, French physicist Claude Pouillet developed and revised Fourier's work on the surface temperature of Earth, creating the first real mathematical treatment of the greenhouse effect. He speculated that water vapour and carbon dioxide could be trapping infrared radiation in the atmosphere, warming the planet sufficiently for it to support living organisms.

Research gathers pace

Using only an air pump, two glass tubes, and thermometers, American amateur scientist, inventor, and women's rights campaigner Eunice Newton Foote was the first to prove the insulating effect of carbon dioxide. After exposing to the Sun two glass tubes – one containing air, the other carbon dioxide – Foote compared the inside temperatures over the course of several hours, noting that they were considerably higher – and "many times as long in cooling" – in the carbon dioxide-filled tube. In 1856, she published her conclusions in the *American Journal of Science and Arts*, stating that carbon dioxide and water vapour both absorb heat

> The atmospheric stratum... exercises a greater absorption upon the terrestrial than upon the solar rays.
> **Claude Pouillet**
> "Memoir on the Solar Heat...", 1838

from the Sun. She also suggested that increasing the amount of carbon dioxide in the atmosphere would change Earth's climate.

In 1859, probably unaware of Foote's earlier paper, Irish physicist John Tyndall devised a complicated laboratory apparatus to show that gases – including carbon dioxide and water vapour – absorb heat. The inevitable consequence of the atmosphere admitting solar energy but not releasing it, he concluded, is a warmer planet.

The hot-house theory

Svante Arrhenius's scientific writing extensively quotes Pouillet's research. In 1896, the Swedish scientist published a paper that used the principles of physical chemistry to estimate how increases in atmospheric carbon dioxide would affect Earth's surface temperature. Although Arrhenius did not use the word "greenhouse" in his 1908 book *Worlds in the Making*, he did refer to the concept advanced by Fourier, Pouillet, and Tyndall as the hot-house theory, because these scientists all "thought that the atmosphere acted after the manner of the glass panes of hot-houses".

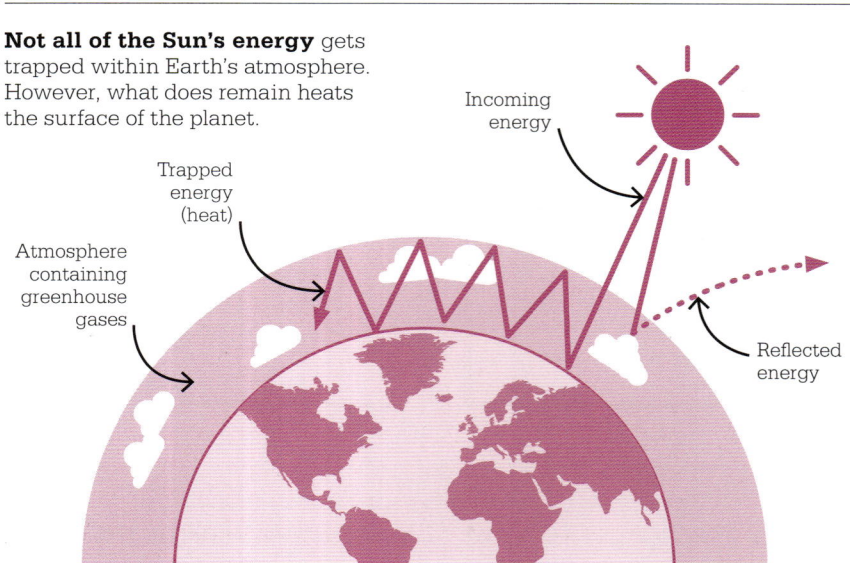

Not all of the Sun's energy gets trapped within Earth's atmosphere. However, what does remain heats the surface of the planet.

The carbon dioxide level in the atmosphere during the Triassic period was five times higher than today. The planet was much warmer, with higher sea levels and altered ecosystems.

Arrhenius went on to indicate that if the level of carbon dioxide (or carbonic acid, as it was then known) fluctuated, so did Earth's air temperature. He calculated that a 50 per cent reduction in carbon dioxide in the atmosphere would lead to a temperature drop of about 4°C (7.2°F); conversely, a doubling of the percentage of carbon dioxide would result in the temperature increasing by the same amount.

Effects on the climate

A close friend of Arrhenius was Swedish meteorologist Nils Gustaf Ekholm, of Stockholm's Central Meteorological Institute. In 1901, he wrote an extensive paper for the *Quarterly Journal of the Royal Meteorological Society* on the variations of the climate through the ages, and the role played by carbon dioxide. The paper included the first reference to a "greenhouse". Ekholm wrote that the atmosphere had a dual role to play with regards to Earth's surface temperature. Firstly, it allowed the light rays of the Sun to penetrate with ease, "like the glass of a green-house"; secondly, it absorbed much of the infrared radiation (or dark rays) emitted from the ground, converting it into heat. He also established that the atmosphere stored heat between the warm ground and the cold space, regulating temperature variations.

Ekholm also recognized that humans could influence the temperature of Earth. He noted that, at rates present in 1899, the burning of coal could eventually double the quantity of carbon dioxide in the atmosphere; and this, he suggested, would cause a very obvious rise in the mean temperature of Earth. However, he did not necessarily consider this a bad thing. By controlling the production and consumption of carbon dioxide, humans could regulate Earth's climate and avoid another ice age. Little did Ekholm realize that anthropogenic emissions of carbon dioxide would subsequently lead to a climate crisis of mammoth proportions. ∎

Svante Arrhenius

Although he originally trained as a physicist, Svante Arrhenius – who was born near Uppsala, Sweden, in 1859 – later became one of the pioneers of physical chemistry.

Arrhenius also investigated the origin of ice ages. In 1896, this led to him using physical chemistry to estimate how increases in atmospheric CO_2 would cause Earth's surface temperature to rise, through what we now know as the greenhouse effect. He also suggested that human-caused CO_2 emissions from the burning of fossil fuels were sufficient to cause what we now call global warming. This revelation ensured that Arrhenius was firmly at the centre of modern climate science.

In 1903, Arrhenius received the Nobel Prize in Chemistry for his work on the conductivities of electrolytes. He died in 1927.

Key work

1908 *Worlds in the Making*

THE COMPOUND CENTRIFUGAL FORCE
THE CORIOLIS EFFECT

IN CONTEXT

KEY FIGURE
Gaspard-Gustave de Coriolis (1792–1843)

BEFORE
1634 Marin Mersenne and Pierre Petit, both French mathematicians, carry out an experiment to prove the rotation of Earth.

1651 Italian astronomer Giovanni Battista Riccioli suggests that a cannonball in flight will be deflected by the rotation of Earth.

AFTER
1856 US meteorologist William Ferrel describes an atmospheric circulation cell partly powered by the Coriolis effect.

1922 According to British physicist Lewis Fry Richardson, weather conditions could be forecast using mathematical models based on physical phenomena, such as the Coriolis effect.

In 1835, Gaspard-Gustave de Coriolis was studying the apparent force at play in rotating machines – a subject of great interest in the Industrial Revolution era. In his paper "On the Equations of Relative Motion of Systems of Bodies", he presented a mathematical formula that could be applied to any spinning object – from water wheels, to Earth itself.

Early experiments

The earliest debates on the effect of Earth's rotation on moving objects centred on whether the planet was rotating at all. For the Greek philosopher Aristotle, the fact that birds ascending into the air were not immediately swept westwards at high speed was proof that Earth does not rotate.

Such thinking persisted until 1543, when Polish astronomer Nicolaus Copernicus suggested that Earth orbits the Sun and rotates once every 24 hours about its axis. In the decades following the publication of Copernicus's theory, several physicists tried to prove the rotation of Earth by measuring any deflection in a falling object.

One such experiment took place in Germany in 1803, when two mathematicians – the German Carl Friedrich Gauss and the French Pierre Simon de Laplace – measured the deflection of iron pebbles dropped into a 90-m- (295-ft-) deep mineshaft. They calculated the expected deflection if Earth was rotating, arriving at 8.8 mm (0.35 in), very close to the observed deflection of 8.5 mm (0.33 in). It was the first measurement of what would come to be known as the Coriolis effect.

Coriolis's cannon

Every point on Earth makes a complete rotation every 24 hours, but because the planet is a sphere, some points must travel farther – and therefore faster. A point on the

This woodcut depicts mathematicians Mersenne and Petit during their 1634 experiment. They fired a ball from a vertical cannon to see if it would fall back down into the barrel.

PHYSICAL GEOGRAPHY 85

See also: Atmospheric circulation and winds 72–73 ▪ Global ocean circulation 140–41 ▪ Extreme weather 142–43

The military… knows all about the Coriolis force and thus introduces the appropriate correction to all missile trajectories.
Neil deGrasse Tyson
Natural History magazine, 1995

equator travels about 40,000 km (25,000 miles) – the circumference of Earth – while a point at one of the poles travels hardly any distance at all. Latitudes between the equator and the poles rotate at intermediate speeds.

As objects move through these regions of varying speed, their paths are deflected by the Coriolis effect. Imagine a cannon at the equator, pointing north. Because of the rotation of Earth, the cannon is moving east at about 1,600 km/h (995 mph). A projectile fired from the cannon will also move to the east at 1,600 km/h. The further north the projectile travels, the faster it moves eastwards, relative to the speed of the ground's rotation below it. If this is not taken into account, the projectile will land to the east of its intended target.

The impact on weather

Weather patterns such as the trade winds show the impact of the Coriolis effect, which is most significant over long distances or at high speeds. Warm air rising near the equator flows towards the poles. In the northern hemisphere, these warm currents are deflected to the east as they move northwards; in the southern hemisphere, to the west as they move southwards. ■

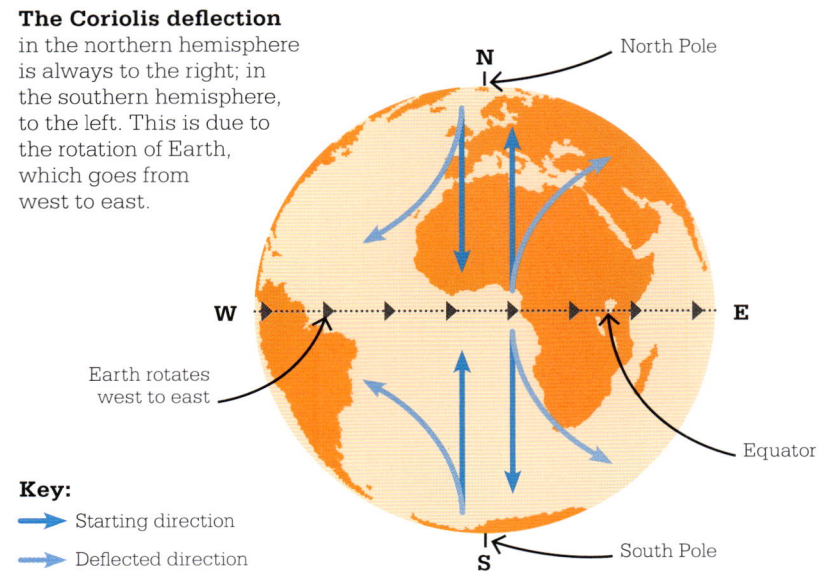

The Coriolis deflection in the northern hemisphere is always to the right; in the southern hemisphere, to the left. This is due to the rotation of Earth, which goes from west to east.

Key:
→ Starting direction
→ Deflected direction

Gaspard-Gustave de Coriolis

The scientist after whom the Coriolis effect is named was born in Paris in 1792, but he grew up in Nancy, eastern France. In 1808, Gaspard-Gustave de Coriolis sat the entrance examination for the renowned École Polytechnique in Palaiseau, a suburb of Paris, placing second of all the students entering that year.

After graduating, he worked with the engineering corps in the Meurthe-et-Moselle district and the Vosges mountains. In 1816, Coriolis accepted a post as a tutor at his alma mater; he continued teaching there until 1838, when he took on the role of director of studies.

The poor health that had afflicted Coriolis since he was a young man became much worse in the spring of 1843, and he died a few months later. His is one of 72 names of noted French scientists and engineers inscribed on the Eiffel Tower.

Key work

1835 "On the Equations of Relative Motion of Systems of Bodies"

THIS GREAT ENGINE SET AT WORK AGES AGO

GLACIATION AND ICE AGES

IN CONTEXT

KEY FIGURE
Louis Agassiz (1807–73)

BEFORE
1787 Alpine rocks in the Jura Mountains lead Swiss geologist Horace-Bénédict de Saussure to propose that they were transported by glaciers.

1815 Swiss hunter Jean-Pierre Perraudin suggests that giant boulders in the Val de Bagnes were left by retreating glaciers.

AFTER
1910–13 Captain Robert Falcon Scott leads the *Terra Nova* Expedition to Antarctica, where his British team carries out important glaciological research.

2022 UNESCO and the World Meteorological Organization (WMO) designate 2025 as the International Year of Glaciers' Preservation.

The glacial theory – that is, the idea that ice cover was previously much more extensive, stretching across vast areas of northern Eurasia and North America – is widely accepted these days. However, when Louis Agassiz first proposed it, in 1837, it was revolutionary.

A radical theory

The presence of large boulders, known as erratics, resting on otherwise flat ground and often far from running water had long puzzled scientists. In the early 19th century, the prevailing view was that they had been deposited by a great flood, such as the one

PHYSICAL GEOGRAPHY

See also: Atmospheric circulation and winds 72–73 ▪ Catastrophism 74 ▪ Isostasy 94–95 ▪ Climatic zones 96–101 ▪ The cycle of erosion 102–07 ▪ Palaeoclimatology 126–27

Louis Agassiz

Born in Môtier, Switzerland, in 1807, Louis Agassiz studied geology and zoology in Paris under the tutelage of Alexander von Humboldt and Georges Cuvier respectively. After returning to Switzerland in 1832, he was appointed professor of natural history at the University of Neuchâtel, where his primary research interest was fossil fish.

While Agassiz was on holiday in the Swiss Alps in 1836, a meeting with Jean de Charpentier and Ignaz Venetz changed the course of his career. He was impressed with their ideas about glaciation and developed them further in his 1840 book *Studies on Glaciers*.

In 1846, Agassiz moved to the US. Two years later, he was appointed professor of zoology and geology at Harvard, where he established the Museum of Comparative Zoology in 1859. Agassiz died in Cambridge, Massachusetts, in 1873.

Key work

1840 *Studies on Glaciers*

---

described in the Bible's Book of Genesis. Known as diluvian theory, this thesis argued that when Noah's flood waters subsided, they left behind erratics; mounds of debris known as moraines; winding ridges of sand and gravel called eskers; and other features – all now known to be associated with glaciation.

Increasingly, though, scientists began to question the diluvian theory. Some argued that erratics and other unusual deposits had been transported by ice. This idea provided a better understanding of landscapes and the dramatic environmental changes produced by ice-sheet formation. The glacial theory also demonstrated that Earth's climate has undergone periodic change.

Early conjectures

Even before Agassiz published his theory, other geologists had considered a glacial origin for erratic boulders. In the first decades of the 19th century, German-Swiss geologist Jean de Charpentier and Swiss naturalist Ignaz Venetz had carried out fieldwork around the active glaciers of the Alps. After examining moraines and other depositional features, both concluded that glaciers had once been larger.

In 1824, Danish-Norwegian scientist Jens Esmark proposed that glaciers and ice sheets had once covered much of Norway and the adjacent ocean. He attributed the origin of moraines and erratics to transportation and deposition by ice, maintaining that if the boulders had been carried by torrents of water, the latter would also have washed away the gravel and sand on which they rested. Esmark also proposed that Norway's fjords had been carved by ice when the glaciers were more extensive.

Swiss geologist Franz Josef Hugi carried out an important experiment in 1827. He marked »

Great sheets of ice, resembling those now existing in Greenland, once covered all the countries in which unstratified gravel is found.
Louis Agassiz
Studies on Glaciers, 1840

The Okotoks Erratic in Alberta, Canada, became dislodged from a mountain during a rockslide. It fell onto a glacier, which moved it to the prairie before melting and leaving it behind.

the rocks of Unteraar Glacier to measure glacial movement, and he was able to confirm that glaciers do flow. In *Principles of Geology* (1830–33), British geologist Charles Lyell also argued that flood waters would not have been capable of carrying large erratic boulders, so these must have been transported on icebergs floating in water. German botanist Karl Schimper, who studied glacial features in his country's Black Forest and Jura regions, proposed the glacial theory in a series of lectures in 1835 and 1836.

Winning over the sceptics

Aware of these scientists' work, Agassiz became convinced that their ideas were correct. In 1837, he presented his own version of the theory at a meeting in Neuchâtel of the Swiss Academy of Sciences. He explained that a global fall in temperature had created a glaciation that had covered Eurasia from the Arctic to the Mediterranean and Caspian seas, and that contemporary glaciers were simply remnants of a once-vast ice sheet.

A change in **Earth's tilt** seasonally varies the amount of **solar radiation** reaching parts of Earth.

Global **temperatures fall**, and polar ice increases.

Since **more solar** radiation is reflected, **temperatures fall** even more.

Greater ice cover increases **surface reflection**.

Ice sheets spread even further.

Agassiz published his findings in 1840, in *Studies on Glaciers*. As well as drawing widespread condemnation from the Christian Church for challenging the diluvian theory, his theory faced criticism from scientists who believed that it contradicted the established view that Earth had been cooling gradually since its origin – so how could it have warmed up again since this widespread glaciation? Gradually, though, influential figures who had previously been sceptics, including Charles Darwin and British geologist William Buckland, were won over.

Solar radiation

From the mid-19th century, the idea that Earth has experienced periods of greater ice cover gradually gained acceptance, but the question of why these ice ages occurred remained unanswered. In 1864, Scottish scientist James Croll proposed that changes in Earth's orbit could increase and decrease the amount of solar energy reaching our planet. When it declined, he suggested, Arctic ice sheets would grow more extensive and reflect more sunlight through a high albedo (surface reflection) value, thereby causing a further fall in temperature and the continued growth of glaciers and ice sheets.

In the 1920s, Serbian astronomer Milutin Milankovitch quantified Croll's hypothesis. He proposed that

The terminal moraine of Alaska's Shakes Glacier sits on the Stikine River. Since the late 17th century, the ice sheet has retreated by about 15 km (9 miles), and it now terminates in Shakes Lake.

The snowball Earth theory suggests that during ice ages such as the Marinoan glaciation, most of the planet was frozen solid, as in this artist's rendition.

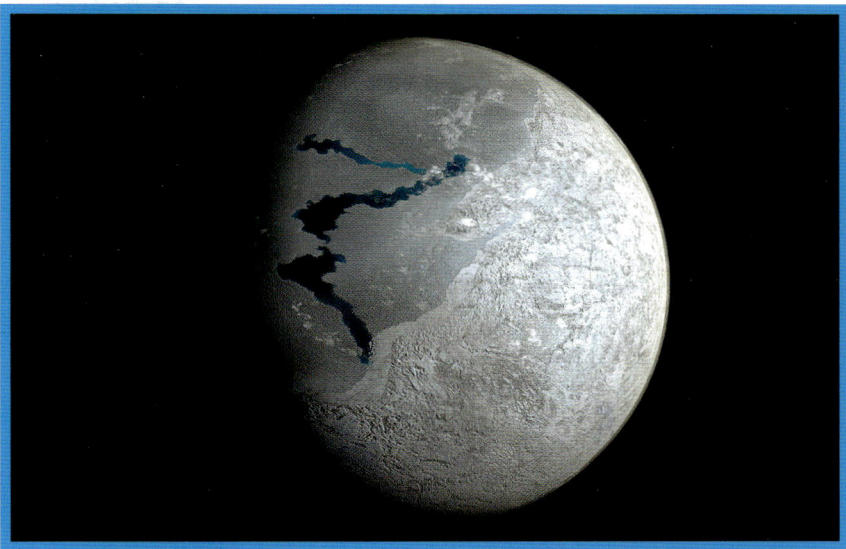

cyclical variations in three elements of Sun–Earth geometry produced predictable changes in the amount of solar radiation reaching Earth. The first of the three variables is eccentricity, or the shape of Earth's orbit, which changes in cycles of 100,000 to 400,000 years. The second is axial tilt with respect to the plane of orbit, which changes over a 41,000-year cycle. The last is precession, the angle at which Earth's axis points, which alters over a 22,000-year cycle. Taken together, these factors result in the Milankovitch cycles, which determine the amount of sunlight reaching different latitudes, so increasing or decreasing global temperatures and ice-sheet extent. The geological record suggests that other factors have also triggered ice ages.

Charting ice ages

Scientists use various methods to chart the fluctuating temperature of Earth's atmosphere and the occurrence of ice ages. Since the 1950s, they have drilled cores from the ice sheets of Antarctica and Greenland. These cores have provided relatively accurate dates for periods of global cooling and warming – glacial and interglacial periods – in the past 800,000 years. Compacted layers of annual snowfall date the ice much like growth rings in a tree. Bubbles of air trapped within the ice give a snapshot of the chemical composition of the atmosphere at any given time, informing palaeoclimatologists when it was colder or warmer.

Other methods of charting Earth's glaciations include analysing nitrogen isotopes in the shells of microscopic marine algae called diatoms, and examining oxygen isotopes in ocean sediments.

It is harder for scientists to date earlier glaciations, but they do know that there have been periods when most of Earth's surface was covered with ice. One of these so-called snowball-Earth episodes occurred during the Huronian ice age, between 2.4 billion and 2.1 billion years ago. During the Cryogenian period (750–600 million years ago), massive volcanic activity is believed to have blotted out sunlight, causing the extreme Sturtian and Marinoan glaciations. Nothing as extensive as these events has occurred since.

The most recent glacial period started about 120,000 years ago and peaked only 20,000 years ago, when temperatures were probably 5°C (9°F) lower than today. For the past 11,700 years, Earth has been in the Holocene interglacial period.

Future glaciations

Since orbital variations can be predicted, it is theoretically possible to forecast their impact on climate. In a 2025 paper, researchers at the University of Cardiff, UK, theorized that the next glacial period would begin in around 11,000 years if no human factors come into play. However, with atmospheric carbon dioxide levels at their highest for 800,000 years due to human activity, greenhouse gas emissions continuing to rise, and rates of glacier loss fast accelerating, this event could be significantly delayed. ∎

> Glaciers… round and polish the rocks that form their foundation.
> **Louis Agassiz**
> *Studies on Glaciers*, 1840

SPECIES IN THE INORGANIC KINGDOM
MINERALOGY

IN CONTEXT

KEY FIGURE
James D. Dana (1813–95)

BEFORE
1546 German mineralogist Georgius Agricola publishes *On the Nature of Rocks*; it becomes a founding document in the study of minerals.

1801 *Treatise on Mineralogy*, by French mineralogist René Just Haüy, also known as Abbé Haüy, addresses crystal structure.

AFTER
2002 The ninth and latest edition of the Nickel–Strunz classification system for minerals is published.

2025 NASA's Perseverance rover on Mars finds carbonate and sulphate minerals that are known to need water to form on Earth.

The surface of Earth is covered in a crust of solid rock. There are several hundred types of rock, each one with a particular composition of several different crystalline minerals cemented together. Mineralogy seeks to classify these minerals so they can be identified and used to further understand the formation of rocks and other processes in Earth's crust.

A founding figure in the field of mineralogy is James D. Dana. In 1837, he published *A System of Mineralogy*, in which he classified minerals in different ways, primarily their physical properties, their crystalline forms,

See also: Karst landscapes 108–09 ▪ The structure of Earth 112–13

and their chemical composition. Eleven years later, in 1848, Dana developed the system further, resulting in the *Manual of Mineralogy*. The first detailed, standardized classification of minerals, it described about 200 minerals, and has been a standard text ever since.

Mineral classes

Dana's *Manual of Mineralogy* also forms the basis of the Nickel–Strunz classification, the modern system that organizes minerals by chemical composition and atomic structure into ten classes, which include sulphides, oxides, carbonates, phosphates, and silicates. It was developed in the 1940s by the German mineralogist Karl Hugo Strunz. Since then, the system has been updated several times with additional contributions from a Canadian colleague named Ernest Nickel.

The Nickel–Strunz classification cannot contain a definitive list of minerals. Estimates range between 4,000 and 6,000 individual mineral types, and more are being added every year. Most of these minerals are extremely rare and are not particularly involved in forming rocks. In fact, 90 per cent of all rocks in Earth's crust are types of silicate. These minerals are »

Atoms arranging themselves in repeating, **three-dimensional patterns** form a **crystal**.

→

Crystals with a definite **chemical composition** and structure form a **mineral**.

↓

The aggregation of different minerals results in the formation of rocks.

←

Geological processes lead to the formation and clustering of **different minerals**.

The Nickel–Strunz classification (2002)

Class 1	Native elements	Metals such as gold, silver, and copper; also, carbon and sulphur.
Class 2	Sulphides and other sulphur-rich minerals	Pyrite (iron sulphide) and ores such as galena, chalcocite, and sphalerite, which are sources of lead, copper, and zinc respectively.
Class 3	Halides	Minerals containing the halogen ions fluoride, chloride, bromide, and iodide. The most abundant is halite, or common salt (sodium chloride).
Class 4	Oxides and hydroxides	Minerals such as hematite, magnetite, and periclase, which contain oxygen.
Class 5	Carbonates and nitrates	Carbonates include calcite and aragonite, the primary constituents of limestones. Nitrates are relatively rare.
Class 6	Borates	Minerals that are rich in boron, including borax.
Class 7	Sulphates; minerals with molybdenum, tungsten, and chromium ions	Sulphates include gypsum, widely used in plaster.
Class 8	Phosphates, arsenates, and vanadates	Phosphates include apatite, a key component of teeth and bones; arsenates and vanadates are oxygen compounds containing arsenic and vanadium respectively.
Class 9	Silicates	Minerals based on the silicate ion, which can form complex arrangements of sheets and chains. The simplest mineral in the class is silica, the main constituent of quartz and sandstones. Others include micas, olivines, and feldspars – all important in the formation of rocks.
Class 10	Mineraloids	Naturally occurring substances that do not have crystalline structures. Volcanic glass, such as obsidian, is one example. Others are amber, tar, and petroleum.

compounds of silicon and oxygen, with a great variety of crystal structures containing a wide range of metallic cations (positively charged ions) in an equally wide variety of proportions.

Visible features

It takes an expert eye to identify a mineral from appearance alone, but that is the starting point of mineralogy. Every mineral is described by its physical properties, the most apparent of which are the colour and lustre of its crystals. However, since crystals transmit and refract light, as well as simply reflecting it, the colour can be misleading. For this reason, in order to be sure of their assessment, mineralogists perform a streak test. This involves scratching the sample on a hard surface, such as a ceramic tile. That creates a streak of powder, the colour of which is often somewhat different from that of the larger sample. For example, pyrite looks golden, but its streak is a greenish-grey.

Crystal colouring can also be the result of impurities in the crystal lattice (the repeating pattern of atoms). For example, an emerald gem only appears green because of the presence of a few chromium ions in a crystal of beryl. Beryl is a colourless silicate made up mostly of beryllium and aluminium ions. Pure beryl has a white streak, and so does the green emerald variety.

Other visible properties are lustre and habit. A crystal's lustre describes how light interplays with its surface. There are several different options. Minerals can be waxy, silky, greasy, earthy, vitreous (meaning glassy), or adamantine, which suggests they are sparkling jewels. All pure metals and several compound minerals, such as pyrite, have a metallic lustre. However, only native metals – that is, those found pure in nature, such as gold and silver – are regarded as minerals.

Mineral samples may also display a habit, which is a characteristic shape taken on by larger samples. There are a large number of habits. For example, native silver is arborescent, or tree-like, while calcite is oolitic, which means it is made of small spheres; and beryl is prismatic, or resembling a prism, with well-defined parallel faces.

Optical mineralogy

To make a definitive identification, it is often necessary to examine minerals under a microscope. This is especially useful when looking at several minerals accreted, or accumulated, into a rock. The rock sample is sliced into ultrathin sections, and a petrographic microscope is able to direct light through the section into the eyepiece. The microscope can reveal how refractive each crystal is and other optical features, such as birefringence, where certain minerals, such a calcite, split light

> The chemical composition of a mineral is the most fundamentally important fact about it.
> **James D. Dana**
> *Manual of Mineralogy*, 1848

James Dwight Dana

Born in 1813 in Utica, New York state, US, James D. Dana worked as a teacher in the US Navy and in a chemistry laboratory after graduating from Yale College, Connecticut, in 1833. Within four years, he had produced the first edition of *A System of Mineralogy*. Later, he served as the geologist for a US expedition that spent four years traversing the Pacific Ocean. For more than a decade afterwards, Dana prepared reports of the findings of the expedition. His knowledge of the US West Coast led him to be widely consulted during the California Gold Rush.

In 1850, Dana was appointed professor of natural history and geology at Yale College, where much of his research focused on Hawaii's volcanic activity. In 1892, his son Edward, also a mineralogist, edited a revision of the *System*. James D. Dana died in 1895.

Key works

1837 *A System of Mineralogy*
1848 *Manual of Mineralogy*

PHYSICAL GEOGRAPHY

Crystal systems

A crystal is built from repeating units of atoms that connect to create a lattice. All true minerals feature this structure: some form crystals several metres (feet) across, while others are barely visible under a microscope. There are seven crystal systems, each defined by the relative lengths and angles of the crystal's axes, imaginary lines that describe the symmetry and shape of the crystal's internal structure.

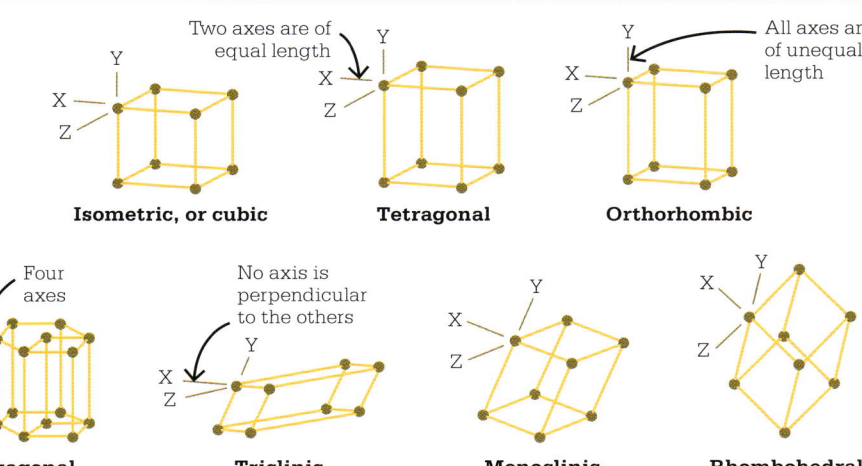

rays, doubling the image. Most revealing is the use of polarized light, which alters the colour and character of particular minerals as an aid to their identification.

Hardness and tenacity

Another important property of a mineral is its hardness, which is most easily assessed using the Mohs scale, invented by German mineralogist Frederick Mohs in 1812. This is a comparative scale, where a mineral is tested by being scratched by 10 reference minerals. The first reference mineral is talc. Since it does not scratch any other mineral, it has the lowest possible hardness of 1. A human fingernail is around 2.5 on the Mohs scale. The hardest mineral of all is diamond, which has a maximum score of 10.

Diamond is famously tough and hard to fracture, but hard minerals are not necessarily strong.

Structure is a key factor in mineral identification. Pyrite crystals, for example, have a cubic structure, while quartz crystals are typically hexagonal with a six-sided pyramid at each end.

The mineralogical property of tenacity describes how likely they are to break up or deform. For example, quartz is a brittle mineral, gold is malleable and bends instead of breaking, while mica forms thin sheets that are surprisingly flexible.

A valuable tool

Classifying minerals allows geologists to understand the properties of rocks and find clues about how and where they formed. However, mineralogy also has practical value as a tool for locating essential natural resources – from ores to building materials. In addition, the discipline plays a key role in the study of industrial applications – for example, naturally occurring crystals can be used in lasers, electronics, and optical devices. Ongoing advances in the field continue to inspire new materials and technologies. ∎

A SOLID CRUST, RESTING UPON A PLASTIC SUBSTRATUM
ISOSTASY

IN CONTEXT

KEY FIGURE
George Biddell Airy
(1801–92)

BEFORE
1743 Swedish scientist Anders Celsius calculates that the sea level in the Gulf of Bothnia, Sweden, is falling at a rate of 1.4 cm (0.55 in) a year.

1802 The Great Trigonometrical Survey of India gets under way. Over the next 69 years, it produces vast amounts of geodesic data.

AFTER
1882 US geologist Clarence Dutton introduces the term "isostasy" and explains how the subsidence of deltas restores the balance between areas of erosion and deposition.

1909 Using seismic waves, Croatian scientist Andrija Mohorovičić discovers the boundary between Earth's crust and mantle, now known as the Moho Discontinuity.

From around the 4th century BCE and the age of classical Greece, scientists have sought to determine the precise shape, size, and density of Earth – a discipline known as geodesy. In the 1730s, when French mathematician Pierre Bouguer conducted geodesic surveys in the Andes for Louis XV, he had expected the great mass of Mount Chimborazo to increase his gravity readings. However, that was not the case, which led Bouguer to conclude that the mountain must have hidden cavities within, reducing its density. Geologists, however, doubted this explanation. More than a century later, in 1855, George Biddell Airy proposed his theory of isostasy as the solution to the gravity anomaly.

Determining Earth's crust

Geodesists had collected large quantities of gravity data, combining survey triangulation with astronomical observation to determine Earth's shape. Airy examined this data to develop his theory that the rocks of Earth's outer crust rest on an inner

Mount Chimborazo was the setting of Bouguer's experiments. At the time, this volcano in Ecuador was thought to be the highest mountain on Earth.

There can be no other support than that arising from the downward projection of a portion of the earth's light crust into the dense lava.
George Biddell Airy
Royal Society paper, 1855

PHYSICAL GEOGRAPHY   95

See also: Geophysical mapping 38–39 ▪ Orogeny 76–79 ▪ Glaciation and ice ages 86–89 ▪ The structure of Earth 112–13 ▪ Plate tectonics 114–21

The exposed marine terraces at Tongaporutu Estuary, on New Zealand's North Island, were formed as a result of isostatic adjustment. Once at sea level, they are now high above the water.

sphere "filled with a fluid of greater density than the crust" and that "beneath mountain ranges, light crustal material replaces heavy lava". As a result, the positive gravitational attraction of mountains and high plateaus is counterbalanced by the presence of less dense rock at depth.

A state of balance

Earth's solid outer shell, or lithosphere, comprises the crust and the topmost layer of the mantle – a layer of rock that makes up the bulk of the planet's interior. The lithosphere rests on the asthenosphere, the thick, pliable section of the mantle.

While buoyancy forces push the lithosphere up, gravity pulls it down. When these forces balance, the lithosphere is said to be in a state of isostasy. This is akin to a floating iceberg. Most of the iceberg is always submerged, and if heavy snow falls, its submerged section will extend even deeper.

When tectonic movement pushes plates together, the lithosphere thickens. The convergence of the Indian and Eurasian continental plates has built the Himalayas, while the subduction of the Nazca oceanic plate beneath the South American continent has created the Andes. Both processes are still ongoing and have produced crust that is about 70 km (43.5 miles) thick.

As mountains erode, Earth's crust gradually rises to maintain isostatic equilibrium. If thick ice sheets cover land, the land is depressed. Conversely, when ice sheets melt, there is a slow post-glacial bounce-back, with the land rising relative to sea level. This explains raised landforms such as wave-cut platforms and cliffs.

In addition to shining a light on Earth's geological history and explaining its topography, isostasy helps scientists to predict the impact of sea level rises produced by climate change. ■

George Biddell Airy

British mathematician George Biddell Airy was born in Alnwick, Northumberland, in 1801. After graduating from Trinity College, Cambridge, in 1826 he became professor of mathematics there. In 1835, Airy was appointed Astronomer Royal, a position that also involved leading London's Royal Greenwich Observatory.

Among Airy's many achievements was explaining how optical devices distorted images of stars, producing bright disks (named after him); establishing a new method for calculating Earth's density; and helping to chart the orbits of planets.

In 1884, Airy was instrumental in establishing Greenwich as the Prime Meridian. He remained at the observatory until 1891, and died the following year.

Key work

1855 "On the Computation of the Effect of the Attraction of Mountain-masses"

TEMPERATURE, PRECIPITATION, AND SEASONAL CYCLE

CLIMATIC ZONES

CLIMATIC ZONES

IN CONTEXT

KEY FIGURE
Wladimir Köppen
(1846–1940)

BEFORE
6th century BCE According to the Ancient Greek philosopher Parmenides, there are five climatic zones based on latitude: one torrid, two temperate, and two frigid.

1842 British botanist Richard Brinsley Hinds investigates the influence of temperature and humidity on vegetation type.

AFTER
1967 Leslie Holdridge updates the bioclimate classification he published 20 years earlier.

2016 The European Space Agency launches Sentinel-3A, the first of a series of four satellites monitoring ocean and land temperatures. Two more launches follow in 2018 and 2025, with Sentinel-3D scheduled for 2028.

No two locations on Earth have precisely the same climate – that is, the same complex combination of temperature, precipitation, seasonality, and wind. However, there are clear patterns of similarity and difference. Ever since humans have travelled, they have attempted to recognize, define, and simplify these patterns.

Köppen's climate zones

Modern climate classification dates from 1884, when Russian-German geographer Wladimir Köppen published a world map and definitions of climate zones, which he modified in 1900, 1918, and 1936. Based on average monthly data for temperature and precipitation, he produced mathematical values to define five major climate types and several subsidiary zones. Rudolf Geiger, a German meteorologist, refined Köppen's classification with amendments in 1954 and 1961, so the system is now known as the Köppen-Geiger classification.

Köppen's five primary zones are tropical (A), dry (B), temperate (C), continental (D), and polar and alpine (E). With the exception of

[Climate] influences the character of native vegetation, soil, drainage, and... the nature of the terrain.
Glenn Trewartha
An Introduction to Climate, 1968

B, which is defined by dryness (combined evaporation and precipitation), each zone is defined by temperature.

Tropical zones have an average temperature of 18°C (64°F) or more in the coolest month. Temperate climates have an average temperature between 0 and 18°C (32–64°F) in the coldest month; they also have at least one month averaging more than 10°C (50°F). Areas that experience at least one month below 0°C and one above 10°C are defined as continental. Polar and alpine climates are

Wladimir Köppen

Born in St Petersburg, Russia, in 1846, Wladimir Köppen spent much of his youth in the Crimean peninsula, where the floral variety awakened his interest in natural sciences. He studied botany at St Petersburg University and completed his doctorate – on the relationship of plant growth to temperature – at Germany's Heidelberg University.

In 1875, Köppen was appointed head of weather telegraphy at the German Naval Observatory in Hamburg, where he remained until his retirement in 1919. He then moved to Austria, where he and Rudolf Geiger co-edited the five-volume *Handbook of Climatology*.

An advocate for the Esperanto language, Köppen also supported reform in land ownership and education. He died in Graz, Austria, in 1940.

Key work

1936 *Handbook of Climatology* (co-edited with Rudolf Geiger)

PHYSICAL GEOGRAPHY

See also: Atmospheric circulation and winds 72–73 ▪ The Greenhouse effect 80–83 ▪ Palaeoclimatology 126–27 ▪ Global ocean circulation 140–41 ▪ Extreme weather 142–43

those that experience an average temperature less than 10°C in every month of the year.

Köppen further divided four of these groups (he excluded polar) into categories based on seasonal precipitation, and into further subcategories based on seasonal temperature. His classification has 31 categories in total.

Climate zones in practice

A few examples from around the world illustrate how the classification works in practice. New York City, US, is classified as Cfa. The "C" denotes that it is temperate; the "f", that precipitation is fairly evenly distributed throughout the year; and the "a" indicates that the average temperature of the warmest month is 22°C (71.5°F) or more.

Kyiv, Ukraine (Dfb), has a humid continental climate with fairly evenly distributed precipitation and warm summers; the temperature of each of the warmest four months is 10°C or above, but the warmest month is less than 22°C.

Lagos, Nigeria, lies in the tropical savannah zone (Aw), with rainfall in the driest month less than 60 mm (2.3 in).

Alternative classifications

The Köppen-Geiger classification is the most commonly used, but there are others. US geographer Glenn Trewartha published an alternative in 1966, revising it in 1980. He redefined mid-latitude climates – Köppen's temperate group C – into three categories: subtropical, which has eight months or more with an average temperature of 10°C or higher; temperate (four to seven months »

Climate-zone classification

Köppen's primary climate zones are loosely latitudinal, with tropical zones around the equator, and polar zones in the Arctic and Antarctic regions. Dry zones are north and south of tropical zones, while temperate and continental zones lie between dry areas and polar zones. Climate zones are further defined by precipitation and temperature variations.

Key:
- Tropical
- Dry
- Temperate
- Continental
- Polar

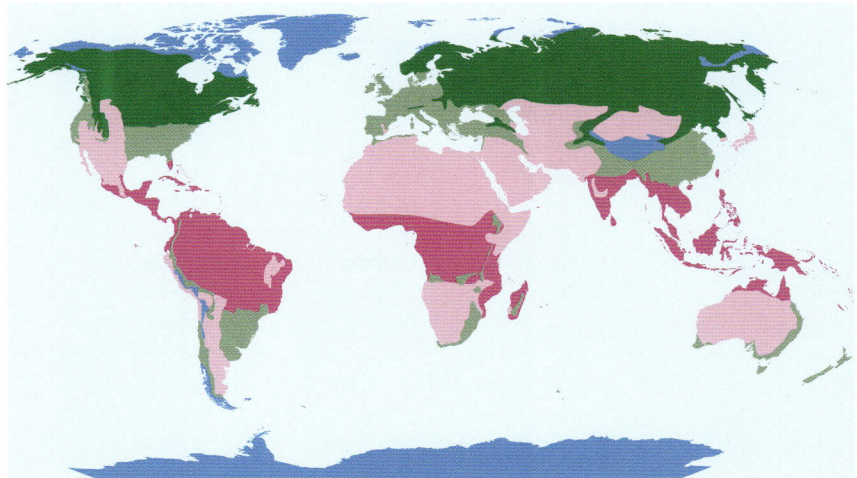

Main climate zones	Precipitation patterns	Temperature variations
A (Tropical)	f (Rainforest) m (Monsoon) w (Savannah, dry winter) s (Savannah, dry summer)	
B (Dry)	W (Arid desert) S (Semi-arid steppe)	h (Hot) k (Cold)
C (Temperate)	w (Dry winter) f (No dry season) s (Dry summer)	a (Hot summer) b (Warm summer) c (Cold summer)
D (Continental)	w (Dry winter) f (No dry season) s (Dry summer)	a (Hot summer) b (Warm summer) c (Cold summer) d (Very cold winter)
E (Polar and alpine)	T (Tundra) F (Ice cap)	

with a temperature of 10°C), and boreal (one to three months). The Trewartha classification better represents biome distinctions within temperate climates and provides a more detailed breakdown of growing season length.

US climatologist Leslie Holdridge adopted a different approach. Published in 1947, his life-zones classification assumes that both soil and climax vegetation – or stable, well-established flora – are mappable once the climate is known. Holdridge's system is based on three main variables: total annual precipitation, biotemperature (considering temperatures only between 0 and 30°C [32–86°F], when photosynthesis occurs), and potential evapotranspiration (PET) ratio – that is, potential evapotranspiration divided by rainfall. PET is the amount of evaporation and transpiration that would occur if water were unlimited.

Holdridge's classification includes 38 categories, such as boreal wet forest, cool temperate steppe, warm temperate thorn scrub, and tropical rainforest. His scheme is useful for modelling how global climate change will affect the ability of plants and animals to survive in different regions in the future.

If you can't measure it, you can't manage it.
María Fernanda Espinosa Garcés
President of the UN General Assembly, on space-based climate data, 2018

Creating a climate

Latitude, altitude, prevailing winds, proximity to water, oceanic circulation, topography, and vegetation combine to produce distinct climates. The nearer a place is to the equator, the more direct sunlight it receives over the course of a year. With increasing latitude, solar radiation arrives at a more oblique angle, spreading its energy and resulting in less heat. Because of Earth's axial tilt, the latitude where the Sun is directly overhead shifts between 23.5 degrees north (boreal summer) and 23.5 degrees south (austral summer), causing the seasons, which are more marked closer to the poles.

If no other factors came into play, temperature would gradually decrease with distance from the latitude of the overhead Sun. The situation is, however, not that simple. Heat is redistributed around the atmosphere by wind circulation cells, which influence climate greatly. For example, southeasterly trade winds blow across subtropical southern Africa. The winds bring warm, moist air – and much rain – from the southern Indian Ocean to Africa's east coast. The winds then blow over the hot, dry interior before reaching the west coast, where they produce little rain. Vilankulos, in Mozambique, on the east coast of Africa, and Walvis Bay, in Namibia, on the west coast, are at the same latitude, yet they receive 780 mm (30.7 in) and 13 mm (0.5 in) of rain annually respectively. Köppen-Geiger defines Vilankulos's climate as tropical savannah and Walvis Bay's as cold desert.

Air temperature falls by about 6.5°C (11.7°F) for every 1,000 m (3,280 ft) rise above sea level. That explains why the summit of Mount Kilimanjaro – despite being close to the equator – is covered with snow and ice, and why mountainous parts of the tropics are relatively cool.

Mountains also affect rainfall. The uplift of a moist airflow on the windward side of a mountain

Namibia's Walvis Bay area sits in the Naukluft Mountains' rain shadow, and the southeasterly trade winds do not carry much rain here. The area does, however, experience frequent fog.

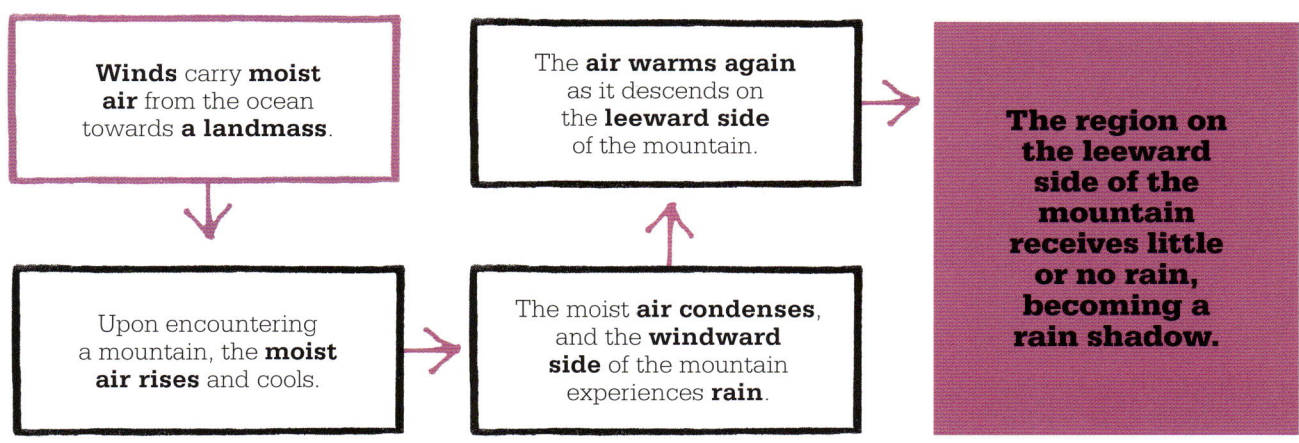

triggers rain. As the air descending on the leeward side warms, it becomes less humid, producing far less moisture. This creates a "rain shadow" – that is, an area of drier weather. As a result, two locations at the same altitude – one on the windward and the other on the leeward side of a mountain – will have different levels of precipitation.

Oceans and vegetation

Water has a higher heat capacity than land, so it can absorb more heat but takes longer to warm and cool. This quality allows it to regulate air temperatures over adjacent land areas, cooling them in summer and warming them in winter. Farther from the ocean, this influence wanes. Despite being at the same latitude, Moscow is 1,900 km (1,180 miles) from the Atlantic Ocean, while Glasgow, UK, is less than 200 km (125 miles) away. The average February temperature is –7.6°C (18.3°F) in the Russian capital, but 3.3°C (37.9°F) in Glasgow; in July, it is 20°C (68°F) and 14.2°C (57.5°F) respectively. Moscow has a warm-summer, humid, continental climate (Dfb), while Glasgow is temperate marine (Cfb). Just as the Hadley, Ferrel, and polar cells redistribute heat around Earth's atmosphere, so surface currents and deep-sea circulation shift warmth around the oceans.

Vegetation also has an impact on climate. Since plants release water vapour into the atmosphere through evapotranspiration, they contribute to a rise in humidity, which can produce cloud cover. This, in turn, reduces the amount of sunlight reaching the ground and increases rainfall. If large tracts of forest are cleared, rainfall will decrease. In addition, temperatures beneath a forest canopy are significantly lower. This is due partly to the shade created, but also to the process of transpiration, which removes heat from the air.

Monitoring changes

Climate classification provides an understanding of the changing distribution of plant and animal life, and how habitable places are for people. It informs decisions on a range of issues – from the best crops to grow, to house construction.

The twin crises affecting climate and biodiversity mean that it is vital to gather data, and technological advances are helping with this task. Computers model the impact of greenhouse gas emissions on temperature, rainfall, ocean currents, and sea levels. Satellites are invaluable tools for tracking sea- and land-surface temperatures, atmospheric composition, and vegetation growth. They also track the development of storms and monitor deforestation and desertification. ■

A cloud of moisture shrouds an area of tropical rainforest. Water evaporating from the leaves of plants gets trapped in the thick canopy, leading to the creation of this misty cover.

STRUCTURE, PROCESS, AND TIME
THE CYCLE OF EROSION

THE CYCLE OF EROSION

IN CONTEXT

KEY FIGURE
William M. Davis
(1850–1934)

BEFORE
1674 French scientist Pierre Perrault studies the discharge of the River Seine, comparing it with the rainfall, evaporation, and transpiration in its drainage basin.

1760s A theory by French naturalist Jean-Étienne Guettard suggests that streams destroy mountains and deposit sediment on floodplains downstream.

AFTER
1914 US geologist Grove Karl Gilbert investigates transport of debris via running water.

2007 A team at the University of Hull, UK, introduces the CAESAR landscape evolution model; it forecasts the impact of erosion and deposition on drainage networks.

The first theory of landscape evolution was advanced by William M. Davis in 1899, when he outlined a model he called the cycle of erosion, or geographical cycle. The site of erosion is the drainage basin, or catchment – that is, the area in which a network of water courses receives surface runoff, as well as throughflow and groundwater flow, which drain below ground through soil and rock, respectively.

The stages of evolution

After recognizing rivers and streams as the main agents responsible for changing landscapes, Davis identified three stages of landscape evolution: youth, maturity, and old age. He suggested that a youthful landscape has the greatest elevation above sea level. It features fast-flowing rivers, channelled along steep-sided V-shaped valleys, which carry the products of eroded bedrock rapidly to the ocean. Because they are so energetic, these rivers do not deposit sediment on their way to the sea, so they do not build floodplains. Interlocking spurs (jutting ridges that appear to overlap) develop due to downward erosion. The pattern of rivers and tributaries here is relatively simple. Davis considered the Himalayas to be an example of a youthful landscape.

As a landscape matures, Davis argued, its overall elevation reduces. At this stage, however, the difference in altitude between hilltops and valley bottoms is at its greatest. This is because the action of energetic flowing water increases the depth of valley

---

> No rocks are unchangeable; even the most resistant yield under the attack of the atmosphere, and their waste creeps and washes downhill as long as any hills remain.
> **William M. Davis**
> "The Geographical Cycle", 1899

William Morris Davis

Considered the father of geomorphology, William M. Davis was born in 1850 to a Quaker family in Philadelphia, US. After graduating from Harvard University, where he studied geography and geology, in 1870 he moved to Argentina, where he worked for three years as a meteorologist at Córdoba's Central Observatory.

From 1878, Davis taught at Harvard: first, geology; then, from 1885, physical geography. Apart from brief periods lecturing at the universities of Berlin and Paris, he remained at Harvard until 1911. After retiring, Davis devoted much of his time to field studies, especially of coral reefs, the origin of which fascinated him. A visiting lecturer at other universities, including Berkeley, Stanford, and Oregon, he also continued to write, publishing more than 500 papers and books between 1880 and his death in 1934.

Key work

1899 "The Geographical Cycle"

PHYSICAL GEOGRAPHY 105

See also: Hydrological cycle 66–69 ▪ Catastrophism 74 ▪ Uniformitarianism 75 ▪ Orogeny 76–79 ▪ Glaciation and ice ages 86–89 ▪ Isostasy 94–95 ▪ Karst landscapes 108–09 ▪ Plate tectonics 114–21

bottoms relatively rapidly, while the tops of the peaks, which are not subjected to such powerful erosional forces, remain largely unaffected. In a mature landscape, many tributaries flow into larger, meandering rivers in U-shaped valleys. Waterfalls, vertical erosion, and interlocking spurs are also features, and rivers begin to develop floodplains in their lower reaches, where they deposit sediment. Here, lateral erosion is greater than vertical erosion. Davis considered the Appalachian Mountains and the Dakota Badlands, both in his native US, to be representative of this mature stage.

In Davis's model, when the landscape approaches the end of the erosion cycle – old age – the valleys have become so wide and gently sloping that the landscape is a generally low-lying, slightly undulating plain, which he called a peneplain. Rivers meander sluggishly across it, and there is virtually no erosion. There may be residual hills that have not been denuded, or worn away; Davis

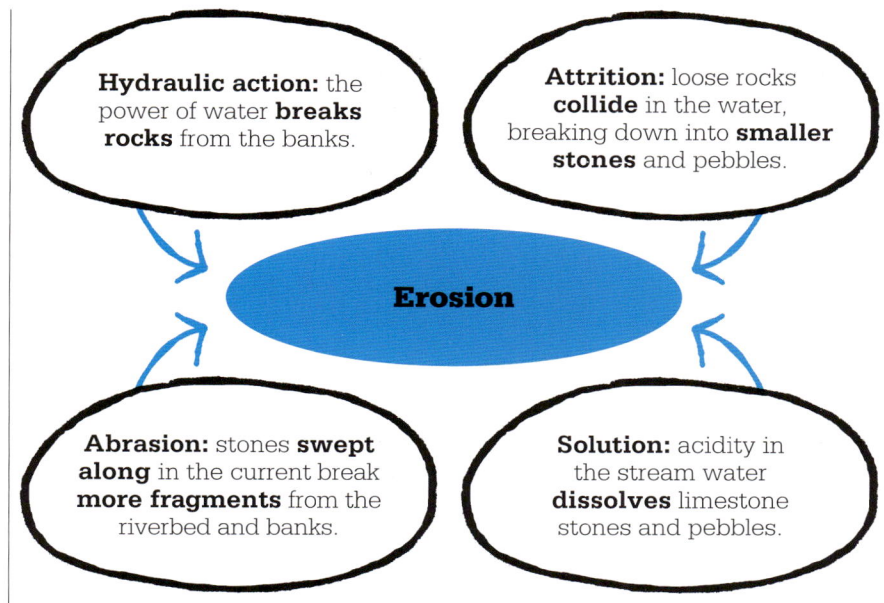

Hydraulic action: the power of water **breaks rocks** from the banks.

Attrition: loose rocks **collide** in the water, breaking down into **smaller stones** and pebbles.

Erosion

Abrasion: stones **swept along** in the current break **more fragments** from the riverbed and banks.

Solution: acidity in the stream water **dissolves** limestone stones and pebbles.

called these monadnocks, after Mount Monadnock, the tallest peak in New Hampshire, US.

A new geomorphology
German geomorphologist Walther Penck was the first to question the Davisian theory, in the 1920s. Whereas Davis believed that uplift and denudation took place alternately, Penck proposed that they must occur at the same time, gradually and continuously.

In the 1940s and 50s, American geomorphologist Arthur Newell Strahler recorded multiple variables, including maximum slope angle, channel gradient, bedrock, climate, vegetation, and soil type. He concluded that

Bhagirathi Gorge, in northern India, is typical of what Davis defined as a youthful landscape. This can be seen in its narrow watercourse and the steep sides of the ravine.

any of these factors can cause a readjustment of the equilibrium angle of a slope – a fundamental aspect of how a landscape erodes.

In 1957, British geologist Lester King proposed a theory of parallel slope retreat for arid and semi-arid environments, with gentle concave slopes, or pediments, in valley bottoms backed by steep scarps bounding upland blocks. As erosion takes place, the scarps undergo parallel retreat rather than any significant reduction in steepness.

The 1960s brought a deeper understanding of plate tectonics as the major driver of mountain-building. When two plates collide, the crust in the zone of contact is not usually raised as a horizontal plane but is distorted, folded, and faulted in a chaotic fashion, causing subsidence as well as lift.

Despite Davis's landscape evolution model being flawed, many of his explanations of river »

106 THE CYCLE OF EROSION

Drainage patterns

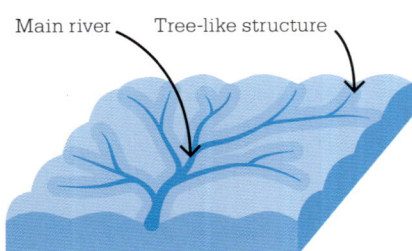
Dendritic — Main river; Tree-like structure

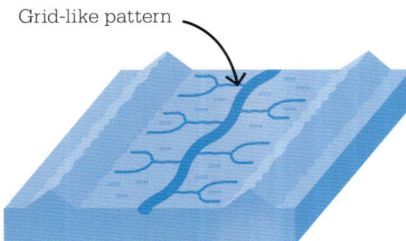
Trellis — Grid-like pattern

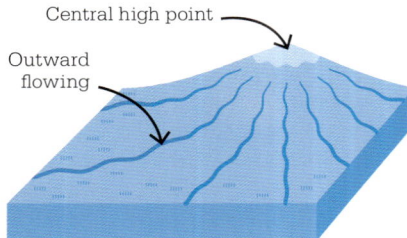
Radial — Central high point; Outward flowing

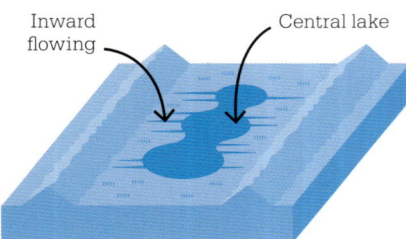
Centripetal — Inward flowing; Central lake

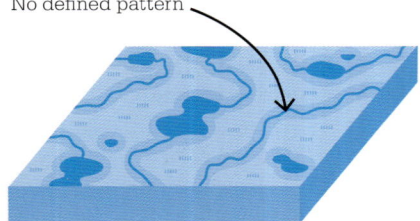
Deranged — No defined pattern

systems remain true. For example, he explained the ways in which rivers adjust to geological structure as they follow the lines of least resistance, such as softer bedrock and fault lines; and how drainage patterns can be superimposed onto older landscapes.

Drainage patterns

There are many different drainage patterns, each giving clues to an area's tectonic, volcanic, and glacial history. The pattern of rivers within a drainage basin is governed mostly by surface geology and structure, slope angle and shape, climate, vegetation, and human activity.

The most common type of drainage basin is the dendritic pattern, which resembles the branching structure of a tree. Small streams join ever larger ones, until they eventually reach a main river. It is typical of areas with no major variations in rock toughness to dictate the direction of stream flow. Examples include the Amazon basin in South America and the Mississippi basin in the US.

A trellis pattern features short tributaries flowing at right angles into roughly parallel larger tributaries, creating a grid. It is typical of drainage basins with alternating bands of soft and more resistant rock. This pattern can be seen in the Brahmaputra basin, which extends across China, India, and Bangladesh; and the Indus basin, which encompasses Afghanistan, China, India, and Pakistan.

Radial drainage is characterized by streams flowing outwards from a volcano or dome. Examples include the Amarkantak Hills in India, from where three rivers originate. By contrast, in a centripetal (or endorheic) basin, streams flow inwards to a central lake, which may be permanent or ephemeral, such as Lake Eyre in Australia. This is a closed system, with no outlet to the ocean. Nepal's Kathmandu Valley is a centripetal basin.

Deranged drainage patterns are common in areas affected by glaciation or volcanity. The presence of moraines, landslides, and lava flows means there is no well-defined pattern to the streams and rivers. Some flow into lakes with no outlet, while others change direction in a chaotic fashion.

Accord or discord?

Drainage patterns may be accordant, meaning that they correlate with the topography and geology of the area, or discordant. Furthermore, there are two types of discordant drainage pattern; these are known as superimposed and antecedent.

In superimposed drainage, a river first forms on a near-horizontal layer of younger, softer rocks with the same resistance. When it eventually erodes down to the older, complex layer of harder rock below, it retains its original pattern, which then appears superimposed on the older rocks. This explains why a river might flow in a narrow gorge through a ridge of hard rock, rather than following a course of less resistance around it.

In antecedent drainage, a river cuts through an area that has been uplifted by tectonic movements. As mountains rise around it, the river maintains its course, forming deep gorges.

PHYSICAL GEOGRAPHY

Horseshoe Bend in the Grand Canyon, US, demonstrates the antecedent drainage system. The Colorado River has maintained its path even after the area was subject to dramatic uplift.

Movements of Earth's crust have an impact on drainage. Uplift tends to re-energize river systems; faulting and folding rearranges them; and subsidence drowns them. Changes in bedrock can also have a dramatic effect. If part of a dendritic drainage pattern erodes through superficial clay sediments to underlying, super-permeable limestone, it may dissolve the latter and forge a subterranean course.

Human effects
For millennia, human intervention has changed river systems, such as when damming streams and creating mill ponds to enhance water flow to power water wheels. In the past 100 years, the scale of these interventions has increased exponentially: people have dammed rivers to generate hydroelectric power and provide fresh water, straightened and dredged water courses for shipping, cleared forests from valley sides for agriculture, embanked rivers to reduce flooding, and built on floodplains. Some of these actions have had disastrous consequences, causing landslides or increasing the risk of flooding in other locations.

Climate change is another factor affecting river systems. If rainfall increases, there is more energy in a drainage basin. There will be more surface runoff and more groundwater, so throughflow will also increase, accelerating erosion and transporting sediment. In turn, this will lead to more sediment being deposited downstream. There may also be more landslides, further accelerating erosion.

Higher rainfall also changes vegetation structure, and this can have a feedback effect. For example, more tree growth will reduce groundwater movement and stabilize valley slopes, thereby reducing erosion.

A warming climate will cause a rising sea level, thus drowning the lowest valleys and creating coastal inlets known as rias. In polar regions, it will cause ice sheets to melt and produce isostatic readjustment, with land masses rising. In terms of real-life impact, much of southern Britain is 200 m (650 ft) higher, relative to sea level, than it was one million years ago. As a result, many rivers have been rejuvenated, so – working to a new base level – their capacity for erosion and sediment transport is a long way from being exhausted.

Studying drainage basins
The so-called drainage basin approach is a valuable tool for geographers researching the way landscapes function at a time of rapid climate change. Combining geomorphology with a quantitative approach, this theory was first proposed in the 1973 book *Drainage Basin Form and Process*, by British geographer Ken Gregory and hydrologist Desmond Walling. It views drainage basins as dynamic systems in which multiple constantly interacting elements – tectonics, geology, soil, vegetation, climate, hydrological cycle, and living organisms, including humans – are considered. Inputs and outputs are quantified, and future change is modelled. ■

The great rivers of high Asia may still be far from completing their prodigious task in 100 million years.
Arthur Holmes
Principles of Physical Geology, 1944

EITHER MECHANICALLY OR THROUGH CHEMICAL CORROSION
KARST LANDSCAPES

IN CONTEXT

KEY FIGURE
Jovan Cvijić (1865–1927)

BEFORE
1642 Published posthumously, *Xiu Xiake's Travels* is the first scientific study into karst features. It relays the Ming Dynasty-era explorer's observations of water-carved landscapes in China.

1689 Johann Weikhard von Valvasor coins the term "karst".

AFTER
1894 Édouard-Alfred Martel, often regarded as the father of speleology, publishes his seminal work *The Abyss*.

1960 German speleologist Herbert Lehmann oversees the production of a series of maps that will later make up the *International Atlas of Karst Phenomena*.

1989 Derek Ford and Paul Williams publish their seminal book *Karst Hydrogeology and Geomorphology*.

Typically, a karst landscape is characterized by the presence of subterranean caverns, sinkholes, caves, and underground streams. In 1893, Jovan Cvijić revealed that these landforms are created by the dissolution of limestone rock by weakly acidic rainwater.

Karstology pioneers

One of the researchers who paved the way for Cvijić was Chinese geographer Xiu Xiake, who between 1613 and 1640 documented hundreds of caves in his country. In 1689, polymath Johann Weikhard von Valvasor explored the flow of underground rivers in the Kras region of his native Slovenia, close to the border with Italy. He observed how, upon reaching a limestone plateau, the river Reka sank below the surface, forming the Škocjan Caves. He defined this topography as "karst", German for Kras.

Another pioneer in the field was French lawyer and cave explorer Édouard-Alfred Martel. In 1893, in Montenegro, he investigated the river Trebišnjica, which flows underground for 90 km (55 miles) – almost half of its full length (187 km/116 miles). Three years later, while exploring the Drach Caves on the Spanish island of Majorca, he discovered what was believed at the time to be the world's largest underground lake, which was named Lake Martel after him.

The Karst process

The late 19th century was a golden age for the exploration of karst regions and caves, with contemporary advances in chemistry aiding comprehension of just how karst landforms came

The whole system is similar as in all fissures in limestone terrain, to the circulation of blood in animals and of sap in plants.
Édouard-Alfred Martel
The Abyss, 1894

See also: The hydrological cycle 66–69 ▪ The cycle of erosion 102–07

The Dydima doline sits on the slope of a limestone mountain in Greece. The weight of accumulated debris probably caused the collapse of an underground cave, creating this vast sinkhole.

to be. Jovan Cvijić was a leading scholar in the field and also the undisputed "father of karst geomorphology"; he carried out extensive research in the karst regions of his native Balkans.

In his 1893 paper "The Karst Phenomenon", Cvijić described karst landforms and processes, noting that falling rain combines with carbon dioxide in the atmosphere to become mildly acidic. When this acid rain seeps into the ground, it dissolves the permeable limestone, sculpting it in a range of different landforms.

Besides cave systems and disappearing streams, karst landscapes include karren, or limestone pavements, such as The Burren, in Ireland; dolines, or sinkholes (hollows in the ground), such as the Devil's Sinkhole in Texas, US; and poljes, or karst fields, which are large flat areas of coalesced sinkholes, such as Livanjsko Polje, in Bosnia and Herzegovina.

Recent developments
Karst Hydrogeology and Geomorphology, a 1989 book by two geomorphologists – Canadian Derek Ford and New Zealander Paul Williams – delves further into the intricacies of karst environments. It discusses the physical and chemical processes involved in karst formation, the classification of cave systems, and the influence of climate and environmental factors on karst development. The authors address practical applications, such as groundwater and water resources, infrastructure development and water management, and the human impact on karst regions. ■

Jovan Cvijić

Born in Serbia in 1865, Jovan Cvijić was a schoolteacher when he published his first paper on karst landscapes in 1889, following excursions in the Serbian countryside. Later, he studied physical geography and geology at the University of Vienna, where one of his tutors was German geographer Albrecht Penck, who encouraged Cvijić to focus on karst landforms.

After receiving his PhD, in 1893 Cvijić embarked on a tour of the Balkans, but his fieldwork was compromised by tense political tensions in the region. In the late 1890s, he developed an interest in human geography, publishing a paper on the anthropogeographical problems of the Balkans in 1902.

As an authority on border delineation, Cvijić was a key contributor to establishing the borders of the Kingdom of Yugoslavia after World War I. He died in Belgrade in 1927.

Key work

1893 "The Karst Phenomenon"

THE CONTINENTS MUST HAVE SHIFTED
CONTINENTAL DRIFT

IN CONTEXT

KEY FIGURE
Alfred Wegener (1880–1930)

BEFORE
c. 1800 Observing geological similarities in South America and Africa, German naturalist Alexander von Humboldt theorizes that the two continents were once joined.

1858 Fossil evidence leads French geographer Antonio Snider-Pellegrini to argue that Europe and North America were once a single landmass.

AFTER
1937 In *Our Wandering Continents*, South African geologist Alexander du Toit proposes that Pangaea split into two supercontinents: Laurasia and Gondwana.

1962–66 Scientists Harry Hess (US), John Tuzo Wilson (Canada), and Frederick Vine, Drummond Matthews, and Dan McKenzie (UK) confirm the theory of plate tectonics.

In 1596, Abraham Ortelius, a Flemish cartographer, commented in his *Thesaurus Geographicus* on the close fit of the east coast of South America and the west coast of Africa. He suggested that the two continents had once been part of the same landmass, which had been broken apart by "a rupture".

In the 19th century, geologists in Uruguay discovered fossils of the freshwater reptile *Mesosaurus*, which lived around 280 million years ago (mya); decades earlier, fossils of the same animal had been found in South Africa too. Since these reptiles were not considered capable of having swum across the Atlantic, their presence on both sides of the ocean was further evidence that the continents were once united.

In 1912, after piecing together several more strands of evidence, Alfred Wegener presented a

Fossils of *Mesosaurus* and of the land-dwelling reptile *Cynognathus* have been found only in southern Africa and South America, corroborating the theory of continental drift.

PHYSICAL GEOGRAPHY

See also: Catastrophism 74 ▪ Uniformitarianism 75 ▪ Orogeny 76–79 ▪ Glaciation and ice ages 86–89 ▪ Isostasy 94–95 ▪ The structure of Earth 112–13 ▪ Plate tectonics 114–21

Alfred Wegener

Born in Berlin, Germany, in 1880, Alfred Wegener had a varied career that encompassed climatology, polar exploration, and geology. After gaining a doctorate in astronomy from the University of Berlin, he worked at Lindenberg's Meteorological Observatory, where he pioneered the use of weather balloons to evaluate upper air conditions.

Through 1906–08, Wegener took part in an expedition to Greenland, where he met fellow German Wladimir Köppen, who was chief meteorologist at Hamburg's Naval Observatory. This meeting inspired Wegener's interest in climatology.

Wegener took part in three further Greenland expeditions. On the last of these, in 1930, he became stranded on the ice sheet 140 km (87 miles) from the nearest supply base. He died in his tent, and his body remained undiscovered for six months.

Key work

1915 *The Origin of Continents and Oceans*

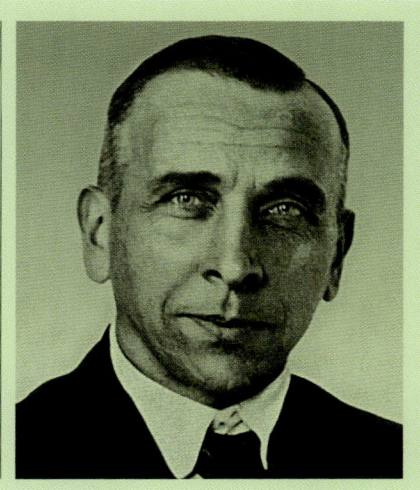

revolutionary new idea to the German Geological Society in Frankfurt. In addition to the jigsaw fit of South America and Africa, and the fossil evidence, he cited strong geological similarities between the Appalachian Mountains of North America and the Caledonian Mountains of Scotland, going on to propose that until around 200 mya all of Earth's landmasses had been joined in one giant continent, which he called Pangaea (from the Greek for "all Earth"). This, he argued, began to break up into smaller units, with the westward drift of South America opening up the Atlantic Ocean.

Explaining the mechanisms

Three years later, Wegener published his theory in *The Origin of Continents and Oceans*. In this book, he also suggested that mountains form when the edge of a drifting continent crumples and folds as it collides with another continent, such as when India converged with Asia to form the Himalayas. Continental drift, as Wegener called this process, continued until the continents arrived at their current position.

Although some in the scientific community supported Wegener's theory, the lack of a mechanism to explain how continents could plough through vast stretches of oceanic crust proved to be a stumbling block to its widespread

The Newton of drift theory has not yet appeared.
Alfred Wegener
The Origin of Continents and Oceans (4th edition), 1929

acceptance. Wegener suggested, incorrectly, that Earth's rotation could be responsible, but this idea failed to find favour, and his theory was largely forgotten.

Mantle convection

British geomorphologist Arthur Holmes offered a possible explanation for the missing mechanism for continental drift in 1919. He suggested that convection in Earth's mantle – with hot rock rising buoyantly towards the surface before cooling and sinking again – could cause the upper mantle to move laterally, like a conveyor belt.

Holmes developed this theory further in 1931, but it, too, was met with scepticism. It was not until the 1960s, when American geologist Harry Hess advanced the thesis of seafloor spreading, that Holmes was proved to be right, and in turn Wegener was shown to be a visionary who had laid the foundation for the modern theory of plate tectonics. ■

AT THAT DEPTH, THERE MUST BE A SUDDEN CHANGE OF MATERIAL MAKING UP THE INTERIOR OF THE EARTH
THE STRUCTURE OF EARTH

IN CONTEXT

KEY FIGURE
Inge Lehmann (1888–1993)

BEFORE
132 CE Zhang Heng, a Chinese scientist, invents an early seismoscope, a device capable of detecting the direction of a distant earthquake by sensing movement in the ground.

1909 Croatian geophysicist Andrija Mohorovičić discovers a discontinuity between Earth's crust and its mantle; this region is now called the Moho after him.

AFTER
1995 US geophysicist Gary Glatzmaier proposes that the inner core rotates faster than the rest of Earth, a phenomenon known as super-rotation.

2024 Australian researchers discover a large doughnut-shaped area of slow-moving liquid metal around Earth's inner core.

To date, humans have not explored very far inside Earth. The deepest borehole descends just over 12 km (7.5 miles), not even deep enough to go through the thinnest section of continental rocky crust, which is about 20 km (12.5 miles) thick. An international effort, Project SloMo, is aiming to drill through 5.5 km (3.3 miles) of oceanic crust to reach the Moho.

To understand what lies beneath the surface, geologists study seismic waves – that is, the ripples and vibrations caused by earthquakes – and the way they reflect and refract as they pass through the different materials within the planet. In 1936, Danish geophysicist Inge Lehmann studied how seismic waves travel and discovered that the centre of Earth consists of two parts: a solid inner core and a liquid outer core.

Detecting waves

Irish geologist Robert Mallet is often credited as the founder of seismology. In the 1840s and 50s, he began using explosives to create artificial seismic waves to help him interpret their behaviours.

As seismographs – the devices that pick up the motion of the ground – became more sensitive,

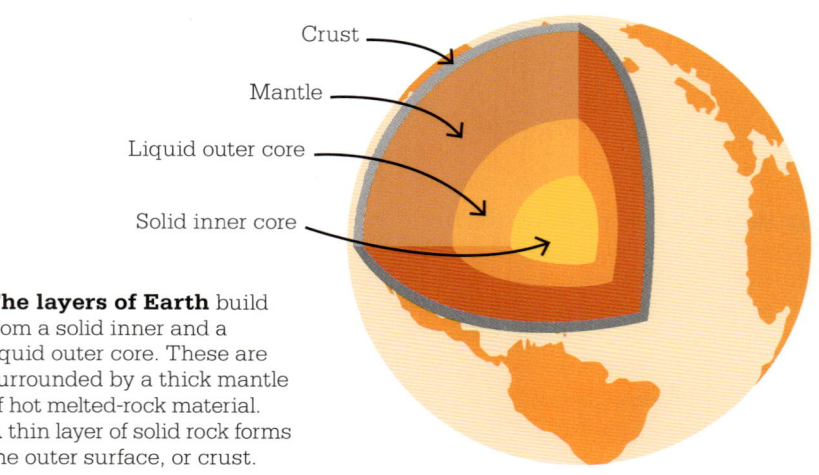

The layers of Earth build from a solid inner and a liquid outer core. These are surrounded by a thick mantle of hot melted-rock material. A thin layer of solid rock forms the outer surface, or crust.

geologists found they could detect waves that had formed on the other side of the world and travelled through the interior of Earth. This offered an opportunity to study the internal structure of the planet.

German geophysicist Emil Wiechert suggested in 1897 that Earth consisted of a mantle of silicate rock with a heavy iron core at the centre. The average density of Earth could only be accounted for if the planet's less dense outer rock was offset by a heavy core.

Types of wave

In 1906, British seismologist Richard Dixon Oldham found that earthquakes form three kinds of seismic wave. Primary waves (P waves) are faster and are therefore detected first; they are pressure waves, formed by compressions and expansions of material, and longitudinal. Secondary, or S waves, have an up-down motion; they are slower and cause more damage due to their side-to-side movement of the ground. Surface waves do not travel far from the earthquake's epicentre but are the most destructive.

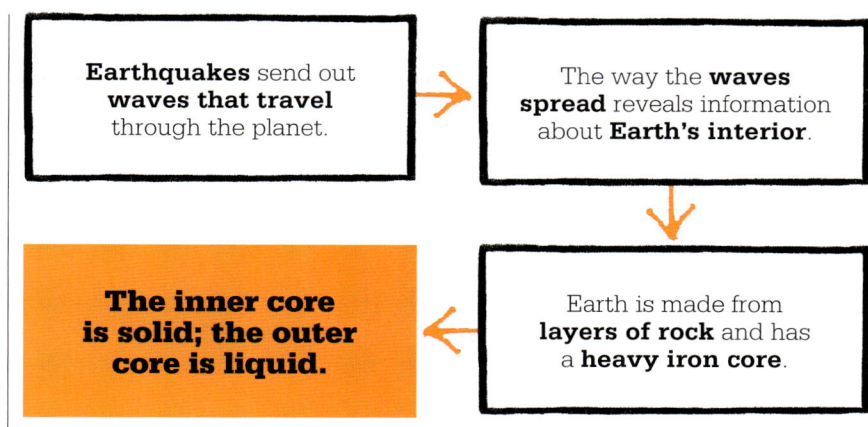

Wiechert's student Beno Gutenberg showed in 1914 that S waves were blocked by material in the centre of the planet, while P waves passed right through. Since S waves cannot move through liquids or gases, this confirmed the existence of a liquid-metal core.

Looking inside

In 1929, a large earthquake struck New Zealand's South Island, and its seismic waves were detected as far away as western Russia and Inge Lehmann's native Denmark.

After several years of analysis of these waves, Lehmann determined that P waves from the quake were reflecting off material deeper inside the core. She proposed that this was evidence that the inner core was a solid sphere inside a liquid outer core. The inner core rotates inside the liquid, and this relative motion is thought to be the primary source of Earth's magnetic field. Lehmann's contribution was one of the final pieces in the puzzle that demonstrated that Earth's interior is built up in layers. ∎

We take it that… the earth consists of a core and a mantle, but that inside the core there is an inner core.
Inge Lehmann
"P", 1936

Different seismic waves spread in different ways. As P waves go through Earth's core, they are refracted by it, while S waves cannot even penetrate the liquid core.

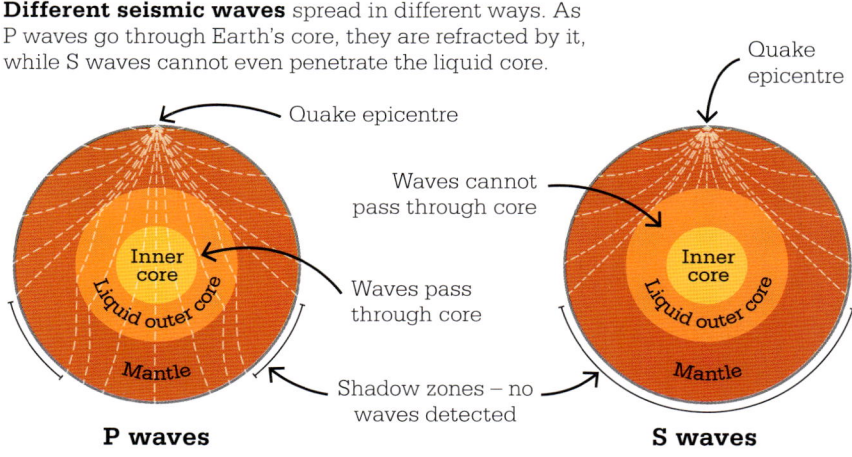

FORCES WHICH DISPLACE CONTINENTS ARE THE SAME AS THOSE WHICH PRODUCE GREAT FOLD-MOUNTAIN RANGES

PLATE TECTONICS

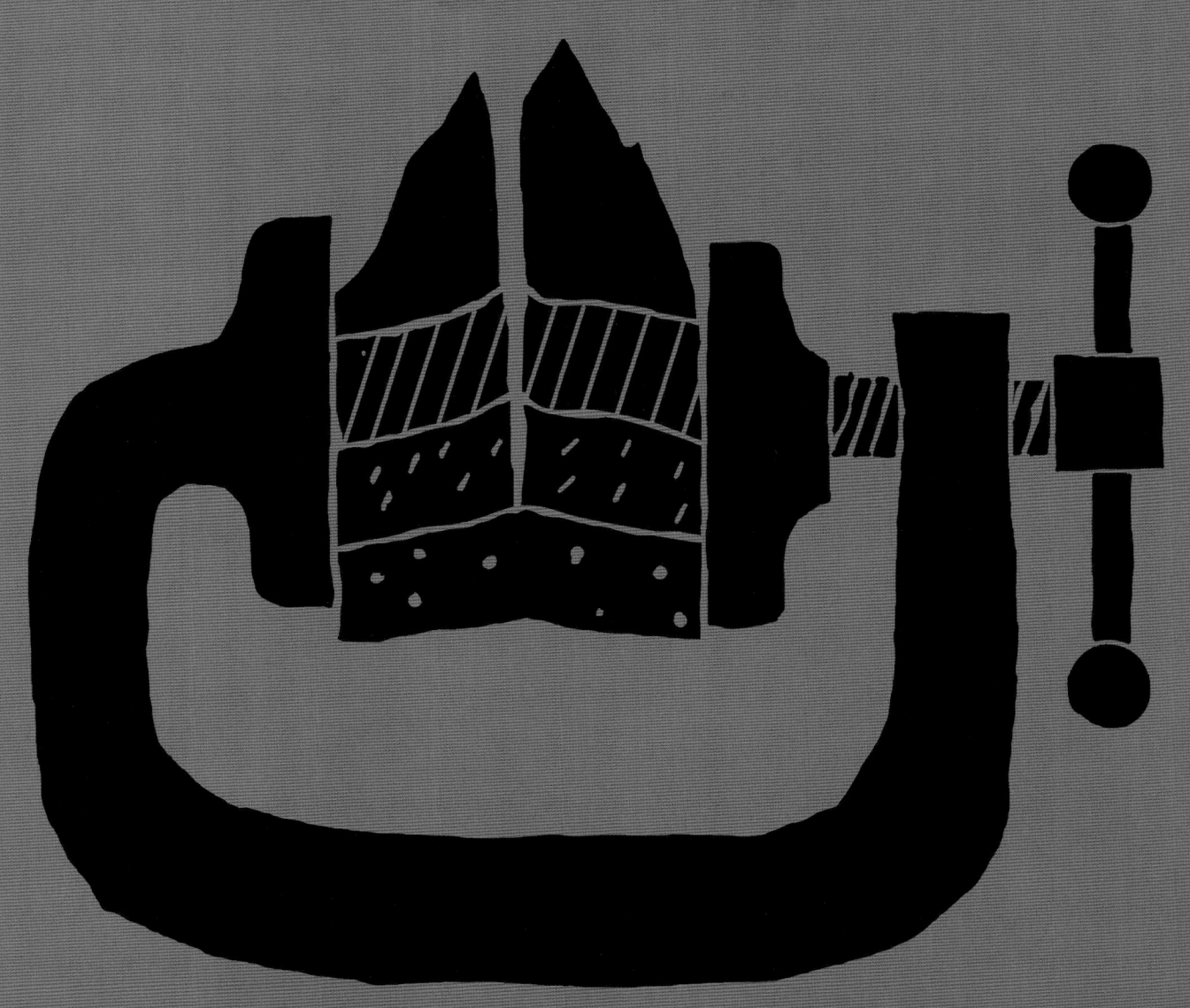

116 PLATE TECTONICS

IN CONTEXT

KEY FIGURE
Harry Hess (1906–69)

BEFORE
1912 Alfred Wegener proposes his theory of continental drift, mentioning a supercontinent that broke apart.

1937 South African geologist Alexander du Toit theorizes that Pangaea broke into two supercontinents: Laurasia and Gondwana.

1953 US oceanographer Marie Tharp uses sonar technology to map the Mid-Atlantic Ridge.

AFTER
2011 The Tohoku earthquake – caused by the Pacific Plate subducting under the Okhotsk plate – creates a tsunami that kills almost 20,000 people.

2025 Swiss and US scientists discover the remnants of submerged plates below the Pacific, in areas with no evidence of past subduction.

By the start of the 1960s, Harry Hess brought together older theories of continental drift and its mechanisms with more recent discoveries to propose a new theory of seafloor spreading. It would do much to advance a final understanding of the action of tectonic plates.

Moving dynamics

In the early 1950s, scientists knew that, throughout geological history, Earth's continents had moved in relation to each other. However, they still did not know the processes responsible for this. Evidence for this continental drift came from various sources: the "jigsaw fit" of continental shelf boundaries; the geological likeness of mountain chains now thousands of kilometres apart; the location of fossil remains; and geomorphological clues provided by glacial erosion in the tropics.

German scientist Alfred Wegener brought these puzzle pieces together in 1912 in his theory of continental drift, but his theory could not explain how or why the continents moved, so

Continents… have drifted like rafts.
John Tuzo Wilson
Speaking at "A Symposium on Continental Drift", 1965

it was considered "case unproven". In 1919, British geologist Arthur Holmes speculated that convection in Earth's mantle could cause plates of the crust to move, but again, solid evidence proved elusive.

Mapping the ocean floor

World War II fostered many technological advances, one of which was more sophisticated sonar equipment to map the ocean floor accurately and at speed. In 1953, US oceanographic cartographer Marie Tharp created maps that showed a striking underwater mountain range running north–south in the

Harry Hess

Born in New York City in 1906, Harry Hess studied geology at Yale, graduating in 1927. After two years spent as an exploration geologist in northern Rhodesia (now Zambia), he went to Princeton and was awarded a PhD in 1932. Six years later, he published his first major paper, on the origin of island arcs. During World War II, Hess saw combat in the Pacific as captain of the USS *Cape Johnson*. The ship was equipped with sonar, which he used to collect ocean floor profiles.

In 1947, Hess took part in research into Caribbean geology and, over the next two decades, wrote papers on ocean geology, geophysics, and mineralogy. He was chair of Princeton's geology department from 1950 until 1966 and, in 1963, president of the Geological Society of America. Hess died in 1969.

Key work

1962 "The History of Ocean Basins"

PHYSICAL GEOGRAPHY 117

See also: Mapping the ocean floor 46–47 ▪ Orogeny 76–79 ▪ Continental drift 110–11 ▪ The structure of Earth 112–13 ▪ Volcanic activity and hotspots 122–25 ▪ Seismology and earthquake magnitude 132–35

Thingvellir National Park in Iceland lies on a rift at the crest of the Mid-Atlantic Ridge, where the North American and Eurasian tectonic plates are slowly moving apart.

middle of the Atlantic Ocean. The presence of this geological feature had been hinted at about a century earlier. In 1855, Matthew Fontaine Maury of the US Navy had described a shallow "middle ground" in the ocean, but at that time nobody had any idea what the significance of this shallower part of the ocean might be.

In fact, the Mid-Atlantic Ridge – as Tharp's discovery came to be known – was later found to run for 16,000 km (10,000 miles) from the Arctic Ocean to sub-Antarctic Bouvet Island, in the South Atlantic. In the late 1950s and 60s, more of the world's ocean topography (bathymetry) was mapped, showing more ridges rising 1,500 m (5,000 ft) or more from the abyssal plain, or the deepest part of the ocean.

Deep trenches were also mapped close to some continental margins, such as the Peru–Chile Trench off the west coast of South America. Others were found close to island arcs far from any continent, such as the Mariana Trench, next to the Mariana Islands, east of the Philippines, in the Pacific Ocean.

Seafloor spreading

Hess's theory suggested that the upwelling of hot material from the mantle reached the surface at ocean ridges, where it cooled and moved away down the flanks on either side. As the seafloor continued to spread, Hess argued, the older ocean floor cooled and »

Heat from the mantle's rising **convection currents** makes the **crust** less dense and **more flexible**.

The **lighter material rises**, the crust fractures, and **hot magma fills its cracks**, spilling onto the surface.

The **magma spreads** in both directions, and **cold seawater** causes it to harden into **new, basaltic oceanic crust**.

As two **plates diverge**, the new oceanic crust moves **further from the spreading zone** and closer to the adjacent continental plate.

Eventually, oceanic crust collides with continental crust, moves beneath it, and is reincorporated into the mantle.

118 PLATE TECTONICS

subsided to the level of the abyssal plain, then was eventually moved in conveyor-belt fashion to ocean trenches. There, it was pushed under lighter continental crust (a process known as subduction) and reabsorbed into the mantle.

Hess theorized that the ocean floor must be oldest near continental margins and youngest near spreading centres. The fact that no oceanic sediments older than the Late Jurassic, about 200 million years ago (mya), have been found is more evidence to support his theory of the "recycling" of oceanic crust.

Magnetic anomalies

Hess still required more evidence for his theory to be given serious consideration, and he did not have to wait long. In 1963, British geologists Frederick Vine and Drummond Matthews – and, independently, Canadian geophysicist Lawrence Morley – reported on their investigations of "magnetic stripes" on the ocean floor. They knew from the work of Japanese geophysicist Motonori Matuyama in the 1920s that Earth's magnetic field periodically reverses polarity, flipping from north to south and back, and that oceanic crust is made of basalt, a rock containing the magnetic mineral magnetite. They were also aware that in 1955, British oceanographer Ron Mason had detected alternating bands of normal and reversed polarity in rocks on the floor of the Pacific Ocean.

When the lava containing magnetite cools, the magnetic minerals line up with Earth's magnetic fields at that time, preserving a record of it.

> Sea-floor spreading is a view about the method of production of new ocean floor.
> **Edward Bullard**
> **"The Emergence of Plate Tectonics: A Personal View", 1975**

Vine and Matthews noticed that there was a symmetrical pattern of alternating magnetic stripes of reversed and normal polarity that had formed on either side of the Juan de Fuca, the mid-ocean ridge they were studying in the northeast Pacific Ocean, opposite the coast of Oregon and Washington states. Their theory, expressed in a 1963 paper called "Magnetic Anomalies over Oceanic Ridges", became known as the Vine-Matthews-Morley hypothesis.

Dating Earth's crust

In another paper published in 1963, Canadian geophysicist John Tuzo Wilson described the four then-current views of Earth's development: a rigid planet either contracting by cooling or expanding by radioactive heating; or a mobile planet whose continents are drifting due to either planetary rotation or convection in the mantle.

By the mid-1960s, studies of terrestrial lava flows had enabled scientists to map the timescale for the last 4–5 million years' worth of polarity reversals. Vine and Wilson applied this to the oceanic "stripes" and dated the crust, confirming that the further from the axis of the ridge it was, the older it was. This was evidence that new oceanic crust was being created at the ridge, and that the seafloor was spreading on each side of it. The axis of the ridge – that is, the line of separation between two plates of oceanic crust – was later named a divergent boundary. Vine and Wilson went

The rugged basalt rock formations at Depoe Bay, along the central coast of Oregon, are the result of an ongoing subduction of the Juan de Fuca Plate beneath the North American Plate.

Tectonic plate boundaries are noted for high levels of seismic and volcanic activity. In spite of this, however, these areas often have high population density because of rich mineral deposits and fertile farming land.

Key:
— Plate boundary
➤ Relative plate movement
plate A major plate

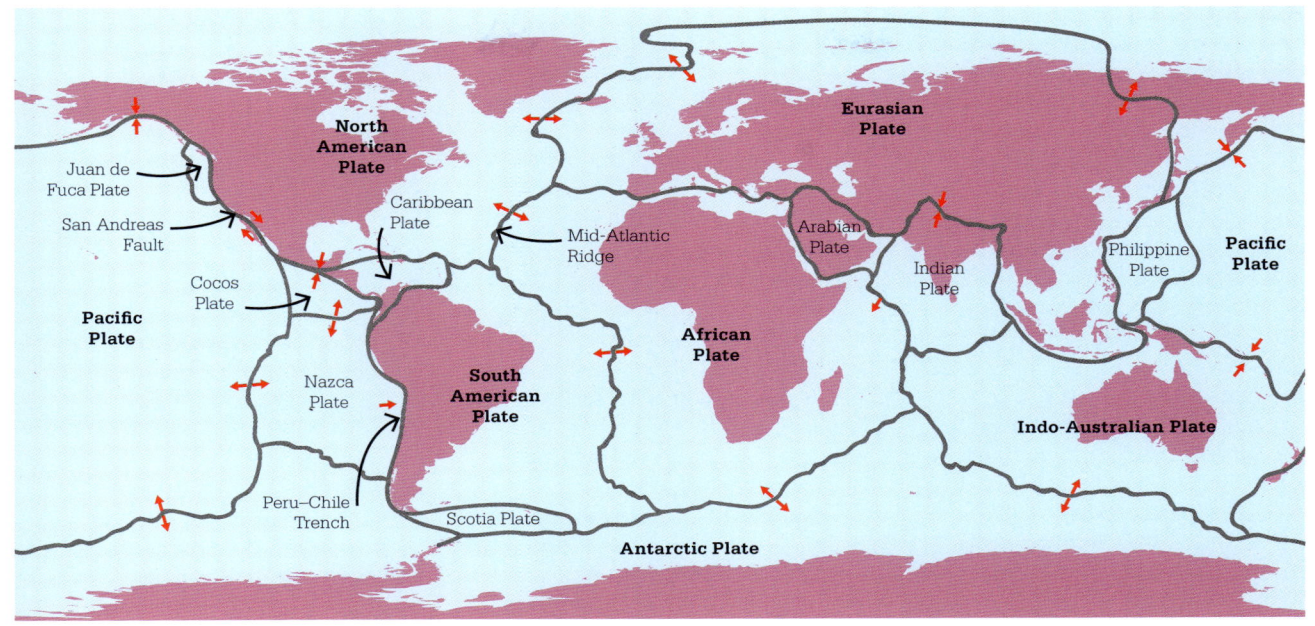

on to calculate that spreading at the Juan de Fuca Ridge, where the Juan de Fuca and Pacific plates diverge, was proceeding at about 3 cm (1.2 in) per year.

Volcanicity had been explained at sites of convergent (or colliding) and divergent (or moving apart) plate boundaries; however, no explanation had yet been provided for volcanoes located at vast distances from the nearest boundary. In 1963, Wilson proposed that plates might move over fixed mantle hotspots, forming chains of islands like those in the Hawaiian archipelago. This was yet more support for the idea of tectonic plate movement.

Two years later, Wilson also identified a third type of plate boundary, which became known as a transform boundary. This allows plates to slide past each other without creating or destroying ocean crust. The best-known example is the San Andreas Fault, between the Pacific and North American plates.

Towards a unified theory

More evidence was mounting that the mantle, so long believed by most scientists to be static, is in constant motion. In 1966, British geophysicist Dan McKenzie suggested in "The Viscosity of the Lower Mantle" that the mantle has two layers, both of which are in motion and control the movement of the tectonic plates above. McKenzie later modelled the generation of magma at divergent plate boundaries and mantle hotspots.

US geophysicist Jason Morgan presented a paper at the April 1967 meeting of the American Geophysical Union in which he drew all the threads together. He covered all the key concepts of plate tectonics, describing seafloor spreading and subduction at divergent and convergent plate boundaries respectively, backing up his arguments with detailed data on the speed of plate movement and seismicity. Morgan published his theory in the *Journal of Geophysical Research* in March 1968. Independently, a few months later, McKenzie published a paper drawing most of the same conclusions.

In the decades since, scientists have constructed a much clearer idea of Earth's structure and the movement of plates. Although they disagree on the number, most believe that Earth's surface is divided into seven major and eight minor plates. The major ones are the Pacific, which is almost »

120 PLATE TECTONICS

entirely oceanic crust, the North American and Eurasian, which comprise mostly continental crust, and the African, Antarctic, Indo-Australian, and South American, which are a mixture of continental and oceanic crust.

Geophysicists consider plates to be made up of the lithosphere, which comprises the crust and the uppermost section of the mantle. Together, these act as rigid plates on the asthenosphere (the next layer of the mantle) below. The rocks of this layer are not molten but are hot and viscous, meaning they can flow slowly over time. Continental plates average 125 km (78 miles) in depth, while oceanic crust is 50–100 km (30–60 miles) thick.

Ridges and rises

Since the 1970s, manned and unmanned submersibles have used instrumentation, including cameras, to evaluate spreading centres along the 60,000 km (37,300 miles) of oceanic ridges and rises. The 1979 RISE project, which investigated the East Pacific Rise, was one of the most successful explorations. Rising 1,800–2,700 m (5,900–8,850 ft) above the Pacific abyssal plain, the East Pacific Rise divides the Pacific Plate to the west and the North American, Rivera, Cocos, Nazca, and Antarctic plates to the east and south.

Near Easter Island, the rate of spreading along the rise is 15 cm (6 in) per year, but in most other places it is much slower. A central zone, with a chain of small volcanoes and hydrothermal vents, is surrounded on both sides by fissured and faulted zones, and there is regular seismic activity.

The Mid-Atlantic Ridge, which separates the North American and South American plates to the west from the Eurasian and African plates to the east, is different in several respects. It is narrower and slower-spreading, and it has a rift valley running almost its entire length. The ridge rises 2,350 m (7,700 ft) above the Atlantic abyssal plains each side. Its rate of seafloor spreading averages 2.5 cm (1 in) per year.

Alvin the submarine took scientists 2,600 m (8,530 ft) below the ocean's surface to investigate the magma source of the East Pacific Rise as part of the RISE research project in 1979.

Investigation into magnetic anomalies in seafloor basalts is providing information about Earth's history. For example, it established that rates of seafloor spreading were much faster 100 mya, so ridges occupied a bigger proportion of ocean basins.

Trenches and subduction

There are 55,000 km (34,000 miles) of convergence zones where denser, relatively thin oceanic lithosphere

Hydrothermal vents

In February 1977, American marine geologist Robert Ballard examined photos taken from cameras 2,500 m (8,200 ft) below the surface of the Pacific Ocean and noticed that hot water was emerging from the seafloor at the Galápagos Rift. It was the first time scientists had seen a hydrothermal vent. Since then, hundreds more vents have been found in all the major oceans.

Vents form when cold seawater percolates into magma-heated subsurface rocks. The water reaches temperatures of up to 400°C (752°F), stimulating chemical reactions that pull in sulphides of copper, iron, and zinc from the rocks, and it emerges through vent openings as a chemical-laden fluid.

Some vents, called black smokers, are tall, chimney-like structures that emit dark, mineral-rich water. Communities of invertebrates live around the vents, despite the intense heat and lack of light. Biologists have so far described around 800 such organisms.

Black smokers and the unique environment they create could have been the catalyst for life on Earth, according to some scientists.

dives beneath lighter, thicker continental lithosphere – or, less commonly, beneath another oceanic plate – at angles of 25 to 75 degrees. An example of the latter is at the 8,000-m- (26,250-ft-) deep Aleutian Trench, which runs for 4,000 km (2,485 miles) in the North Pacific. Here, the Pacific Plate plunges beneath the North American Plate at an angle of about 45 degrees. The resultant heat has generated intense volcanic activity, creating the Aleutian Arc, a chain of 40 volcanoes. Scientists believe that this zone of convergence has been active for 50–55 million years.

The Peru–Chile Trench runs for 5,900 km (3,700 miles) and is more than 8,000 m (26,250 ft) deep in places. It marks where the Nazca Plate is subducting under the South American Plate, a process that has continued for at least 140 million years. The movement of the Nazca Plate is unusual in that, after diving in the usual way, it becomes near-horizontal ("flat-slab" subduction) at some locations, before plunging again. Its current rate of horizontal movement averages 7.7 cm (3 in) per year.

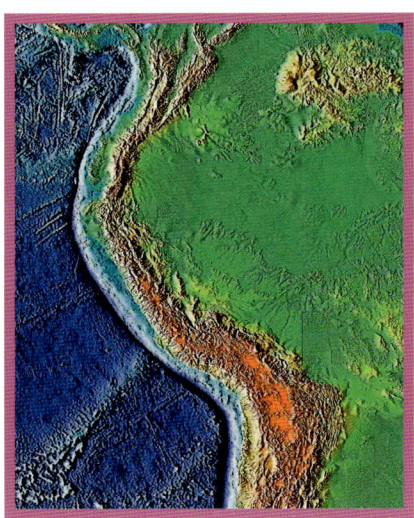

Subduction recycles Earth's crust by destroying old oceanic lithosphere and returning it to the mantle. Subduction zones are the locations of the most powerful earthquakes, known as megathrust earthquakes.

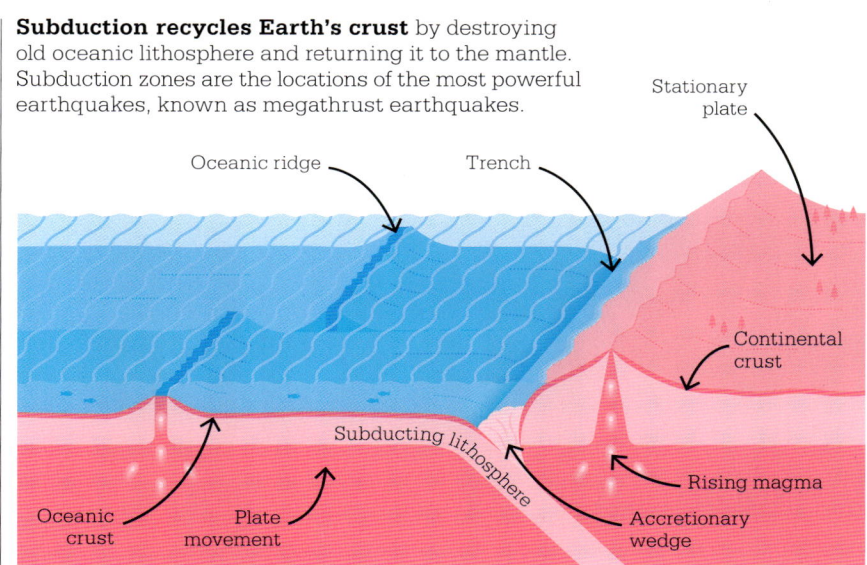

Subduction has also produced the Andes Mountains, which rise to almost 7,000 m (23,000 ft), locating them 15,000 m (49,200 ft) above the deepest points in the Peru–Chile Trench in some places.

Ongoing research

The plate tectonics theory has given, and continues to provide, deep insights into Earth's history. One more recent discovery is the role of plate tectonics in the deep carbon cycle, whereby carbon within subducting ocean plates is taken into the mantle and later released back into the atmosphere as CO_2 via volcanic hotspots.

Scientists now use GPS technology, satellite imaging, and computer modelling to study tectonic activity and the structure of Earth. Earthquakes, tsunamis, and volcanic eruptions are all

The Peru–Chile Trench, also known as the Atacama Trench, is the subject of ongoing scientific research into plate tectonics and the dynamics of earthquakes and tsunamis.

products of the constant shift between plates – forces with great destructive power, as seen in the devastating earthquake in Sumatra, Indonesia, in December 2004. Caused by subduction of the Indo-Australian Plate beneath the Burma plate, which resulted in a sudden uplift of the seafloor, the earthquake registered at 9.2–9.3 on the moment magnitude (M_W) scale, and caused a tsunami that killed more than 220,000 people. ∎

The earth's surface is considered to be made of a number of rigid crustal blocks.
W. Jason Morgan
"Rises, Trenches, Great Faults, and Crustal Blocks", 1968

A THERMAL ANOMALY IN THE MANTLE
VOLCANIC ACTIVITY AND HOTSPOTS

IN CONTEXT

KEY FIGURE
John Tuzo Wilson (1908–93)

BEFORE
1920s British geologist Arthur Holmes proposes mantle convection as the driving force behind continental drift.

1935 In "On the Activity of Deep-Focus Earthquakes", Kiyoo Wadati provides seismological evidence for plate subduction.

AFTER
1996 The National Oceanic and Atmospheric Administration in the US establishes a monitoring station on an underwater volcano in the North Pacific.

2025 US geologists establish that the Louisville hotspot in the southern Pacific Ocean helped to form the 120-million-year-old Ontong-Java Plateau, in the mid-Pacific.

The converging and diverging movements of Earth's plates result in tectonic activity on the edges of those plates. However, while these geological shifts account for most volcanic activity, they do not explain the presence of volcanoes far from plate boundaries, such as those of the Hawaiian archipelago, which is in the middle of the Pacific Plate. In 1963, John Tuzo Wilson addressed this apparent anomaly by suggesting the hotspot theory.

Defining hotspots
Hotspots are areas fed by stationary superheated plumes rising from unusually hot regions

PHYSICAL GEOGRAPHY 123

See also: Orogeny 76–79 ▪ Continental drift 110–11 ▪ The structure of Earth 112–13 ▪ Plate tectonics 114–21 ▪ Seismology and earthquake magnitude 132–35

A deadly tsunami caused by a volcanic eruption on the Indonesian island of Anak Krakatau in December 2018 killed more than 400 people.

of the mantle. They are thought to have a bulbous "head" fed by a long, relatively narrow "tail" rising from the depths of the mantle. Upon reaching the lithosphere, the plume head spreads into a mushroom shape, or diapir.

As magma rises through the lower mantle and asthenosphere in these plumes, its nature changes from viscous to molten as pressure decreases near the boundary of the lithosphere. The molten material breaks through the rigid plate to reach the surface, creating a hotspot where an outpouring of low-viscosity basalt builds up to form a volcano.

Classes of volcanoes

Volcanoes are vents or fissures from which magma, or molten rock, and other material emerge as lava flow, explosive bursts of ash and gas, and fountains of incandescent spray. Magma may reach the surface through a central "pipe", around which lava and ash accumulate to form a volcano; or it may emerge through long fissures, with low-viscosity basaltic lava spreading in a wide flow.

All volcanoes are classified according to their level of activity (active, dormant, or extinct) and their structure. There are four main structural classes. Composite (or strato-) volcanoes – such as Mount Etna, Italy, and Mount St Helens, Washington State, US – have the classic, tall cone shape, constructed from cooled, hardened lava and ash. They are known for erupting explosively because their magma is too viscous to allow gases to escape easily.

Cinder cones are steep-sided and usually relatively small, built from erupted ash, cinders, and larger rock fragments. Dormant Paricutin, Mexico, and active Cerro Negro, Nicaragua, are examples.

Shield volcanoes – such as Kilauea and Mauna Loa on Hawaii's Big Island – resemble a shield lying on the ground, with broad, gentle slopes built up from cooling low-viscosity basaltic lavas. Finally, lava domes are bulbous mounds that build up where viscous lava »

As a **tectonic plate passes** over a **hotspot**, magma breaks through the ocean floor.

The **build-up of lava** creates an **undersea volcano**, which eventually emerges **above sea level**.

The **volcanic island** erupts until **plate movement** takes it away from the hotspot.

A new volcano begins to form above the latest hotspot position, with the original volcano becoming extinct as it moves away.

It is difficult to find a source for so much heat.
Edward Bullard
"The Flow of Heat Through the Floor of the Atlantic Ocean", 1954

124 VOLCANIC ACTIVITY AND HOTSPOTS

solidifies near the vent – one formed at Mount St Helens after its massive 1980 eruption; Chaiten, Chile, is a currently active example.

Tectonic discoveries

As much as three quarters of Earth's volcanic activity takes place underwater, and the study of such volcanoes began in the late 19th century. Dredged ocean-floor samples collected by Britain's HMS *Challenger* expedition of 1872–76 included rocks of igneous origin far from land, suggesting some kind of underwater volcanism. Later, seafloor mapping showed oceanic ridges, rift valleys, and volcano-shaped features.

Several discoveries made from the late 1930s onwards helped to revive German scientist Alfred Wegener's earlier theory of continental drift. In a series of papers published between 1935 and 1954, Japanese seismologist Kiyoo Wadati and his American colleague Hugo Benioff showed that earthquake sources are deeper as one moves away from an oceanic trench to a section of continental crust. The plane on which the deepening earthquake epicentres lie indicates the boundary between overriding and subducting plates – evidence for plate tectonics and an explanation for volcanism at convergent plate boundaries.

In 1960, American geologist Harry Hess described volcanism at divergent plate boundaries, and how this led to seafloor spreading.

Different hotspots

Three years later, Wilson proposed that if a crustal plate moved over a fixed hotspot, it might create an active volcano over the "rising vertical current". If this process continued, the volcano would rise above sea level and form an island. However, as the crustal plate continued to move, the volcano would be starved of its hot magma

Eruptions of Castle Geyser, in Yellowstone National Park, happen every 16 to 17 hours, producing vertical columns of hot water that reach a height of 27 m (90 ft).

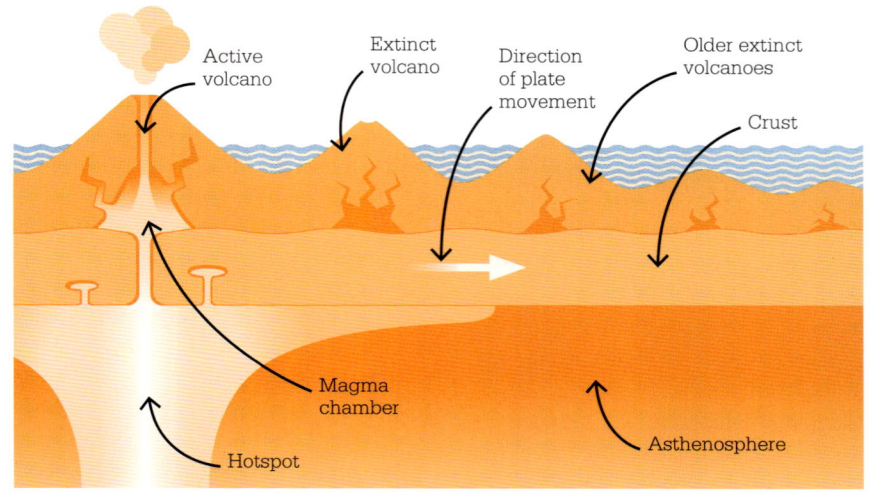

and become extinct. New volcanoes would build as the crust shifted across the hotspot, forming a chain of islands over time.

Not all hotspots are located in ocean basins. The best-known continental example is the one situated below Yellowstone National Park in Wyoming, US, which is renowned for its geysers, hot springs, and regular earthquakes – all indicators of volcanic activity. Yellowstone is now directly above a hotspot that can be tracked across Idaho, Montana, Nevada, and Oregon as the North American Plate moved in a west–southwest direction over the past 17 million years.

One example of the Yellowstone hotspot's legacy are the Steens Basalt flows of Oregon, which cover about 50,000 sq km (19,300 sq miles). They are up to 1 km (0.6 mile) thick, and were extruded 15–16 million years ago (mya). Another evidence is the line of giant volcanoes, the oldest of which – McDermitt Caldera, on the border of Oregon and Nevada – last erupted 16 mya.

Today scientists know that hotspots begin with concentrated radioactive decay deep in Earth's mantle. Most also agree that there are 40–50 hotspots around the globe, although some believe that number is much higher.

Volcanic archipelagos form when tectonic plates move over hotspots. Scientists estimate the size of the hotspot that formed the Hawaiian Islands to be around 320 km (200 miles) across.

In 2003, a team of French geophysicists suggested that most hotspots do not meet the criteria for a deep origin near the core–mantle boundary. Hotspots that do not originate at the lower–upper mantle boundary produce island chains but no large volcanic province, and these include the Samoa, Tahiti, and Cook islands in the Pacific Ocean. There is only convincing evidence for seven deep-origin "primary" hotspots; these include Hawaii, Reunion (in the Indian Ocean), Louisville (in the southwest Pacific), and Yellowstone (in the US).

Underwater study

A detailed study of marine volcanic environments is now possible using autonomous underwater vehicles (AUVs) to take hot hydrothermal fluid samples and measure temperature, depth, magnetism, light scattering, ocean-floor movements, and other variables. They can also film volcanic activity.

Permanent monitoring takes place at some sites, including on the Axial Seamount, the youngest volcano on the Cobb–Eickelberg undersea mountain range, 480 km (300 miles) off the Oregon coast in the North Pacific. Axial is unusual because, as well as having a hotspot origin, it is on the Juan de Fuca Ridge, a spreading centre. An observatory established on its flank in 1996 recorded its eruption two years later. Axial erupted again in 2011 and 2015.

The importance of research

Tracking the movement of hotspot volcanoes can provide many clues about Earth's geological history. For example, the Hawaiian–Emperor seamount chain is now known to extend for 6,200 km (3,850 miles) from southeast of the Hawaiian Islands to near the Kamchatka Peninsula in Russia. At the current position of the hotspot, a volcano is building from the ocean floor in the vicinity of Hawaii's Big Island but has yet to break the surface, and there are already several active volcanoes in the Hawaiian archipelago itself.

Atolls running northwest from the archipelago are the remains of formerly active volcanoes in the seamount chain that date from between 7 and 28 mya. Going even further back in time, the Emperor seamounts, the summits of which are now below sea level, date from 39 to 85 mya.

With the global climate changing rapidly due to the increased level of greenhouse gas emissions, studying volcanicity has never been more important. Volcanoes release large quantities of these gases into the atmosphere, and technological advances enable scientists to collect more data on this. A better understanding of the seismicity of volcanic zones, both terrestrial and marine, will help to forecast earthquakes and tsunamis. ∎

The hot spots and their volcanic trails… mark the passage of the plates.
Kevin C. Burke & John Tuzo Wilson
"Hot Spots on the Earth's Surface", 1976

John Tuzo Wilson

Prominent 20th-century geologist John Tuzo Wilson was born in 1908 in Ottawa, Canada. After becoming the first graduate in geophysical studies at any Canadian university (Toronto, 1930), he worked with the Geological Survey of Canada from 1936 to 1939.

During World War II, Wilson served with the Royal Canadian Engineers, then taught geophysics at the University of Toronto from 1946 to 1974. He was one of the world's leading exponents of the revived theory of continental drift. In addition to his work on mantle plumes and hotspots, he explained the role of transform faults in the movement of tectonic plates and the cycle of ocean opening and closing.

Wilson died in 1993, and the Wilson Mountains of Antarctica are named after him.

Key work

1963 "Evidence from Islands on the Spreading of Ocean Floors"

WHEN YOU LOOK DEEPER INTO THE ICE, YOU GO BACK IN TIME
PALAEOCLIMATOLOGY

IN CONTEXT

KEY FIGURE
Claude Lorius (1932–2023)

BEFORE
1088 Observing fossilized bamboo in a region too cold for the plant to grow, Chinese scientist Shen Kuo concludes that the area must have had a warmer climate in the past.

1840 Swiss zoologist and palaeontologist Louis Agassiz drills into the ice of a mountain glacier to take temperatures; he also places flow markers to track how the ice moves.

AFTER
2020 In the UK, a study by Loughborough University shows that climate change has caused increased amounts of organic carbon to be trapped in lakes.

2025 An ice core drilled at Dome C, a mound of ice in eastern Antarctica, contains a section that is 1.2 million years old, the oldest ever extracted.

Earth's climate is linked to **atmospheric gases**. → Evidence from **ancient ice** reveals the levels of **greenhouse gases** in the distant past.

↓

Periods with **high greenhouse gas** concentrations also have **warm climates**. ← Growth rings and other **fossil indicators** capture periods of **warm and cold climates** in the past.

↓

Evidence from ancient climates proves that rising levels of greenhouse gases now will lead to a warmer climate in future.

To prove that the climate is changing and to understand how this might affect the planet, it is necessary to have a record of the long history of Earth's climate. This is the job of palaeoclimatology. An important part of the climate record is the relative abundance of greenhouse gases – such as methane and carbon dioxide – in the atmosphere through the years. This significant information indicates the varying impacts of the greenhouse effect in the past. The data is collected by drilling ice cores – a technique pioneered in the 1970s by French climatologist Claude Lorius.

Ancient bubbles
Lorius went on many polar expeditions as a young researcher. In 1965, after a day's drilling, he used some spare samples collected

PHYSICAL GEOGRAPHY 127

See also: Steno's law of superposition 70–71 ▪ The Greenhouse effect 80–83 ▪ Glaciation and ice ages 86–89 ▪ Climatic zones 96–101 ▪ Climate change 232–39 ▪ Climate modelling 268–71

Scientists collect ice samples near the Rothera Research Station, in the Antarctic Peninsula. The Antarctic ice sheet is believed to preserve up to 800,000 years of Earth's climate history.

from deep down in the ice to cool a glass of whisky. He was intrigued to see bubbles sparkle out of the ice as it melted. Lorius realized that the frozen chunk contained samples of air that had been trapped when the ice had formed, and those gases were only just being released as it melted.

The polar regions are covered in ice, yet somewhat surprisingly, the precipitation of snow or rain there is very low. As a result, new ice forms slowly and in small amounts, but it stays frozen for a long time as it is steadily buried deeper under deposits of younger ice. In keeping with the law of superposition, the deeper one drills, the older the ice becomes, and the more ancient the air trapped within it.

Core drilling

Typically, the first 60 m (200 ft) of an ice sheet is compacted snow. The solid ice below that is sampled using a rotating cutting tool that removes a ring, or annulus, of ice from around the core. The process is necessarily slow and gentle. Any sudden shocks could crack the core, and excessive frictional heat from the drilling might melt the ice and allow contaminants to get in. Cores are removed from the borehole every few metres (feet). The longest boreholes are 3,000 m (9,850 ft) deep. The combined cores of this length contain ice laid down over hundreds of thousands of years.

Climate history

The concentrations of gases extracted from ice of different ages show how the composition of the atmosphere – including its greenhouse gases – fluctuates over the course of millennia. Palaeoclimatologists then cross-refer this atmospheric record with other evidence of ancient climates, such as cores of sediments from the sea floor and lake beds. These represent a record of what kinds of wildlife communities were in the waters through the years, as well as showing up periods of drought. The growth rings seen in fossilized tree trunks and corals indicate climate conditions over the lifetime of the organism. Thick rings show warm years ideal for growth. Collecting corals and trees from a wide span of time creates a record to match that of ice cores.

Taken together, this evidence proves that when the concentration of greenhouse gases in the air is high, Earth has a warmer climate as indicated by the biological records from the same period. This is an important fact for understanding the climate impact of the current increase of greenhouse gases. Moreover, palaeoclimatologists have shown that the concentration of greenhouse gases is rising faster now than at any time documented in the historical record. ▪

The paleoclimate record shouts out to us that... the Earth's climate system is an ornery beast which overreacts even to small nudges.
Wallace Broecker
"Cooling the Tropics", 1995

SOIL HAS A PROFILE
PEDOLOGY

IN CONTEXT

KEY FIGURE
Hans Jenny (1899–1992)

BEFORE
1697 German chemist Georg Stahl promotes the idea that soil fertility is dependent on decayed organic matter.

1883 Vasily Dokuchaev produces the first scientific system of soil classification.

1935 US geologist Curtis Marbut's soil classification underlines the importance of climate and vegetation.

AFTER
1984 Peter Birkeland, a US geologist, uses soil as a tool to date landscape forms and interpret past climates.

2025 A Scottish study reveals that treated sludge used to feed crops often contains contaminants that are harmful to soil and human health.

Soil is one of Earth's most important materials, and human life would be untenable without the range of functions it provides. Soil acts as a growing medium for plants (as much as 95 per cent of global food production is dependent on soil); it filters and cleans our water; it helps mitigate natural hazards such as flooding; and it is the largest terrestrial store of carbon, therefore helping to offset the effects of climate change. In addition, a quarter of all terrestrial species live in it.

Understanding this important resource is crucial. In 1941, Swiss-American soil scientist Hans Jenny

PHYSICAL GEOGRAPHY 129

See also: The cycle of erosion 102–07 ▪ Biomes and ecological zones 136–39 ▪ Desertification 228–29

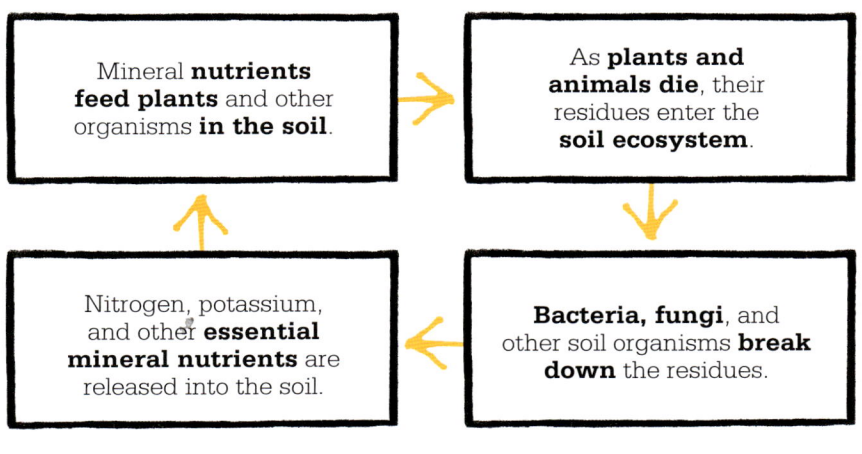

| Mineral **nutrients feed plants** and other organisms **in the soil**. | → | As **plants and animals die**, their residues enter the **soil ecosystem**. |

| Nitrogen, potassium, and other **essential mineral nutrients** are released into the soil. | ← | **Bacteria, fungi**, and other soil organisms **break down** the residues. |

Worms boost soil fertility by burrowing, which improves aeration and drainage, and breaks down organic matter into nutrient-rich castings.

introduced mathematical models to describe and predict soil formation, showing that soil properties can be predicted if the environmental conditions are known. By so doing, he transformed soil science into a systems-based discipline.

Understanding soil

Since the development of agriculture some 12,000 years ago, soil management has been a crucial factor in the advancement of human civilization. The shift from hunter-gatherer societies to an agrarian way of life changed the course of human history, but it also altered natural nutrient cycling within soils.

Nutrients cycle naturally from soil to plants and animals, and then back to the soil, primarily through decomposition. Agriculture has a profound effect on this cycle, with intensive cultivation and harvesting of crops depriving the soil of nutrients. In order to maintain fertility for sufficient crop yields, the soil typically requires occasional boosts. Early humans soon learned to add animal manure, ash, and other nutrients to improve fertility – a function that is carried out today by chemical fertilizers alongside organic nutrient sources. »

The soil system is an open system; substances may be added to or removed from it.
Hans Jenny
Factors of Soil Formation, 1941

Hans Jenny

Born in Basel, Switzerland, in 1899, Hans Jenny earned a PhD in agricultural chemistry from Zurich's Federal Institute of Technology in 1927. After emigrating to the US, he spent eight years as assistant professor at the University of Missouri. In 1936, he moved to the University of California at Berkeley, where he taught pedology, eventually becoming head of the department of soils. In 1941, Jenny published *Factors of Soil Formation*, which is widely regarded as a seminal book on soil science.

The Hans Jenny Pygmy Forest Reserve, in California's Mendocino County, is named in his honour, after Jenny conducted a long-term study of the soils in this rare coastal landscape. Along with his wife Jean, he also did much to preserve many areas of wildland in California and the western US.

Jenny retired in 1967 but continued to work and conduct research until his death in 1992.

Key work

1941 *Factors of Soil Formation: A System of Quantitative Pedology*

130 PEDOLOGY

As early as the 2nd millennium BCE, the Greeks recognized that different soils had different properties. They chose plants appropriate to each soil, and developed the concept of soil profile. The naturalist Theophrastus wrote what was probably the first classification for soils around the 4th century BCE, and other ancient civilizations built up a similar store of knowledge based on observation.

Early soil science

A scientific understanding of soil and its formation came only in the 19th century. In 1809, German agronomist Albrecht Thaer proposed that soil fertility was determined largely by the amount of organic matter, or humus, present in it.

Thaer's theory influenced fellow German Friedrich Fallou, who coined the term pedology (from the Greek *pedon*, meaning "soil") for the scientific study of soil. He described soil as having its own structure and composition, observing the presence of several different layers, which he suggested developed under environmental influences, or weathering.

Dokuchaev's horizons

In 1883, Russian scientist Vasily Dokuchaev, widely regarded as the founder of classical soil science, suggested that soil was more than just weathered rock or a medium for growing plants. He regarded it as a complex entity with its own structure and function.

Dokuchaev identified five key factors that control soil formation, or pedogenesis. They are climate, such as temperature and moisture; parent material, or the underlying geological foundation; organisms, or vegetation, animals, and microbes in the soil; relief, or land topography; and time. Dokuchaev explained that these factors interact, leading to the creation of five different layers, or horizons, in the soil profile. The uppermost layer is the O horizon, consisting of organic matter in various stages of decomposition. Below it are the A horizon, or topsoil, which is rich in humus; the B horizon, or subsoil, with an accumulation of clay particles and minerals washed down from the higher layers; and the C horizon, or parent material. Underneath all these layers is the R horizon, made up of consolidated bedrock.

The Jenny Equation

Building on Dokuchaev's insights, Hans Jenny set out the conceptual framework of pedogenesis in his 1941 book *Factors of Soil Formation*. The Jenny Equation is $S = f(cl, o, r, p, t, …)$, where S represents soil formation and f is the relationship, or function, between the various soil-forming factors: cl for climate; o for organisms in the soil; r for relief, or topography; p for the parent material; and t for time. The "…" allowed for new discoveries to be added later.

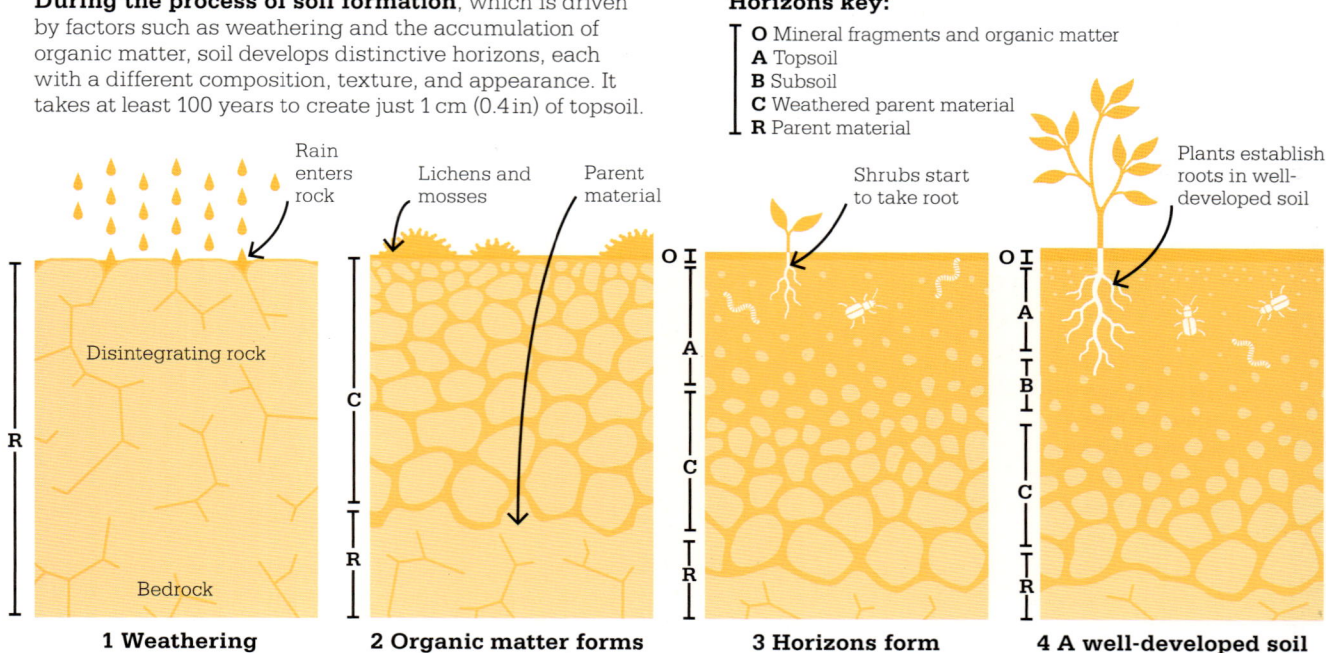

During the process of soil formation, which is driven by factors such as weathering and the accumulation of organic matter, soil develops distinctive horizons, each with a different composition, texture, and appearance. It takes at least 100 years to create just 1 cm (0.4 in) of topsoil.

Horizons key:
- O Mineral fragments and organic matter
- A Topsoil
- B Subsoil
- C Weathered parent material
- R Parent material

1 Weathering 2 Organic matter forms 3 Horizons form 4 A well-developed soil

PHYSICAL GEOGRAPHY

Intensive farming practices have a negative impact on soil health. Heavy machinery leads to compaction and the collapse of underground habitats vital for the cycle of soil nutrients.

Jenny explained how soil formation begins with the physical and chemical breakdown of rocks through the process of weathering. Physical disintegration breaks down the rock into ever smaller pieces, and eventually into sand, silt, and clay particles. Heating and cooling of exposed rocks creates physical stress and cracking, and abrasion by water, ice, or wind also contributes to the breakdown.

The composition of weathered rock has a bearing on soil chemistry and fertility. Limestone and basaltic lava have a high content of soluble minerals essential for plants and produce fertile soil in humid climates, while soils developed over sandstone can be acidic and unsuitable for agriculture.

Soil-forming processes

Living organisms also play a key role in pedogenesis. The cycling of nitrogen and carbon through the soil is almost completely due to plants and animals. Through the process of decomposition, organisms add nutrient-rich humus that increases fertility and affects soil structure. Surface vegetation binds the soil, protecting the upper layers from erosion by wind and water.

Climate is an important factor in soil formation, and soils differ widely from one major climatic zone to another, as they undergo different soil-forming processes. A process common to soils found in tropical and subtropical environments, for example, is laterization, whereby soils acquire a distinctive reddish coloration. This is caused by rapid weathering due to high temperatures and heavy rainfall; the large amounts of water moving through the soil remove minerals from the soil, leaving only the iron oxides that give it its unique red hue. Decomposition of the vegetation is almost the only source of nutrients.

Another soil-forming process is podzolization, which is associated with coniferous forests in humid, cold climates, and results in a grey-white sandy sub-layer. This is due to the combined effect of heavy summer rain and decomposing coniferous litter, which creates acidic conditions that strip soluble compounds from the topsoil.

Calcification of soil occurs when water loss by transpiration is greater than the amount of rainfall, resulting in the upward movement of dissolved salts from the groundwater. This can result in the formation of a hard layer of calcium carbonate in the subsoil. Salinization is similar to calcification, but it takes place in much drier climates, and the salt deposits occur at or very near the soil surface.

Future risks

Erosion, unsustainable farming, and deforestation are contributing to alarming rates of topsoil loss. Topsoil takes centuries to form, and without it, food production, water filtration, and carbon storage are severely compromised. Understanding soil science can help develop regenerative practices and restore degraded land. ∎

The Nation that destroys its soil destroys itself.
Franklin D. Roosevelt
"Letter to all State Governors on a Uniform Soil Conservation Law", 1937

A METHOD OF SEGREGATING LARGE, MODERATE, AND SMALL SHOCKS

SEISMOLOGY AND EARTHQUAKE MAGNITUDE

IN CONTEXT

KEY FIGURE
Charles F. Richter (1900–85)

BEFORE
1923 Harry Wood and John Anderson, two US seismologists, develop an advanced seismometer to measure ground vibrations.

1931 Wood and another US seismologist, Frank Neumann, adapt Mercalli's classification to produce the Modified Mercalli Intensity Scale.

AFTER
1995 In the aftermath of the Kobe earthquake, the Japanese government develops three national seismic monitoring networks.

2023 A National Seismic Hazard Model by the US Geological Survey shows that almost 75 per cent of the US could experience potentially damaging earthquakes.

In the late 19th century, various attempts were made to classify the magnitude of earthquakes before Italian volcanologist Giuseppe Mercalli developed an earthquake intensity scale in 1902. Refined in 1931 as the Modified Mercalli Intensity (MMI) scale, this was based on subjective factors, such as observed effects on people and the damage done to buildings and other structures.

In 1935, American seismologists Charles F. Richter and Beno Gutenberg took advantage of technical advances to propose a very different earthquake scale, which would become the global standard for more than

PHYSICAL GEOGRAPHY 133

See also: Orogeny 76–79 ▪ Continental drift 110–11 ▪ The structure of Earth 112–13 ▪ Plate tectonics 114–21 ▪ Volcanic activity and hotspots 122–25 ▪ Early warning systems 296–303

Charles Francis Richter

Known for the scale named after him, Charles F. Richter was born in Ohio in 1900; when he was nine, his parents divorced, and he moved to California with his mother. He graduated in physics from Stanford University and had started to work on a doctorate at the California Institute of Technology (Caltech) when his interest in earthquakes led to a job at a new seismological laboratory in Pasadena. In 1937, he returned to Caltech, where he remained until retiring in 1970.

Richter helped make seismology a major area of scientific study. Living in earthquake-prone California, he recognized the need for advance warnings, mapped the US according to earthquake susceptibility, and advocated more robust construction techniques to minimize building collapse. With Beno Gutenberg, he wrote two major seismology books. Richter died in 1985.

Key work

1958 *Elementary Seismology*

---

four decades. They were partly guided by the pioneering work of Japanese seismologist Kiyoo Wadati, who had shown in a paper published in 1928 that the amplitude of a seismic wave – that is, its maximum height – decreases at a constant rate as it moves away from the earthquake's source.

New calculations

Commonly known as the Richter scale, the scale devised by Richter and Gutenberg works on the premise that earthquakes generate seismic waves that pass – and vibrate – through rock, mud, sand, and buildings. In 1906, British seismologist Richard Dixon Oldham had established that there are three kinds of seismic wave. Primary (P) waves are the fastest, compressing and expanding the ground. Secondary (S) waves are slower, and they shake the ground side to side or up and down, often causing stronger movement. Surface waves do not travel far from the earthquake's epicentre, but they cause the strongest shaking.

Richter and Gutenberg used a sensitive, state-of-the-art Wood–Anderson torsion seismometer that comprised a continuously unwinding roll of paper with a pendulum and marking device suspended above. Any ground vibrations caused the pendulum and the trace on the paper to deviate. The stronger the vibration, the greater the deviation recorded on the seismograph.

The Richter scale classified an earthquake by measuring the maximum amplitude, as shown on the seismograph; it also took account of an earthquake's distance from the epicentre – the point on Earth's surface directly above the hypocentre, or the underground origin of the tremor.

On the scale, every whole-number increase represents a tenfold growth in the amplitude of motion recorded by a seismometer. As an example, a magnitude-4 earthquake will produce 10 times the amount of ground shaking – and, potentially, damage to structures – as a magnitude-3 quake.

The San Francisco quake

In a sense, the origins of Richter's work lay almost 30 years earlier, in the notorious 1906 San Francisco earthquake and subsequent firestorm that together destroyed 80 per cent of the US city and killed at least 3,000 people. »

Fire caused more damage than ground tremors in the 1906 San Francisco earthquake. Many buildings were made of wood, and ruptured gas mains fuelled the flames.

134 SEISMOLOGY AND EARTHQUAKE MAGNITUDE

Like most earthquakes, the San Francisco quake had been caused by movement at a tectonic plate boundary. Stresses had built up at the San Andreas Fault – a crack in Earth's crust that marks the boundary between the Pacific and North American plates – until one side of the fault suddenly slipped about 8 m (26 ft) in relation to the other. Slippage occurred along 477 km (296 miles) of the fault, whose hypocentre was 11.7 km (7.3 miles) below the surface, and created powerful waves that shook the rocks above and caused widespread damage.

Although there had not been a similarly destructive event in California since the 1906 San Francisco earthquake, the state experienced thousands of tremors annually. To better understand and evaluate the risk, a seismological observatory was established in Pasadena in 1921. As a young

A tsunami that struck Japan in 2011 killed close to 20,000 people, injured 6,000 more, and caused the near destruction of many coastal towns and cities in the Tohoku region.

man, Richter worked there, and he and Gutenberg developed their scale primarily to measure tremors in southern California.

Richter also confirmed that the degree of shaking intensity in an earthquake varies according to several factors, the most important one being the distance from the tremor's hypocentre. Generally, closer means more intense; however, the amount of shaking – and possible resulting damage – also depends on the nature of the substrate. If the ground is mud or sand, it will be subject to more shaking than if it is solid bedrock. It follows that structures built on "soft" ground may be damaged more than those on rock, even if the

latter are closer to the hypocentre. Construction methods and materials are also factors, something Richter later advised on.

Sub-ocean hypocentres

Many earthquakes occur beneath the ocean. On 22 May 1960, the most powerful earthquake ever recorded had its hypocentre 160 km (100 miles) off the coast of Chile, where the Nazca plate subducts the South American plate. Registering about 9.5 on the Richter scale, the initial quake caused widespread damage to buildings, but most casualties were caused by the 25-m- (82-ft-) high tsunami that struck the coast 15 minutes later. The energy discharged by the earthquake was such that when the tsunami hit the Hawaiian Islands, 10,000 km (6,200 miles) away and 15 hours later, it was still 11 m (36 ft) high.

On Boxing Day 2004, the category 9.2–9.3 Indian Ocean earthquake, centred 160 km (100 miles) off the west coast of northern Sumatra, generated a devastating 30-m (100-ft) tsunami that killed 230,000 people.

Whereas normal ocean waves have a relatively short wavelength, tsunamis – also known as seismic sea waves – possess wavelengths of up to 200 km (125 miles) in the deep, open ocean. There, they travel at up

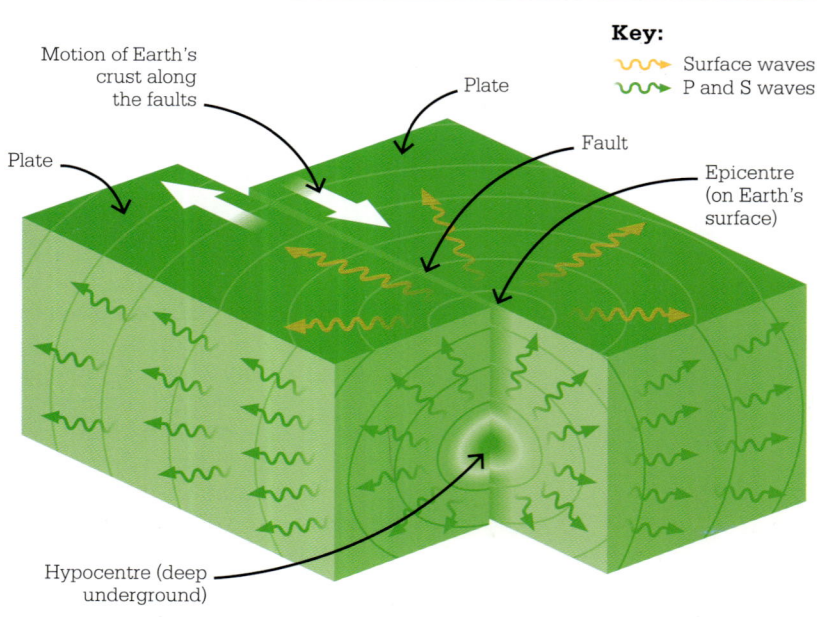

Earthquakes occur when tectonic plates move along fault lines. The energy released takes the form of seismic waves, which spread from the centre of the tremor through the ground, causing it to shake.

PHYSICAL GEOGRAPHY 135

Earthquake frequency and destructive power

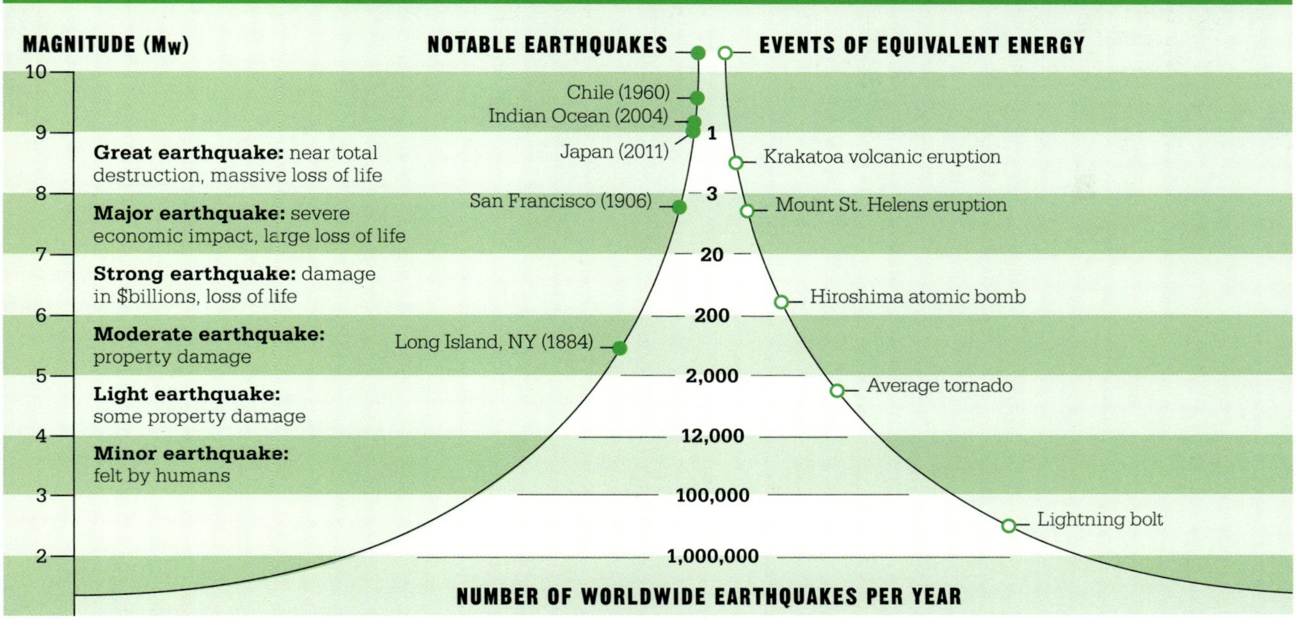

MAGNITUDE (M_W)	NOTABLE EARTHQUAKES		EVENTS OF EQUIVALENT ENERGY
10			
9	Chile (1960) / Indian Ocean (2004)	1	Krakatoa volcanic eruption
8 — **Great earthquake:** near total destruction, massive loss of life	Japan (2011)	3	Mount St. Helens eruption
7 — **Major earthquake:** severe economic impact, large loss of life	San Francisco (1906)	20	
6 — **Strong earthquake:** damage in $billions, loss of life		200	Hiroshima atomic bomb
5 — **Moderate earthquake:** property damage	Long Island, NY (1884)	2,000	Average tornado
4 — **Light earthquake:** some property damage		12,000	
3 — **Minor earthquake:** felt by humans		100,000	Lightning bolt
2		1,000,000	

NUMBER OF WORLDWIDE EARTHQUAKES PER YEAR

to 800 km/h (500 mph) but have a small amplitude, so they are barely noticeable far from land. However, as they move into shallower water, they undergo wave shoaling: the wave is compressed and slows, but its amplitude greatly increases.

Moment magnitude

In the late 1970s, seismologists Hiroo Kanamori from Japan and Thomas Hanks from the US developed the moment magnitude (M_W) scale, which is now the standard earthquake measure. This approach is based on the total energy released by an earthquake, and it takes into account factors such as the fault's total displacement and the force required to move it. As with the Richter scale, each increase of one whole number corresponds to a tenfold increase in the amplitude of the motion. For smaller earthquakes, estimates of moment are similar to Richter magnitudes, but only M_W can accurately measure events of magnitude 8 and above.

With 21st-century technology, constellations of Earth-orbiting satellites that are linked to ground stations – collectively known as global navigation satellite systems – can rapidly detect the extent of tectonic plate activity. Data from this is then fed into models that predict tsunami activity and alert warning systems so that at-risk communities may be evacuated.

After the 2011 Tohoku earthquake, with a hypocentre 130 km (80 miles) east of the Honshu coast, Japan's advanced tsunami-warning system was triggered within just three minutes. The first tsunami did not strike the Japanese coast until 20 minutes after the quake, allowing many people to escape the danger zone. Similar warning systems are now in place in many nations around the Pacific Ocean, the most seismically active region on Earth. After a quake, satellite-based synthetic aperture radar (SAR) instruments can detect potentially quake-damaged buildings and assess the extent of flood water if a tsunami has struck, to guide where relief efforts should be focused. ■

Then the mud began to fall from the wall, and the bookshelf at my side fell down.
Kiyoo Wadati
"Born in a Country of Earthquakes", 1989

THERE IS NOTHING LARGE OR SMALL IN NATURE
BIOMES AND ECOLOGICAL ZONES

IN CONTEXT

KEY FIGURE
Heinrich Walter (1898–1989)

BEFORE
1799 Alexander von Humboldt travels to South America and studies the way plant populations vary at different latitudes and altitudes.

1916 Frederic Clements describes how plant communities grow in predictable stages.

1926 Vladimir Vernadsky publishes *The Biosphere*.

AFTER
2007 Argentinian ecologist Sandra Díaz explores the ways in which societies can value and support ecosystems.

2008 Erle Ellis and Navin Ramankutty introduce the concept of anthromes, which are biomes shaped by humans.

In 1875, Austrian geologist Eduard Suess coined the term "biosphere" to describe the part of Earth where life exists. In reality, the biosphere extends vertically for about 20 km (12.5 miles) – from the depths of the oceans, to the tops of mountains, while most life exists within a 6-km (3.7-mile) comfort zone.

The biosphere consists of different biomes, which are areas classified according to the species – particularly the types of vegetation – that inhabit it. In the 1970s, Heinrich Walter developed a classification system that divided Earth's surface into nine major biomes (or, as he

PHYSICAL GEOGRAPHY

See also: Climatic zones 96–101 ▪ Natural resource management and conservation 222–27 ▪ Desertification 228–29 ▪ The Anthropocene 256–59 ▪ Monitoring biodiversity 306–07

Heinrich Walter

Best known for his biome classification system, which emphasizes the link between climate and vegetation, Heinrich Walter was born in 1898 in Odessa (then Russian Empire, now Ukraine). After studying botany in his home town, he completed his doctorate at the University of Jena, in Germany, where he was a student of Alsatian botanist Ernst Stahl.

From 1923, Walter worked at the University of Heidelberg, becoming a professor of botany in 1927. In the meantime, he had married Erna, the daughter of botanist Heinrich Schenck, and together they travelled extensively, collecting plant specimens and meticulously recording their findings.

Walter died in Stuttgart, Germany, in 1989.

Key work

1973 *Vegetation of the Earth and Ecological Systems of the Geo-Biosphere*

No chemical force on Earth is more constant than living organisms taken in aggregate, none is more powerful in the long run.
Vladimir Vernadsky
The Biosphere, 1926

called them, zonobiomes), based on the annual cycle of temperature and precipitation.

Biogeography
The discipline concerned with the geographic distribution of living things and the factors that affect it is called biogeography, and German explorer and naturalist Alexander von Humboldt is credited with laying its foundations. Between 1799 and 1804, during an expedition to South and Central America, he observed that changes in altitude were linked to shifts in climate zones, similar to the way climate changes from the equator to the poles. Humboldt saw nature as a unified whole formed by dynamic interactions between organisms and their environment, climate, geography, and human activities.

South Sister, in Oregon, US, has dense forests on its lower slopes. At 2,300 m (7,500 ft), alpine shrubs replace trees, and above 3,000 m (9,800 ft), areas of bare rock and snowpack appear.

This concept was reprised by Soviet scientist Vladimir Vernadsky. In his 1926 book *The Biosphere*, Vernadsky used the term to describe the total mass of living organisms on Earth and their interactions with their environment. He suggested that the biosphere was a dynamic system in which energy and nutrients were constantly processed and recycled. Vernadsky's research was influential in showing the link between biology and geography and helping to understand large-scale ecologies. He also founded the discipline of biogeochemistry, which is the study of chemical processes relating to living things and their environment; this was a significant influence on the development of biomes.

Defining biomes
The concept of biomes first emerged in the 1916 book *Plant Succession* by American botanist Frederic Clements, who used the term as a synonym for a biotic community, or a population of living organisms. In the late 1930s, while working with American »

138 BIOMES AND ECOLOGICAL ZONES

Walter's equatorial biome covers areas of humid tropical rainforest, with a thick canopy, a middle layer of plants that thrive in shaded conditions, and a dark forest floor, or decomposition zone.

zoologist Victor Shelford, Clements refined the term to mean a biotic community on a large geographic scale, or all the organisms – plants and animals – that inhabit a specific region.

The variety of organisms present in a particular biome depends on a number of factors, and these can be split into two broad categories. Abiotic (or non-living) factors are things such as temperature, latitude, moisture, and type of soil present; while biotic factors refer to interactions with living organisms, such as predation and competition for resources.

A biome is not the same thing as an ecosystem. While a biome describes a particular geographic area, an ecosystem describes the interaction of living and non-living things within a particular environment. It is therefore possible for several ecosystems to exist within a single biome.

Biome classification

Scientists disagree on the question of how many biomes there are. Some list five: forest, grassland, tundra, desert, and aquatic (often divided into freshwater and marine), while others count as many as 11 terrestrial biomes and four aquatic ones.

Heinrich Walter's classification scheme, which includes nine biomes, is built upon the earlier Köppen climate classification system, which categorizes climates by temperature and precipitation patterns. Walter's system is

Walter's classification comprises nine zonobiomes defined by temperature and precipitation. The map also illustrates a tenth area, known as the orobiome, which represents mountain areas.

Key:
- Equatorial
- Tropical
- Subtropical
- Mediterranean
- Warm temperate
- Nemoral
- Continental
- Boreal
- Polar
- Orobiome

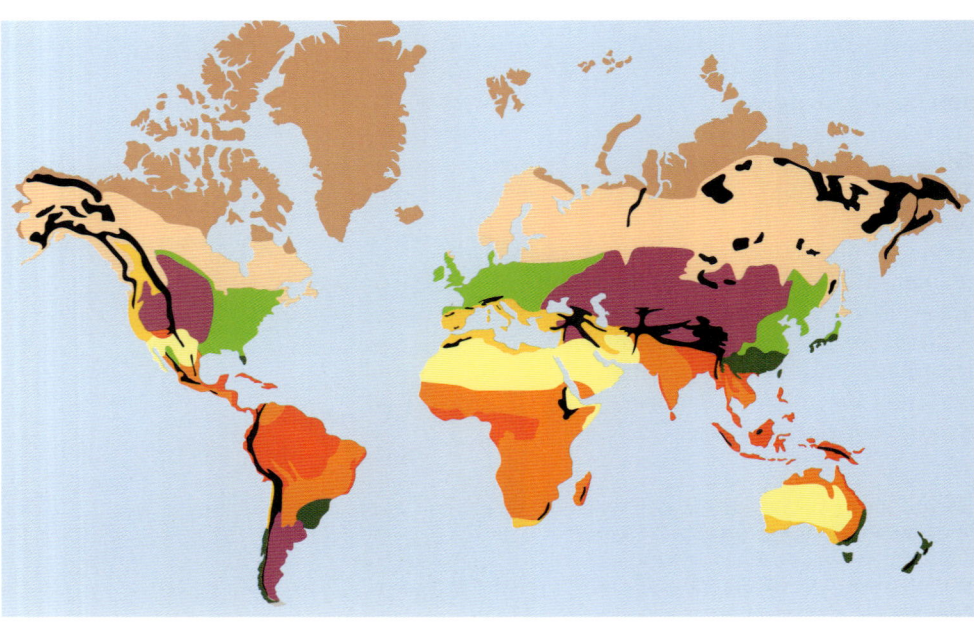

essentially a biological interpretation of the Köppen classification, and it shows how climate influences the distribution of plant and animal life in a particular area. For example, Köppen's tropical rainforest climate, which is marked by fairly constant temperatures averaging 18°C (64°F) and the absence of a dry season, corresponds to Walter's equatorial biome, with shaded, humid tropical rainforest. At higher latitudes, the subtropical arid climate zone receives little or no precipitation, and the typical vegetation types here are desert scrub and succulent plants, which thrive in a perennially water-stressed environment.

Other biomes in Walter's classification include tropical, featuring rainy summers and dry winters; Mediterranean, with dry summers and rainy winters; warm temperate, with mild winters and warm, rainy summers; nemoral, with four months of -10°C (14°F) temperatures and freezing winters; continental, with warm summers and cold winters; boreal, with a cool summer and a cold season lasting about six months; and polar, with fewer than 30 days of temperatures above 10°C (50°F) and long, cold winters.

Biomes and ecozones

The same biome can occur in areas that are geographically distinct but with similar climates – for example, temperate deciduous forests exist across the mid-latitude regions of North America, Europe, Asia, and Australia. Equally, the same biome can be home to different species in different areas. The temperate deciduous forests of Europe tend to be dominated by hornbeams, oaks, and beech, while those of North America have maples, American chestnuts, and hickory. Although classified under the same biome, these forests fall in different ecozones.

Ecozones, which are the broadest of the biogeographic divisions, map out Earth's surface into areas within which organisms have been evolving in relative isolation, separated by barriers such as mountain ranges, deserts, or oceans over long periods of time.

Human-altered biomes

An obvious limitation to the biome classification systems in use today is that they mostly fail to take into account the influence of human activity on the biosphere. Sustained direct human intervention within ecosystems – from urbanization and pollution, to deforestation and agriculture – has fundamentally changed the development of biomes.

In 2008, environmental scientists Erle Ellis and Navin Ramankutty proposed that Earth's terrestrial ecosystems should be considered as anthropogenic biomes, or anthromes, because more than

Delhi's Old District is a densely built area with heavy traffic. Sustained human intervention in this urban anthrome has resulted in poor air quality and a rise in temperature.

A new ethic… sees humanity as part of the biosphere and its faithful steward, not just the resident master and economic maximizer.
Edward O. Wilson
"Vanishing Before Our Eyes",
Time magazine, 2000

three quarters of ice-free land on the planet has been altered "as a result of human residence and land use". Ellis argued that we cannot manage ecological patterns or changes if we do not understand how humans have reshaped Earth's ecosystems. His anthrome classification system has 19 terrestrial biomes, such as urban settlements, villages, croplands, and populated woodlands, and only two biomes that are classified as wildlands, defined as areas that do not have human settlements or agriculture. ∎

GREAT OCEAN CONVEYOR
GLOBAL OCEAN CIRCULATION

IN CONTEXT

KEY FIGURE
Wallace Broecker (1931–2019)

BEFORE
1851 In a bid to map ocean currents, US naval officer Matthew Fontaine Maury drops sealed bottles at various locations and lets them drift.

1902 Swedish oceanographer Vagn Walfrid Ekman theorizes that wind-driven currents affect ocean waters to a depth of about 100 m (330 ft).

AFTER
2000 The international Argo programme launches 4,000 floats to map the temperature and salinity of the oceans.

2021 The Intergovernmental Panel on Climate Change (IPCC) predicts that climate change will probably alter the thermohaline circulation in the next 75 years.

2025 The Arctic experiences its lowest-ever sea-ice extent for the month of February.

In the early 1960s, American researchers Henry Stommel and Arnold Arons established that both abyssal (deep-ocean) and surface currents were part of a vast flowing system known as thermohaline circulation. The term "thermohaline" derives from the Greek words *therme*, meaning heat, and *halos*, meaning salt or salinity. These two factors affect the density of seawater, and they are also the drivers of the circulation process in the ocean.

More than two decades later, in 1987, US geoscientist Wallace "Wally" Broecker showed that this circulation system connected all of Earth's oceans, slowly mixing their water and bringing a fresh supply of oxygen and nutrients to ocean life closer to the surface. This process

---

The **surface** of the ocean moves in **currents** driven by the **wind**.

As **currents at higher latitudes** cool, cold temperatures, evaporation, and **increased salinity** make **surface seawater denser**.

Denser seawater sinks, pulling in warmer, less dense water **from other regions** to replace it.

Wind, Earth's rotation, and changes in seawater density create a global system of ocean circulation.

PHYSICAL GEOGRAPHY   141

See also: Mapping the ocean floor 46–47 ▪ The hydrological cycle 66–69 ▪ Atmospheric circulation and winds 72–73 ▪ The Greenhouse effect 80–83 ▪ The Coriolis effect 84–85

also has a significant impact on the global climate by distributing energy around the planet. Broecker dubbed thermohaline circulation the "great ocean conveyor".

Conveyor mechanics
There is no starting point to the conveyor belt, which runs in a loop. Wind-driven surface currents push warm water from the equator towards the poles. In the North Atlantic, the climate is cold and windy, with low rainfall. These factors increase evaporation from the surface, resulting in colder, saltier, and denser water that sinks to the bottom to form the North Atlantic Deep Water (NADW) mass.

Another cold and salty water mass, the Antarctic Bottom Water (AABW), develops in the Southern Ocean. Its formation is due to rapid cooling and to processes such as sea-ice formation, which makes water denser by expelling salt from the ice, and the evaporation caused when retreating sea ice exposes surface water to the wind.

> The discussion of today's climate and its… potential future changes is often framed in the context of the ocean's thermohaline circulation.
> **Carl Wunsch**
> "What Is the Thermohaline Circulation?", *Science*, 2002

Thermohaline circulation brings warm surface water from the equator to the poles. Here it cools, becoming denser and sinking deep into the ocean, where it is redirected towards the tropics.

Key:
➡ Warm surface current
➡ Cool deep current

These deep-water masses spread out along the sea floor, heading into the deep ocean basins. As the cold water travels towards warmer regions, it gradually heats up, rising to the surface – a process known as upwelling. Other upwellings are caused by the action of wind and currents at the surface, which draw up deep water. Driven by ocean currents, deep cold water moves northwards, eventually warming in the Indian and Pacific oceans, before finding its way into the Atlantic, where the cycle begins again.

The conveyor moves at a few centimetres per second, considerably slower than other ocean currents. It is estimated that it takes 1,000 years for water to complete a full circuit.

Impacts on climate
The complex circulation of air and water that creates the Southern Oscillation climate system – also known as El Niño – involves warmer surface water in the Pacific Ocean moving in the opposite direction to the colder, deeper water; this creates an upwelling of nutrient-rich water in the east, while the water in the west sinks to the bottom.

Both El Niño and the North Atlantic Oscillation, an area of see-sawing atmospheric pressure, are now at risk of being disrupted. Fluctuations in the temperature and salinity of the oceans caused by climate change will affect the great ocean conveyor over the coming decades and lead to unpredictable impacts on these climate systems. ∎

Currents colliding off the coast of Hokkaido, northern Japan, create an eddy around the edges of which phytoplankton, which is the basis of the marine food web, blooms.

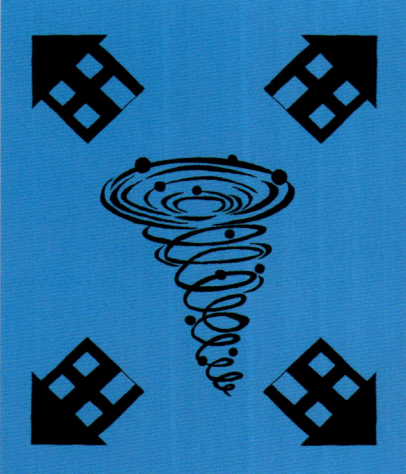

THERE WAS NO TYPE OF BUILDING CODE THAT WAS OF ANY VALUE…
EXTREME WEATHER

IN CONTEXT

KEY FIGURE
Herbert Saffir (1917–2007)

BEFORE
1450 Italian architect Leon Battista Alberti invents the anemometer, a device that catches the air in spinning cups to measure wind speed.

1805 British Royal Navy admiral Sir Francis Beaufort develops a scale for estimating wind speeds from observations of their effects on sea and land.

AFTER
1980 US meteorology professor Fred Sanders coins the term "weather bombs" to describe storms as powerful as tropical cyclones that tend to form in higher latitudes.

2025 A survey by NASA's Grace satellite shows that extreme weather events are becoming more frequent, more severe, and longer-lasting, probably due to human-caused climate change.

Hurricanes and some other **extreme-weather events** are characterized by **high winds**.

The high winds are **dangerous** and cause **damage to buildings**.

The authorities must **give warnings** of this extreme weather **to reduce the dangers**.

Intensity scales based on the damage caused by the extreme weather help the public understand what to do.

Extreme-weather events such as hurricanes are classified according to the damage they have the potential to cause. This knowledge allows the public to act accordingly when warned in advance.

Hurricanes are classified according to five categories – 1 to 5, with 5 being the most powerful, featuring winds in excess of 250 km/h (155 mph). This classification uses the Saffir–Simpson Hurricane Wind Scale, which dates back to the 1970s. Its creators were Herbert Saffir, an American structural engineer who was investigating how best to build low-cost housing in areas prone to hurricanes, and Robert Simpson, the head of the US National Hurricane Center at the time.

Building a warning system
The National Hurricane Center had been set up in Florida in 1965 to track tropical storms as they headed due west across the Atlantic Ocean towards North America, and to give warnings when the storms reached hurricane size and threatened coastal communities. The Central Pacific Hurricane Center in Hawaii assumed a

PHYSICAL GEOGRAPHY

See also: Atmospheric circulation and winds 72–73 ▪ Seismology and earthquake magnitude 132–35 ▪ Crisis mapping with crowdsourcing 304–05

Hurricane Dorian, seen here in a satellite image, was an exceptionally destructive Category 5 storm that struck the Caribbean region and the Atlantic coast of North America in 2019.

similar role for the West Coast in 1970, although hurricane-force storms in this part of the world are known as typhoons and cyclones.

However, despite the forecasting and warning infrastructure, there was no simple way to communicate the severity of the incoming storms. Saffir proposed a scale based on the damage caused to structures and natural habitats by the winds and the storm surge of ocean water that rises up over the land.

Intensity scales
There is no such thing as a safe hurricane. A Category 1 storm, with wind speeds between 119 and 153 km/h (74–95 mph) will always cause some damage to buildings. Category 3 creates devastating damage, destroying roofs and walls, while a Category 5 storm is catastrophic, leading to the total destruction of buildings.

Why does the United States have 75 per cent production of all the global tornadoes?
Ted Fujita
Interview with the American Meteorological Society, 1988

Around the same time as the hurricane scale was being created, Japanese-American meteorologist Ted Fujita was developing a similar system for classifying tornadoes. Tornadoes are smaller and shorter-lasting than hurricanes, but they have faster winds and can cause a great deal of damage with very little warning.

Like the Saffir–Simpson scale, the Fujita scale was organized by wind speed and potential damage. A tornado with an intensity of F0 has winds below 117 km/h (73 mph) and only causes light damage. By contrast, an F5 whirlwind has wind speeds in excess of 419 km/h (260 mph) and causes what the scale calls "incredible" levels of damage – in effect, total destruction.

In 2007, the scale was upgraded to the Enhanced Fujita (EF) scale. An EFU ("U" for unknown) intensity was added to indicate "dust devils" that cause no surveyable damage at all, and the wind boundaries between the ratings EF0 to EF5 were simplified. The actual wind speeds in a tornado are largely estimates, so the enhanced intensity ratings are more focused on the potential for damage.

Frequency of events
The Saffir–Simpson hurricane wind scale and Enhanced Fujita scale have been retrospectively applied to storms that predated the systems' creation. This has made it possible to analyse the relative frequency of high-wind events and their corresponding intensities.

There is no evidence that the number of hurricanes per year is rising, but the intensity of those that do form has increased over recent decades, and so has their destructive power. As for tornadoes, their frequency in Tornado Alley – the central region of the US most prone to such events, due to its atmospheric setup – is dropping, but whirlwinds are becoming more common elsewhere, especially in the southernmost states. ■

HUMAN
GEOGR

APHY

INTRODUCTION

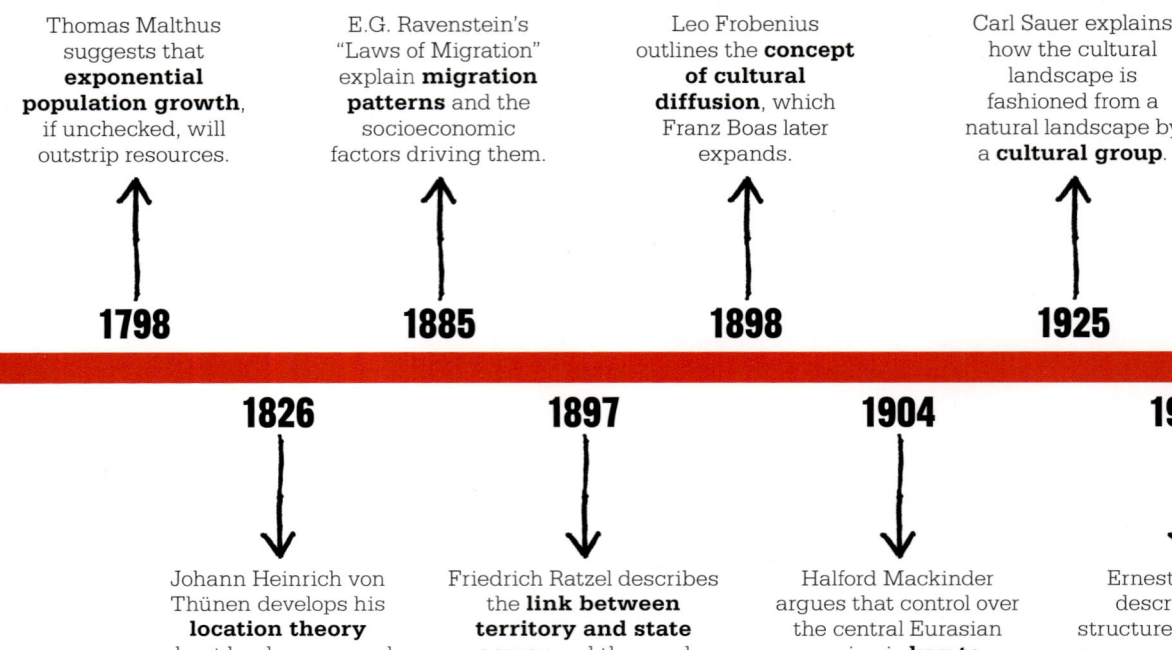

For millenia, geography was viewed largely as the study of location and the physical features of the world, now known as physical geography. However, from the end of the 18th century, the relationship between people and the places they inhabit grew into a distinct field of scientific study: human geography. Using the same principles of reason, rationality, and systematic examination, it evolved alongside, and was influenced by, other social sciences, including anthropology, sociology, politics, and economics.

The prediction by British economist Thomas Malthus that a population, if left unchecked, would exhaust available resources became the basis of geographic population studies from 1798. In the early 19th century, German economist Johann Heinrich von Thünen's theory of land use laid the foundations of economic geography. Over time, the evolving fields of population and economic geography began to include studies of human migration, such as those by E.G. Ravenstein, who examined both its causes and patterns.

Politics and culture

In the 1890s, German geographer Friedrich Ratzel's theory that nation states resembled living organisms gave rise to a new research area – geopolitics. British geographer Halford Mackinder added to the field in the early 1900s with his Heartland theory that emphasized the strategic importance of central Eurasia. During World War II, American political scientist Nicholas Spykman developed the Rimland theory in the 1940s, which noted the tactical significance of the coastal areas of Eurasia.

From the early 20th century, geographers also began to analyse the relationship between human culture and environment. Influenced by sociology and anthropology, theorists explored how environment shapes human behaviour and, conversely, how people tailor their surroundings to their needs. Examining human interaction with the environment, American geographer Harlan Barrows talked of "geography as human ecology". Later thinkers, such as German-English geographer Carl Sauer, described how culture affects landscape, and German-American anthropologist Franz Boas built on earlier ideas that cultural traditions spread through "diffusion".

HUMAN GEOGRAPHY 147

Warren Thompson's **demographic transition model** outlines the stages of population growth and economic development.

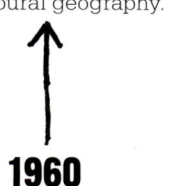

Kevin Lynch's *The Image of the City* studies how humans mentally map cities, introducing the idea of behavioural geography.

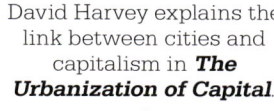

Waldo Tobler's **first law of geography** proposes that everything is related but nearer things are more closely related.

David Harvey explains the link between cities and capitalism in *The Urbanization of Capital*.

1929 — **1960** — **1969** — **1985**

1947 — **1960** — **1970s** — **1993**

The United Nations establishes its Statistics Division to gather population data and monitor **socioeconomic development**.

Walt Rostow publishes *The Stages of Economic Growth*, which presents a model of **development processes**.

Yi-Fu Tuan suggests a humanistic approach to geography, coining the word **"topophilia"** for people's emotional connection to places.

Gillian Rose's *Feminism and Geography* examines how **gender influences the use of geographical space**.

Behavioural geography, which emerged in the 1960s, examined how human decision-making is influenced by perceptions about space. In the 1970s, Chinese-American geographer Yi-Fu Tuan's humanistic geography emphasized the links between emotion and place, and Doreen Massey examined how local and global processes are mutually influential. From the 1980s, she and fellow British geographers Gillian Rose and Linda McDowell also explored the links between gender and space.

Urban living

As a result of industrialization in the West, urban populations expanded and rural models of land use were superseded by a growing field of urban studies. Building on German economist Johann Heinrich von Thünen's theory of agricultural land use, American sociologist Ernest Burgess developed the "concentric zone model" (1925) that provided a systematic explanation for the internal structure of cities. In the mid-20th century, spatial interaction theory – how people, materials, services, and information move between locations – influenced American geographer Waldo Tobler and others looking to model and predict urban patterns.

The power of economics

Economics has remained a potent influence on human geography. In 1929, American demographer Warren Thompson revealed direct links between a nation's birth and death rates and its economic stage of development. American economist Walt Rostow, in 1960, created a "stages of growth" model to show how developing nations could progress in a capitalist world. This was countered in the 1970s by American sociologist Immanuel Wallerstein, who posited that, in a capitalist world-economy, powerful nations exploit those nations on the periphery.

Scholars such as British-American geographer David Harvey have described how capital shapes urban spaces, and spatial justice research continues to explore how location is a crucial factor in determining a person's quality of life. Meanwhile, in the modern digital age, the "online space" has become essential for communication, commerce, and access to information. This provides new spaces and cultures for the study of human geography. ∎

FAMINE SEEMS TO BE THE LAST, THE MOST DREADFUL RESOURCE OF NATURE
MALTHUSIAN THEORY

IN CONTEXT

KEY FIGURE
Thomas Malthus (1766–1834)

BEFORE
1795 In *Sketch for a Historical Picture of the Progress of the Human Mind*, French philosopher Marquis de Condorcet describes a continually improving society based on scientific progress.

AFTER
1965 Ester Boserup's *The Conditions of Agricultural Growth* suggests that population growth stimulates agricultural innovation.

1968 In *The Population Bomb*, US biologist Paul Ehrlich defines overpopulation as a population unable to sustain itself without depleting non-renewable resources.

1981 US economist Julian Simon makes the cornucopian claim that humans will always find a way to produce the necessary resources.

The 18th century was an age of growing prosperity for much of Europe, with trading economies replacing the last vestiges of feudalism. It was a time of great optimism, with the promise of societal progress, and philosophers such as Jean-Jacques Rousseau and William Godwin describing the potential perfectibility of human society.

In reaction to this, Thomas Malthus formulated his view. He believed that increasing prosperity would lead not to a Utopia, but to catastrophe. He reached this conclusion by observing the effect of prosperity – especially an increase in food production – on population. In times of prosperity, he noted, the greater supply of food prompted population growth, which resulted in a greater strain on resources, rather than the anticipated higher standard of living.

A pessimistic scenario
Taking his observations further, Malthus argued that population growth is an inevitable consequence of increased prosperity, and this, in turn, leads to an increase in demand for food production. In his view, if increased supplies were constantly being eaten up by a growing population, an ideal, prosperous society was simply not possible. He also noted that the rate of population growth far outstrips the rate of increase in

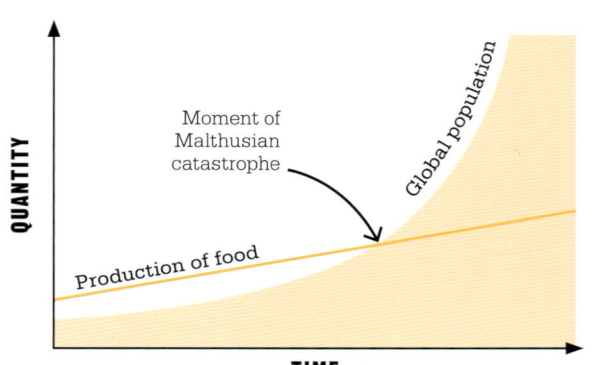

Malthus believed that, left unchecked, population multiplies over time, while food production increases by the same amount each time. The mismatch between population growth and food supply leads to the Malthusian catastrophe.

the resources necessary for a basic standard of living. Left unchecked, population grows exponentially, while food supply only increases arithmetically, and there will come a point when demand exceeds supply.

Malthus argued that this pessimistic scenario (the so-called "Malthusian catastrophe") can be mitigated only by checks on population growth. He identified natural positive checks – such as famine, disease, and war – that limit population in times of scarcity, especially among the lower social classes; he also advocated restraints such as celibacy and chastity as preventative checks.

A divisive theory

Expounded in *An Essay on the Principle of Population* (1798), Malthus's theory sparked a controversy that continues to divide thinkers. In the late 1850s, British naturalists Charles Darwin and Alfred Russel Wallace were struck by his ideas, which informed their notions of natural selection, but others were more sceptical, especially as the benefits of increased productivity came to be felt during the Industrial Revolution.

The optimistic view that people respond to the increased demands of population growth with innovation persisted through the 20th century, supported by huge advances in technology. In 1965, for example, Danish economist Ester Boserup argued that the increased demand brought about by population growth is what drives progress, in particular in more intensive agricultural production.

Malthus's theory has maintained a following – for example, the environmental movement has

Industrial-scale agriculture has increased production; however, larger harvests have come at a considerable cost to biodiversity and soil health.

highlighted the damaging effects of overexploiting natural resources. In a world of finite supplies, the dangers of overpopulation have re-emerged as an issue. However, fertility rates have fallen in most countries, and the UN forecasts that the world's total population will peak at 10.3 billion. While this eases the long-term pressure on food systems, food production challenges linked to climate disruption and resource scarcity remain. ∎

Thomas Malthus

The son of a prosperous land-owner, Thomas Malthus was born in Surrey, UK, in 1766. He was educated at Warrington Academy, Lancashire, and Jesus College, Cambridge. Devoutly religious, he took orders to become a Church of England curate in 1789.

Despite being exposed from an early age to Enlightenment ideals (his father had numbered the philosophers David Hume and Jean-Jacques Rousseau among his friends), Malthus had a bleaker view of social progress, which he illustrated in his 1798 *Essay on the Principle of Population*.

In 1805, Malthus was appointed professor of history and political economy at the East India Company College, Hertfordshire, and went on to become a prominent figure in both the Political Economy Club and the Statistical Society until his death in 1834.

Key work

1798 *An Essay on the Principle of Population*

THE NATURE OF THE INDUSTRY DETERMINES ITS LOCATION
LOCATION THEORY

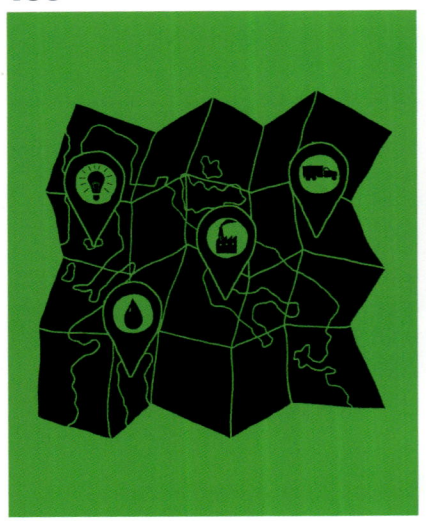

IN CONTEXT

KEY FIGURE
Johann Heinrich von Thünen (1783–1850)

BEFORE
1776 Scottish economist Adam Smith describes the role of free markets in shaping the economy in *The Wealth of Nations*.

1817 According to British economist David Ricardo's theory of rent, differences in the quality of land affect its economic use.

AFTER
1933 German geographer Walter Christaller explains the distribution of cities and towns in his central place theory.

1964 William Alonso, an Argentine-American planner and economist, applies von Thünen's model to urban land.

2006 US economist Ann Markusen explores how creative workers shape the places they inhabit.

Location theory explores the siting of various economic activities in specific places. This field of study has its origins in early 19th-century rural Germany, where economist Johann Heinrich von Thünen sought to explain the pattern of agricultural land use around a central market town.

Born into a wealthy land-owning family, von Thünen observed that proximity to markets appeared to be a determining factor in what sort of goods were produced and where. From this, he developed a theory, later backed up by thorough data-gathering, which he explained in his 1826 book *The Isolated State*.

Concentric rings

Von Thünen took as a hypothetical example a city with a marketplace that is self-sufficient and has no contact with the outside world. Around it is farmland on a uniform plain, which is surrounded by wilderness. Farmers in this "isolated state" transport their products to the central market on carts.

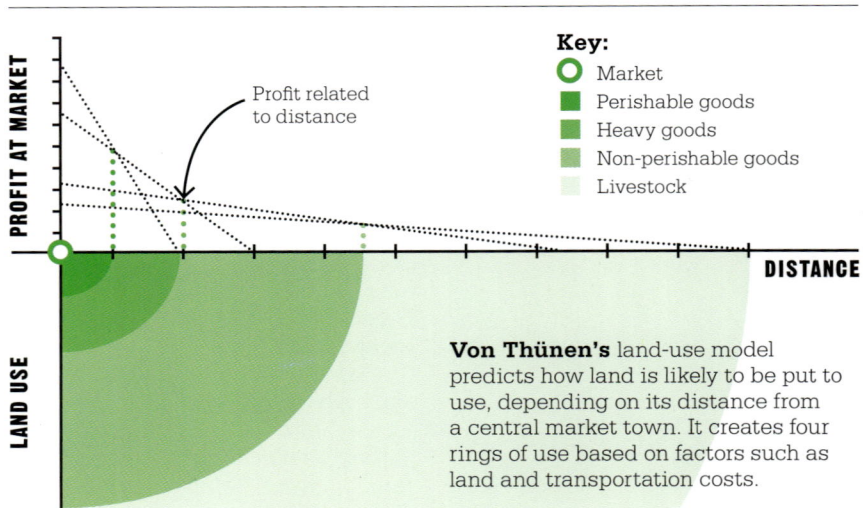

Von Thünen's land-use model predicts how land is likely to be put to use, depending on its distance from a central market town. It creates four rings of use based on factors such as land and transportation costs.

HUMAN GEOGRAPHY 151

See also: Malthusian theory 148–49 ▪ Migration theory 152–53 ▪ Organic theory 154–55 ▪ Rank-size rule 160 ▪ Primate city rule 161 ▪ Geography as human ecology 162-63 ▪ Central place theory 182–83

Farmers' markets such as this one in Esslingen, southern Germany, illustrate von Thünen's theory, attracting local producers of vegetables, dairy, meat, and other perishable goods.

What happens, von Thünen proposed, is that different types of land use develop in concentric rings around the market. In the ring closest to the centre, farmers would produce goods that are perishable, but have a high market value – such as dairy products, vegetables, and fruit, which can be most intensively farmed to counterbalance the relatively greater cost of land near the city.

Heavy goods that are difficult or costly to transport, such as timber, would be produced in the next ring. Beyond that, non-perishables such as grain; then, in the outer ring, livestock, which need no transport and can be walked to market.

Land use principles

Von Thünen's model was a simplification, and advances in transport and technology have diminished the importance of proximity to markets. However, his theory set the agenda for the subsequent study of land use and location. The principle that producers will choose locations in order to maximize their profits, and minimize expenses such as transport, material, and labour costs, had been established, and it could be applied to areas beyond agricultural land use.

Industrialization, for example, led to a reappraisal of von Thünen's theory, with the emphasis shifted to the sources of raw materials,

From ring to ring the staple product, and with it the entire farming system, will change.
Johann Heinrich von Thünen
The Isolated State, 1826

workforce availability, and transportation to market. According to German economist Alfred Weber, who developed his location triangle theory in 1909, these factors are key when deciding the optimum location for a production site.

Location theory is still relevant today, even after innovations such as air transport and the internet, helping firms to weigh up the pros and cons when choosing a suitable business location. ■

Johann Heinrich von Thünen

For many generations, the von Thünens had owned and farmed land near the town of Jever, in the Friesland region of northern Germany. Johann Heinrich von Thünen was born on the family estate in 1783.

Von Thünen went to agricultural college in Holstein, where he formulated the ideas of his location theory. After further study at the University of Göttingen, he married and bought the Tellow estate in Mecklenburg in 1810. As well as running the estate, von Thünen used the opportunity to gather information to support his theory of land use and location, which he published in 1826 in *The Isolated State*. The 1850 version of his magnum opus also included an additional section devoted mainly to an exposition of his wage theory. Von Thünen died at the Tellow estate in September 1850.

Key work

1826, 1850 *The Isolated State*

MIGRATION MEANS LIFE AND PROGRESS
MIGRATION THEORY

IN CONTEXT

KEY FIGURE
Ernst Georg Ravenstein
(1834–1913)

BEFORE
1776 Scottish economist Adam Smith argues in *The Wealth of Nations* that people should be free to move where job opportunities are.

1870s British epidemiologist Dr William Farr remarks that migration appears to go on "without any definite law".

AFTER
1920 A book on the Polish diaspora to the US marks the beginning of the study of migration.

1966 US sociologist Everett S. Lee's "A Theory of Migration" identifies push and pull factors in the movement of people.

2008 According to American sociologist Douglas Massey, Mexican migration to the US is driven as much by labour demand as it is by poverty.

In the 19th century, in the wake of the Industrial Revolution, Britain underwent a period of radical change. The creation of mills and factories triggered a wave of urbanization, creating centres of economic activity with unprecedented employment opportunities in big cities. The workforce responded by moving from agricultural communities to where the work was.

Ernst Georg Ravenstein pioneered the study of such migrations. Using data from censuses between 1871 and 1881, he attempted to identify patterns of migration in Britain in order to understand its causes and predict future movements of populations.

Absorption and dispersion

It may seem obvious today that migration in 19th-century Britain was driven by economic factors – from rural areas with few opportunities for work to thriving industrialized urban centres. It was only in the 1880s, however, that anyone thought to analyse the data and identify patterns of movement. Ravenstein undertook that research and presented his conclusions to the Journal of the Statistical Society of London as his "Laws of Migration". Originally seven in number, they were later expanded to 11.

Ravenstein observed that certain areas of the country, which he called "counties of absorption", attracted more people than they lost, while in other areas – "counties of dispersion" – more people left than were taken in. A simple comparison of the areas of absorption and dispersion showed him that migration was, to a large extent, from rural to urban areas, providing him with one of his fundamental laws of migration.

Cotton mills, such as this one in the British county of Lancashire, were among the factories attracting rural workers to industrial towns with employment opportunities.

HUMAN GEOGRAPHY 153

See also: Malthusian theory 148–49 ▪ Location theory 150–51 ▪ Rank-size rule 160 ▪ Primate city rule 161 ▪ Geography as human ecology 162–63

Untenable living conditions in rural areas force **workers to become mobile**. → **Migrants** move from agricultural areas of **dispersion** to industrial areas of **absorption**.

Migration flows along well-defined geographical channels. ← **Vacancies** left in rural areas **are filled** by migrants from more **remote districts**.

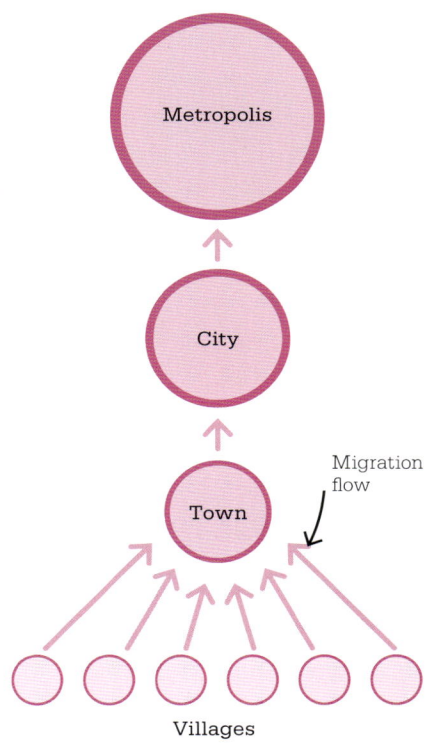

People tend to migrate in stages, moving first from their village to a small town, then a city, and finally a metropolis. Ravenstein called this progression "step migration".

By mapping the movement of people, Ravenstein was able to show that there were discernible currents of migration, and although most migration was over a short distance, vacancies left by departing migrants were filled by others, creating flows that stretched across the country.

Other migration laws

Given the rise in employment opportunities in urban areas, compared with the dearth of jobs and poor working conditions in agriculture, the pattern is far from surprising. However, as Ravenstein examined the detail of his data, less obvious "laws" began to emerge, such as that women tend to be more migratory within a country, but men more migratory over long distances.

Ravenstein's laws were initially dismissed as merely finding a pattern for the particular internal migration in Britain. Later studies in other European countries, however, proved them to be useful guidelines, and they were also found to be applicable to models of international migration. ■

Ernst Georg Ravenstein

Best remembered today for his 11 laws of migration, which helped establish the field of migration studies, Ernst Georg Ravenstein was primarily a cartographer and a historian of geography.

Born in Frankfurt am Main, Germany, in 1834, Ravenstein was apprenticed to German cartographer August Heinrich Petermann, but in 1855 he moved to London to take up a post in the topographical department of the British War Office. He became a naturalized British citizen, anglicizing his name to Ernest George, and spent most of the rest of his life in London.

Among Ravenstein's published works were atlases, as well as histories of cartography and geography. He was also interested in demography, estimating the population of the world as it was then and into the future, and developing his theory of migration. He died while visiting Germany in 1913.

Key work

1885 "Laws of Migration"

THE STATE IS AN ORGANISM
ORGANIC THEORY

IN CONTEXT

KEY FIGURE
Friedrich Ratzel (1844–1904)

BEFORE
1859 *On the Origin of Species*, by British naturalist Charles Darwin, helps to shape the idea that strong societies dominate weaker ones.

1864 British anthropologist Herbert Spencer coins the expression "survival of the fittest". A supporter of social Darwinism, he applies natural selection to human societies.

AFTER
1902 Peter Kropotkin, a Russian geographer, proposes that the fittest are not necessarily the strongest, but those who manage to cooperate with each other.

1926 Karl Haushofer popularizes the concept of *Lebensraum* in his native Germany. After World War II, he dies by suicide rather than stand trial at Nuremberg.

In the same way that a living organism must seek out a source of food to grow and survive, so must a nation acquire land to expand and thrive. German geographer Friedrich Ratzel proposed this concept, which became known as organic theory, in his 1897 book *Political Geography*.

The organic theory of the state emphasizes the close relationship between a nation's geographical space and resources and its political power. The more territory a country controls, the more it can sustain and preserve its integrity.

Everything that wants space on our planet earth must draw on the finite amount… of its surface.
Friedrich Ratzel
"Lebensraum", 1901

The implication is that a nation that does not conquer new territories risks failing, because other nations will be doing just that; thus, stronger countries naturally grow at the expense of weaker ones.

Origins of *Lebensraum*

In his 1860 review of Charles Darwin's *On the Origin of Species*, German geographer Oscar Peschel coined the term *Lebensraum* (living space) to denote the specific natural region in which a particular people emerged and developed. Although Peschel only used the term within a geographic and anthropological context, Ratzel later applied the word *Lebensraum* to describe the physical territory a nation needed for its survival.

Organic theory posits that a political entity must seek out *Lebensraum*, conquering all the territory that it possibly can in order to secure space, resources, and power. At a time when industrialized nations such as Britain, France, and Germany competed for global influence, Ratzel's theory, which provided intellectual justification for colonialist expansion, soon became popular.

See also: Environmental determinism 156 ▪ Heartland theory 158 ▪ Rimland theory 159 ▪ Territoriality theory 202–03

A German marching band crosses the border into Czechoslovakia after the 1938 annexation of the Sudetenland region. This was Adolf Hitler's first step in his pursuit of *Lebensraum*.

A pupil of Ratzel's, Swedish political scientist Rudolf Kjellén saw modern political states as being similar to organic systems that could flourish and then decay. His 1900 book *Introduction to Sweden's Geography* contains the first mention of the term "geopolitics", which he used to describe the problems and conditions within a state that arise from its geographic features.

The impact on politics

Kjellén's theories, also expressed in his 1916 book *The State as a Lifeform*, found fertile ground in Germany. In 1911, General Friedrich von Bernhardi interpreted the concept of *Lebensraum* as a racial struggle for living space. He believed that Eastern Europe was a necessary acquisition in order to protect Germany's racial supremacy and to defend the country against cultural stagnation. Organic theory was also on the syllabus at the Institute for Geopolitics in Munich, established in 1922 by German professor and diplomat Karl Haushofer.

Adolf Hitler was personally mentored by Haushofer and eagerly embraced ideas about competing nations. In *Mein Kampf*, published in 1925, Hitler upheld *Lebensraum* and the geopolitics of "inevitable expansion" as the philosophic basis for the German colonization of Eastern Europe.

After World War II, organic theory was largely discredited for the role it had played in the rise of fascism. Today, geopolitics is concerned more with diplomacy and economic influence than with the acquisition by force of new territory. ∎

Friedrich Ratzel

The son of the Grand Duke of Baden's household staff manager, Friedrich Ratzel was born in Karlsruhe, Germany, in 1844. After studying zoology, he began a period of travel during which he grew increasingly interested in geography. Letters he wrote about his field work in the Mediterranean led to a job as a travelling reporter for a Cologne-based newspaper.

During a trip to North America and Cuba in 1874–75, Ratzel studied the influence of people of German origin in the US. On his return, he started teaching geography at the Technical High School in Munich.

Ratzel explained his theory of environmental determinism in *Anthropogeography*, published in two volumes in 1882 and 1891, while *Political Geography*, published in 1897, introduced his ideas on the meaning of *Lebensraum*. Ratzel died in Bavaria, Germany, in 1904.

Key works

1882, 1891 *Anthropogeography*
1897 *Political Geography*

MAN IS A PRODUCT OF THE EARTH'S SURFACE
ENVIRONMENTAL DETERMINISM

IN CONTEXT

KEY FIGURES
Friedrich Ratzel (1844–1904),
Ellen Churchill Semple (1863–1932)

BEFORE
1377 Arab polymath Ibn Khaldun suggests people are sedentary or nomadic because of soil, food, and climate.

1785 In *Notes on the State of Virginia*, US statesman Thomas Jefferson proposes that tropical climates encourage laziness.

AFTER
1920 British geographer Thomas Griffith Taylor's "stop and go" determinism is a middle ground between environmental determinism and possibilism.

1997 In his book *Guns, Germs, and Steel*, US scientist Jared Diamond argues that environmental factors do have a relevance in determining which civilizations succeed and which fail.

German geographer Friedrich Ratzel developed the theory of environmental determinism in his two-volume book *Anthropogeography* (1882 and 1891). He argued that, like all living organisms, human populations face multiple environmental challenges – from physical geography to climate. Ratzel concluded that the environment is responsible for shaping human behaviour and social development. Echoing Darwin's *On the Origin of Species*, he claimed that the adaptability of a nation was essential to its ability to thrive.

In *Influences of Geographic Environment* (1911), US geographer Ellen Churchill Semple reiterated that the environment is the primary influence on human behaviour. This trend continued with Ellsworth Huntington, another US geographer, who pointed to climate as the key factor moulding human society.

Racist agendas
Environmental determinism was the dominant theory in geography in the West in the early 20th century; against a backdrop of global imperialism, many of its ideas were appropriated to support an array of racist agendas, especially in Nazi Germany.

By the 1920s, critics such as US geographer Carl Sauer argued for a more nuanced approach, which took into account the agency of individuals and societies to develop despite environmental constraints. ■

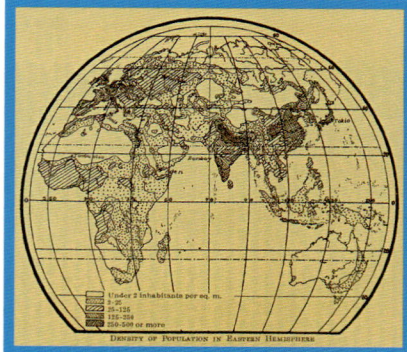

Semple's 1911 map of the eastern hemisphere showed how environmental conditions such as climatic conditions affect population distribution.

See also: Location theory 150–51 ■ Organic theory 154–55 ■ Possibilism 157 ■ Geography as human ecology 162–63 ■ Cultural landscape theory 172–75

HUMAN GEOGRAPHY

THERE ARE NO NECESSITIES, BUT EVERYWHERE POSSIBILITIES
POSSIBILISM

IN CONTEXT

KEY FIGURE
Paul Vidal de la Blache
(1845–1918)

BEFORE
64 BCE The Greek geographer Strabo writes that people can escape the physical constraints of their environment.

1748 Montesquieu, a French judge and philosopher, argues that humans possess free will and can choose from different opportunities.

AFTER
1922 French historian Lucien Febvre first uses the word possibilism in *A Geographical Introduction to History.*

1925 In *The Morphology of Landscape,* US geographer Carl Sauer argues that human activity is the main driver of landscape change.

2021 British architect Michael Pawlyn suggests adopting a possibilist mindset to deal with environmental challenges.

In 1911, French geographer Paul Vidal de la Blache questioned the then-dominant theory of environmental determinism. He argued that surroundings do not have to dictate human behaviour; instead, people can modify their environment to meet their own needs. Vidal de la Blache believed that social conditions trumped environmental circumstances, and his views were later dubbed possibilism.

Moulding the environment

For millennia, human initiatives to mitigate challenging environments have enabled civilization to flourish in the harshest conditions. Where extreme aridity limits agriculture and potentially holds back civilization, people practise irrigation. In regions where heavy rainfall could wash away topsoil and crops on steep slopes, communities construct terraces.

In 1952, British-born Australian geographer Oskar Spate introduced the concept of "probabilism" as a compromise between environmental determinism and possibilism. However, possibilists have always understood that environmental constraints do exist. While shelters made from snow could be built in the tropics using refrigeration, igloos are found at high latitudes where snow is abundant.

Possibilism helps to provide an understanding of how human society has developed and how it must develop further in the face of a rapidly changing climate. ■

Man, as master of the possibilities, is the judge of their use.
Lucien Febvre
A Geographical Introduction to History, 1922

See also: Location theory 150–51 ▪ Organic theory 154–55 ▪ Geography as human ecology 162–63 ▪ Cultural landscape theory 172–75

WHO RULES THE HEARTLAND COMMANDS THE WORLD-ISLAND
HEARTLAND THEORY

IN CONTEXT

KEY FIGURE
Halford Mackinder
(1861–1947)

BEFORE
13th century The Mongol Empire controls much of Central Asia, or what would be defined as the heartland.

1890 Alfred Thayer Mahan, a US naval officer, stresses the importance of maritime power in *The Influence of Sea Power upon History*, a study of naval warfare.

AFTER
1997 Zbigniew Brzezinski, a Polish-American diplomat, revives heartland ideas in the context of US strategy to prevent any one power from dominating Eurasia.

2022 Russia launches a full invasion of Ukraine, a former Soviet republic on the edge of the heartland that has been moving towards closer ties with NATO.

One of the most influential geopolitical ideas of the 20th century, the heartland theory was proposed in 1904 by British geographer Halford Mackinder. He believed that global power depended on control of a central area of Eurasia, which he called the pivot area and, later, the heartland. He saw this landlocked region, stretching from Eastern Europe into Central Asia, as the geographic pivot of history. In his words: "Who rules East Europe commands the Heartland; who rules the Heartland commands the World Island; who rules the World Island commands the world." The "world island" was Mackinder's term for the combined Eurasian and African landmasses.

Continental power

At the time, sea power had long dominated geopolitics, thanks in part to the British Empire's naval strength. Mackinder challenged this view, warning that railways and land-based transport were making continental control more important than naval supremacy. He argued that the heartland – rich in resources and hard to invade due to its geography – could become the base of a future world empire.

The theory gained attention during both World Wars and the Cold War, as powers vied for influence in Eastern Europe and Central Asia. Nazi Germany's push for *Lebensraum* in Eastern Europe, and Soviet dominance in the same area after World War II, gave new life to Mackinder's ideas and helped shape Western strategies. ∎

The Heartland offers the basis of an all-powerful militarism.
Halford Mackinder
Democratic Ideals and Reality, 1919

See also: Location theory 150–51 ▪ Organic theory 154–55 ▪ Rimland theory 159 ▪ Territoriality theory 202–03

WHO CONTROLS THE RIMLAND RULES EURASIA
RIMLAND THEORY

IN CONTEXT

KEY FIGURE
Nicholas J. Spykman
(1893–1943)

BEFORE
1897 In *Political Geography*, German geographer Friedrich Ratzel puts forward the organic theory, likening a country to a living organism.

1904 Halford Mackinder's heartland theory emphasizes the importance of seizing control of central Eurasia.

AFTER
1947 US president Harry S. Truman's domino theory states that changes in the political structure of one country tend to spread to neighbouring countries.

2013 China proposes the development of the Maritime Silk Road, a shipping route that will link the country with Asia, Africa, and Europe via the Indian Ocean and the Mediterranean Sea.

In 1942, US political scientist Nicholas J. Spykman adapted what British geographer Halford Mackinder had identified as the inner crescent surrounding the heartland, and reimagined it as the rimland. Spykman argued the maritime-oriented rimland powers on the periphery of Eurasia had greater trading opportunities with the outside world by virtue of having open access to sea. Therefore, the influence of land-oriented heartland powers could be kept in check by the rimlanders, which included Western Europe, the Middle East, and Asia's Pacific nations. Spykman concluded that Mackinder had overestimated the importance of the heartland, stating that whoever controlled the rimland could dominate the world.

Cold War significance

After the end of World War II, the USSR was largely in control of Mackinder's heartland, having expanded its influence into Eastern Europe. During the Cold War years, Spykman's rimland

The US intervention in the Vietnam War (1955–75) aimed to prevent the spread of communism from the country's north to the republican south.

theory was especially significant as the US sought to prevent Soviet expansion into coastal zones. The formation of the North Atlantic Treaty Organization (NATO) in 1949 by the US and its West European allies can arguably be seen as a rimland challenge to heartland domination

US foreign policies supporting Eurasian rimland allies in an effort to prevent the spread of heartland communism was a direct application of Spykman's ideas, resulting in conflicts in, for example, Korea and Indochina. ■

See also: Organic theory 154–55 ■ Heartland theory 158 ■ Territoriality theory 202–03

A RELATIONSHIP BETWEEN SIZE AND RANK THAT IS BOTH QUITE PRECISE AND QUITE SIMPLE
RANK-SIZE RULE

IN CONTEXT

KEY FIGURE
George Zipf (1902–50)

BEFORE
1925 Canadian-American sociologist Ernest Burgess sets out the concentric zone model for the development of cities.

1929 The theory of demographic transition by US sociologist Warren Thompson investigates the link between population growth and economic development.

1933 German geographer Walter Christaller's central place theory explores the hierarchy of settlements.

AFTER
1939 Mark Jefferson, an American geographer, proposes the primate city rule.

1990 After decades of Seoul-centred industrialization, the Korean capital's primacy weakens, and the country's urban system moves closer to a rank-size distribution.

The rank-size rule states that a city's population is inversely proportional to its rank. This relationship was first observed in 1913 by German physicist Felix Auerbach, but it was US linguist and statistician George Zipf who popularized it in 1935.

Zipf's law

While studying the frequency of words used in different languages, Zipf discovered that some words appeared often, whereas others were used more rarely. He established that the second most common word occurred approximately half as often as the first, and the third most common word a third as often. As a result of his study, this inverse relationship became known as Zipf's law.

When applied to urban studies, this concept states that the second-largest city will have a population that is approximately half the size of the largest city, the third-largest city will have a population roughly equal to one third the size of the largest city, and so on.

The number of cities in the first largest country is twice... that in the second largest.
Bin Jiang et al, 2015

The rank-size rule can be applied in urban planning and development. For example, understanding the distribution of city sizes can help with decisions about where to build transportation networks.

Zipf's law works well in many countries, such as China and the US, but not in all. India has a few megacities – such as Mumbai, Delhi, and Kolkata – that are significantly larger than its other cities. Similarly in Canada: Toronto, the largest city, is closely followed by Vancouver and Montreal. ■

See also: Migration theory 152–53 ■ Primate city rule 161 ■ Demographic transition model 176–81 ■ Central place theory 182–83

HUMAN GEOGRAPHY

A COUNTRY'S LEADING CITY IS DISPROPORTIONATELY LARGE
PRIMATE CITY RULE

IN CONTEXT

KEY FIGURE
Mark Jefferson (1863–1949)

BEFORE
1826 Johann Heinrich von Thunen, a German economist, produces his isolated state model, which explores patterns in land use.

1935 American linguist and statistician George Zipf formulates the rank-size rule.

AFTER
1950s In post-World War II Japan, Tokyo's full primacy is challenged by other industrial cities like Osaka and Nagoya.

1961 French geographer Jean Gottmann argues that megalopolises – such as the region stretching from Boston to Washington – weaken the concept of a primate city.

2021 Nigeria's Kano State government promotes Kano's development to help reduce Lagos's primate-city dominance.

In some countries, one city dominates over all others in its cultural and economic influence and as a population centre. In 1939, American geographer Mark Jefferson coined the term "primate city" to describe this phenomenon. A primate city, according to Jefferson, is one that is at least twice as large and twice as important as the next-largest city.

Primate cities may be, but are not always, capital cities. They dominate their country and are the national focal point in terms of commerce and culture.

These factors, combined with plentiful employment opportunities, exert a strong pull factor, drawing more people to the city. The net result is that the primate city grows even larger and more dominant over the smaller ones.

A national focal point
Some geographers now define a primate city as one that is larger than the populations of the second- and third-largest cities combined. In the UK, for example, capital city London has four times as many inhabitants as Birmingham, the country's second city; Paris is five times more populous than Lyon; and Columbo in Sri Lanka, is 45 times bigger than Kandy.

Not all countries have primate cities. India, China, and the US are examples of countries with several cities of comparable size and no single dominant city. ■

Bangkok is a primate city with an estimated population of 10.7 million, while Thailand's second city, Chiang Mai, is home to 1.2 million people.

See also: Location theory 150–51 ■ Migration theory 152–53 ■ Rank-size rule 160 ■ Cultural landscape theory 172–75

GEOGRAPHY BORE MANY CHILDREN
GEOGRAPHY AS HUMAN ECOLOGY

IN CONTEXT

KEY FIGURE
Harlan H. Barrows
(1877–1960)

BEFORE
1866 Ernst Haeckel, a German zoologist and naturalist, coins the term "ecology" to describe the interactions between the living world and its environment.

c. 1882 German geographer Friedrich Ratzel advocates environmental determinism, proposing that human activities are shaped by physical geography.

AFTER
1925 Carl Sauer, a US geographer, advances the idea that humans both adapt to and transform their environment.

1960s US urban planner Kevin Lynch and Australian-American geographer Reginald Golledge look at how human behaviour shapes the physical world.

As the 19th century came to a close, it seemed that the entire surface of Earth had been, or would very soon be, thoroughly mapped. However, geographers' attention then turned to the relationships between the landscapes that were described by physical geography and the humans who inhabited those landscapes.

The actions of people

Harlan H. Barrows believed that simply describing what exists in the landscape was not enough. He asserted that the actions people take to adapt and change their environments in order to survive was just as important a focus of study for geography. He made this case in a speech to the Association of American Geographers in 1923, arguing that "geography is the science of human ecology".

In 1936, Robert E. Park, an American sociologist whose work heavily influenced thinking on human–environment interactions, used the term "human ecology" to describe the study of the relationships between humans

Harlan Harland Barrows

Born in Armada, Michigan, in 1877, Harlan H. Barrows taught history and geography at the Ferris Institute in Big Rapids, Michigan, before graduating from the University of Chicago in 1903. He attended graduate school at the same university's newly formed department of geography – the first in the US – and became an assistant professor in 1908. He was later made an associate professor of geography, in 1910, and was named a professor in 1914.

In 1922, Barrows was elected as president of the Association of American Geographers. From 1933, he served as a consultant to the administration of President Franklin D. Roosevelt. Barrows stepped down as chair of the University of Chicago geography department in 1942 and was named emeritus professor. He died in Highland Park, Illinois, in 1960.

Key work

1923 "Geography as Human Ecology"

HUMAN GEOGRAPHY 163

See also: Location theory 150–51 ▪ Environmental determinism 156 ▪ Possibilism 157 ▪ Cultural landscape theory 172–75 ▪ Behavioural geography 194–95

Scientists and scholars use overlapping ideas from human and physical geography – as well as concepts from ecology, sociology, and anthropology – to explore and better understand the complex relationships between people and their environments.

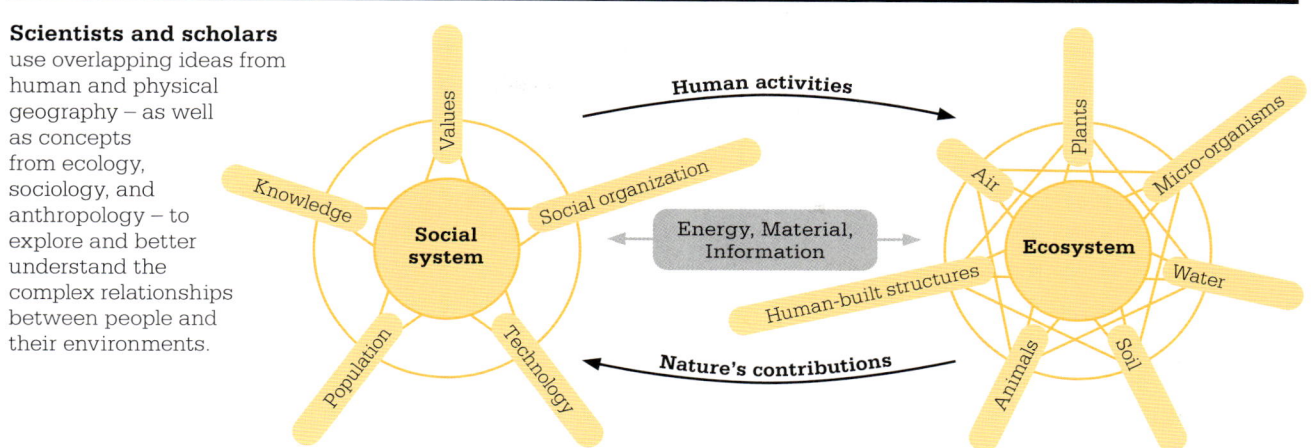

and their environment, whether that be the natural environment or the built environment.

A complex relationship

Human ecology explores not only the influence humans have on their environment, but also the way the environment influences human behaviour. Barrows maintained that geographers should study the deliberate interactions people have with the environment rather than looking for passive influences, and that human–environment relationships were complex. Knowledge of the natural world had to be balanced with awareness of social and economic pressures.

This way of thinking countered the 19th-century theory of environmental determinism, which suggested that human cultures were influenced almost entirely by their environment, failing to consider the fact that in many cases people make positive efforts to mould their environments to suit their needs, rather than adapt themselves to their environment – a perspective that came to be known as possibilism.

Population dynamics is a key focus of human ecology studies, looking at how population size and density affect the availability of resources and the stresses placed on the environment – for example, deforestation. It also examines how humans adapt to a range of different environments through technological innovation or cultural practices. At the same time, the theory accounts for the constraints placed on people by environmental factors.

The physical landscape – its topography and climate – is the stage for human activities, but humans can contribute to a certain amount of scene-setting within the space, reshaping that landscape over time as required. This is accomplished through the social and economic structures created by people, such as urban planning and systems of agriculture, to manage their relationships with the environment. ■

Humans change their environment to solve problems and access resources. One example is the building of dams to supply water, support farming, and generate power.

DIFFUSION OF CUSTOMS OVER ENORMOUS AREAS

CULTURAL DIFFUSION

CULTURAL DIFFUSION

IN CONTEXT

KEY FIGURE
Franz Boas (1858–1942)

BEFORE
1890 In *The Law of Imitation*, French sociologist Gabriel Tarde uses the term "diffusion" in a social scientific sense.

1897 German ethnologist Leo Frobenius identifies several *Kulturkreise*, or culture circles, showing similar traits spread by diffusion or invasion.

AFTER
1911 Grafton Elliot Smith, an Australian-British anatomist, proposes that all human civilizations have a single origin (hyperdiffusionism).

1962 Canadian philosopher Marshall McLuhan coins the term "global village".

2018 Indian-Canadian academic Annamma Joy observes how cultural diffusion is transforming patterns of consumption.

Humans have been grouping together to enhance their chances of survival ever since they first appeared on Earth. Every human society has its own culture – a collective wealth of shared language, traditions, and values. However, cultures are not fixed. When a group comes into contact with another group of a different culture – through means such as migration, communications, and trade – both may borrow and adapt each other's cultural traits. This process of evolution through contact is at the core of the theory of cultural diffusion, and one of its main proponents was Franz Boas.

Developing a concept

Preceding the idea of cultural diffusion was the concept of cultural evolution – that is, the theory of a linear progression in a society's culture. British anthropologist Edward Burnett Tylor theorized that cultures, just as species, progressed along the same path through three stages of development: from savagery to barbarianism, and then on to civilization. Within this pattern of unilineal, or strictly linear,

Survivals are early cultural practices that persist despite being outdated. Examples include blood-letting, which continued even after more effective medical treatments became available.

evolution, Tylor viewed European societies as the most advanced on the evolutionary ladder. He argued that people in different locations were equally capable of progressing through the various stages of cultural development, and that some societies were simply at different stages of evolutionary development. The parallel evolution of different cultural traditions across different societies was

Franz Boas

Pioneering anthropologist Franz Boas was born in the Westphalia region of Germany in 1858. He studied geography, mathematics, and physics at the universities of Heidelberg and Bonn; he then went to the University of Kiel, where he was awarded a PhD in physics in 1881.

Two years later, keen to investigate the relationship between the objective world and people's subjective experiences of it, Boas rekindled his interest in geography. He travelled to Baffin Island, Canada, to work with the Inuit. In 1886, on his way back from Canada, he decided to stay in New York, becoming an editor of the magazine *Science*. His first teaching position was at Clark University, Massachusetts, in 1889. In 1899, he became professor of anthropology at New York's Columbia University.

Boas was one of the founders of the American Anthropological Association. He died in 1942.

Key work

1911 *The Mind of Primitive Man*

HUMAN GEOGRAPHY 167

See also: Possibilism 157 ▪ Geography as human ecology 162–63 ▪ Cultural landscape theory 172–75 ▪ Humanistic geography 200–01 ▪ The urbanization of capital 204–05

The existence of any pure race with special endowments is a myth, as is the belief that there are races… foredoomed to eternal inferiority.
Franz Boas
Race and Democratic Society, 1945

A society comes into **contact** with another via activities such as **trade, media, and migration**.

→

People in the receiving society are **exposed to** the other group's **cultural traits**.

↓

Integration happens when new cultural traits are fully normalized within the receiving society.

←

These **traits** – beliefs, practices, foods – slowly **start to filter** into the receiving society.

explained, Tylor believed, by the fact that people are all basically the same, with the same needs.

According to Tylor, simpler contemporary societies resembled ancient societies, and proof of cultural evolution could be gleaned through the presence of what he called "survivals" – that is, traces of earlier customs or beliefs that persisted despite being outdated.

Although he was an evolutionist, Tylor was open to ideas of diffusion. In *Researches in the Early History of Mankind and the Development of Civilization* (1865), he proposed the concept of diffusion to explain why similar cultural elements appeared in different geographical groups, and how these elements changed over time within the same group. He suggested that shared traits across different cultures were not always necessarily the result of independent invention, as claimed by evolutionists, but could be the result of simple diffusion following contact between different groups.

Boas's theories

In the early 20th century, diffusion began to challenge evolution as the mechanism for explaining cultural differences. A leading opponent of evolutionism was Franz Boas, who was interested in how the physical environment influenced human perception, thought, and culture. At the time, geographers were divided over the root causes of cultural variation. Many believed that the physical environment played a key role, while others argued that the diffusion of ideas through human migration was more important.

Boas stressed the importance of fieldwork and trustworthy data collection. In his view, the cultural »

This Chinese New Year celebration in London is a result of relocation diffusion. Thanks to migration, people outside China can experience Chinese festivals, traditions, and foods.

168 CULTURAL DIFFUSION

evolutionists' fixed stages of progress were based on insufficient evidence gathered by "armchair anthropologists" who relied on third-hand accounts.

Historical particularism
Boas promoted cultural relativism – that is, the belief that each culture should be understood on its own terms, rather than evaluated by Western standards, such as "progress". He also introduced the concept of historical particularism: to truly understand a culture, one should refrain from attempting to fit it into a universal evolutionary model. Cultural traits, he argued, should be analysed in light of their own historical development, beginning with the circumstances of their initial introduction and continuing to the stage when their original source becomes obscure.

The cultural make-up of a people, Boas suggested, results from the cumulative effects of diffusion, weaving together diverse cultural threads into new patterns. Over time, this process leads to the spread of technologies, practices, and ideas across large areas. Boas also pointed out that similarities between cultures in different parts of the world did not necessarily mean that these cultures had gone through the same developmental stages. Instead, they might have borrowed from one another, or coincidentally come up with similar solutions to the same challenges. Boas believed that this kind of borrowing was a common practice between neighbouring cultures and an important tool for understanding why cultures are the way they are.

How culture spreads
Cultural diffusion helps to explain why different geographic regions share similar cultural elements. Cultural traits that originate in a specific area, known as the cultural hearth, spread outwards into a larger cultural region, which means that the geography of culture is in constant flux.

Cultural diffusion occurs in several different ways, some of which were identified in the 1960s by Swedish geographer Torsten Hägerstrand. Migration is one of the most important vectors of diffusion. When people move from their country to another, they tend to bring their cultural practices – such as their language, cuisine, and traditions – with them, and these gradually seep into the receiving culture. This is known as relocation diffusion.

Expansion diffusion occurs when ideas spread to new places while staying strong in their cultural hearth. This process can take three different forms: contagious, stimulus, and hierarchical. Contagious diffusion arises because people have a tendency to imitate one another – for example, when they see something on television or the Internet. In this context, ideas and behaviours spread almost like a virus, with celebrities and influencers serving as catalysts for this dissemination – for example, when a dance challenge posted by a teenager on platforms such as TikTok suddenly goes viral.

When the underlying principles of a concept are embraced but adapted to the local culture, it is a case of stimulus diffusion. This can be observed in the way national foods are interpreted abroad – for example, thin-based Italian pizza evolved into Chicago's deep-dish pie in the US. Similarly, large US fast-food chains adapt

Types of cultural diffusion

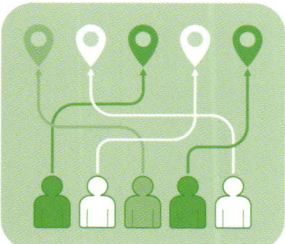

Relocation diffusion
People migrate from one place to another, taking their culture with them.

Contagious diffusion
Cultural traits spread fast from person to person, regardless of social status or location.

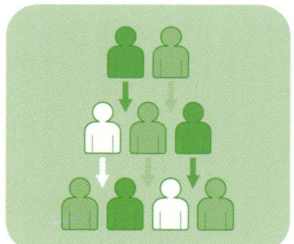

Hierarchical diffusion
Values spread from people or places of influence to those under influence in a top-down approach.

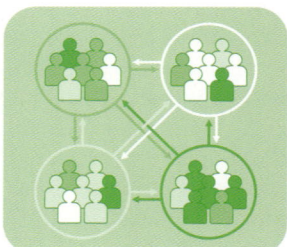

Stimulus diffusion
Cultural values spread to other nations, which modify and adapt them to fit with their own culture.

Tokyo Disneyland adapts the Disney experience for the Japanese market – with nods to local events such as cherry-blossom season – in an example of stimulus diffusion.

their menus to local tastes and practices, with McDonald's offering rice-based dishes in Asian markets, and halal meat in Muslim countries.

Hierarchical diffusion describes the way fads and fashions trickle down from large urban centres to smaller towns, or from early adopters to the wider population. Such is the case with many social media platforms, which initially appeal to a young audience but eventually become mainstream among older generations, too.

Types of culture

There are two main categories of culture: folk and popular. Folk culture is typically more localized, often rooted in rural or isolated communities, and shaped by local social practices and environmental factors. Transmission of folk culture – which includes regional dances, traditional crafts, and folk songs – occurs mostly by contagious or relocation diffusion.

By contrast, popular culture is often created with the intention of generating revenue. It includes entertainment (films, TV shows, music), consumer goods, and fashion. Popular culture generally originates in highly developed economies such as North America, Europe, and East Asia, spreading to regions where people are able to pay for it. Hierarchical diffusion, facilitated by mass media and the Internet, is the main vector for its transmission.

Culture can also be classified as material or non-material. Material culture refers to the physical artefacts associated with a group of people, such as their architecture, clothes, and crafts. Non-material culture – language, customs, and rituals, for example – can be further divided into mentifacts and sociofacts. Mentifacts are ideas and values, such as religious beliefs and moral codes; while sociofacts are forms of social organization – from family structures to political systems.

The role of global cities

Global cities have a significant impact on the diffusion of culture. These are major urban centres that are highly connected in terms of both transport infrastructure and telecommunications. Vitally important within the economy of their own country, they also play a key role on the global stage, serving as hubs for international commerce and finance.

New York, London, Paris, Tokyo, and Hong Kong are major cultural centres. All of them are home to important financial institutions, attracting large amounts of investment, both domestic and from abroad; much of the population is highly educated and well paid. This means that there is a ready market for expensive items, such as luxury clothing and high-end accessories. These global cities therefore draw people who are capable of designing, producing, and distributing these exclusive objects.

Since global cities are often also major tourist destinations, visitors encounter cultural artefacts that they later carry home and »

The study of the present surroundings is insufficient: the history of the people… must be considered.
Franz Boas
***The Principles of Ethnological Classification*, 1887**

spread via contagious diffusion. Cultural elements thus work their way out from global cities to cities and towns in other regions, eventually appearing in smaller communities as they spread outward by hierarchical diffusion.

A globalized culture

Disseminated by travel, trade, and media, cultural ideas spread across national boundaries, influencing people worldwide. This widespread diffusion can result in cultural homogenization, whereby once-diverse cultural practices and traditions become standardized. As predominantly Western cultural products and behaviours permeate virtually every corner of the world, local traditions and customs are supplanted by a globalized culture of profitable mass-market products, leading to the loss of cultural diversity and the marginalization of Indigenous and minority cultures.

Appadurai's flows

In 1990, Indian-American anthropologist Arjun Appadurai identified five "-scapes", or "dimensions of global cultural flow": ethnoscape, technoscape, financescape, mediascape, and ideoscape. None of these acts in isolation, but they all interact with one another, shaping the globalization process.

The ethnoscape, or the flow of people across boundaries, includes both those travelling out of necessity, such as refugees and economic migrants, and people travelling for leisure. Tourism, one of the fastest-growing commercial sectors, creates many opportunities for cultural diffusion. When people

Global cities, such as Hong Kong, act as diffusion hubs by way of their tourist appeal. After being exposed to urban cultural elements, visitors take those influences back to their communities.

from developed parts of the world travel to countries in the developing world, cultural exchanges are skewed towards the dominant culture and may lead to the phenomenon of cultural commodification – that is, turning traditional customs into marketable commodities in order to appeal to foreign visitors. Such is the case with luaus in Hawaii, which have morphed from a community event to a spectacle to entertain tourists.

The technoscape refers to the flow of technology. This is evident in the electronics or automotive industries, with products designed in one country, manufactured or assembled in another, and sold worldwide. High demand for new products in wealthy countries can lead to challenging working conditions in poorer nations, which struggle to keep up with production demand. As innovation accelerates, the flow of technology also

Central [to] global culture today is… the mutual effort of sameness and difference to cannibalize one another.
Arjun Appadurai
Modernity at Large, 1996

increases, changing the way people think of themselves, their jobs, and their place in a globalized world.

Appadurai defined the flow of money across political borders as the financescape. Computerization has accelerated the global transfer of money, with the result that the effect of transactions in the world's major stock markets can be felt around the world almost immediately. Banking transfers, online payment systems, and cryptocurrency fall under this flow.

The flow of information via mass and electronic media platforms is called mediascape. Thanks to digital communications, information can be shared across borders instantly. For example, people can livestream global news reports while travelling to work, or watch a cricket match being played in India from anywhere in the world.

Finally, the ideoscape refers to the flow of ideas. This can be on a small scale, with individual people posting their ideas on social media, or on a larger scale, such as when a country seeks to impose its political ideologies on its neighbours.

Cultural dynamics
People respond in different ways when exposed to another culture. Some individuals adopt the values and behaviours of the culture with which they have come into contact, a process called acculturation. This can be a two-way exchange, with both cultures influencing each other while still retaining their original identity to some extent – for example, when immigrants adopt the language and some of the host country's customs while still adhering to their own cultural traditions and practices.

Assimilation happens when individuals or groups adopt the values of another culture, losing their own cultural identity in the process. In some instances, cultural assimilation can be carried out forcefully, with the dominant culture deliberately suppressing a minority culture – one example is the Chinese government's suppression of the Uyghurs, a predominantly Muslim ethnic minority.

Cultural appropriation occurs when one culture adopts the customs of another for its own gain. These appropriations can be exploitative or disrespectful of the minority culture, such as when Caucasian people get tribal tattoos from Polynesian cultures, divorcing them from their spiritual significance. To combat this, a culture may try to defend its right to keep its customs within its community and prevent others from benefiting from its traditions.

Digital culture
In the 21st century, digital media is one of the main gateways for cultural exchange. Nearly 70 per cent of the global population – around 5.5 billion people – have access to the Internet, producing and consuming content that embodies their cultural world views and influences those of others.

Technology can break down barriers to the diffusion of cultural information, but it can also reinforce cultural biases by providing online spaces for communities that share the same values. Promoting digital literacy and ensuring equal access to technology helps to avoid these digital cultural ghettos so that the Internet becomes a tool to expand people's horizons instead of narrowing them. ■

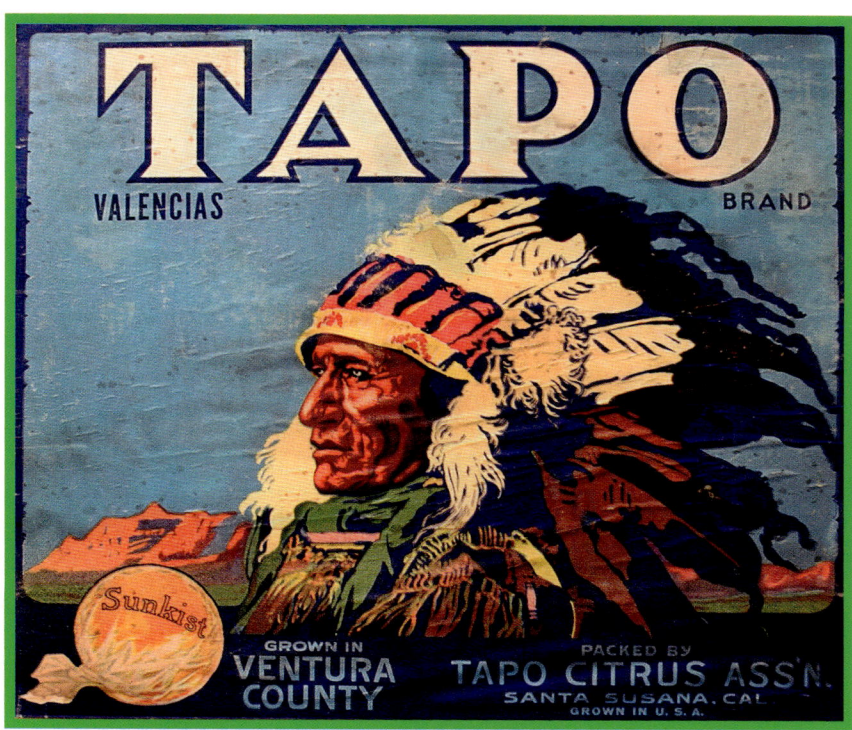

An Indigenous American chief wearing his feathered war bonnet appears in a 1930s advertisement for California-grown oranges in a glaring example of cultural appropriation.

CULTURE IS THE AGENT, THE NATURAL AREA THE MEDIUM
CULTURAL LANDSCAPE THEORY

IN CONTEXT

KEY FIGURE
Carl Sauer (1889–1975)

BEFORE
1897 Friedrich Ratzel writes about a "landscape modified by human activity".

1923 US geographer Harlan H. Barrows defines geography as the science of human ecology.

AFTER
1939 Carl Troll, a German geographer, coins the term "landscape ecology".

1974 Chinese-American geographer Yi-Fu Tuan popularizes the word "topophilia" to describe the emotional connection people have to places.

1981 US landscape ecologist Richard Forman defines his discipline as the study of the structure, function, and dynamics of landscapes.

Proposed by US geographer Carl Sauer in 1925, the theory of cultural landscape examines the ways in which, over the course of time, humans have reshaped natural environments to suit their needs. Sauer suggested that human culture has an influence on – and, in turn, is subject to the influence of – the landscape in which it develops.

Among the forces moulding cultural landscapes are the social, religious, and technological aims and limitations of their human inhabitants, as well as the physical characteristics of the environment. As a result, there is no singular cultural landscape, but rather

HUMAN GEOGRAPHY 173

See also: Location theory 150–51 ▪ Organic theory 154–55 ▪ Environmental determinism 156 ▪ Geography as human ecology 162–63 ▪ Central place theory 182–83 ▪ Behavioural geography 194–95 ▪ Humanistic geography 200–01

> The cultural landscape is fashioned from a natural landscape by a culture group.
> **Carl Sauer**
> "The Morphology of Landscape", 1925

multiple perspectives that reflect the complex interactions between peoples and environment.

Schlüter's landscapes

The relationship between the natural environment and human intervention within it started to come under scrutiny in the late 19th century. Some geographers, especially in Germany, called this new discipline landscape science, and they attempted to precisely define regions, settlements, village types, and agricultural systems within their countries. One of those scholars, Otto Schlüter, was a pupil of fellow German and geographer Friedrich Ratzel, who had been particularly interested in the interactions between people and their environment.

Schlüter first used the term "cultural landscape" in his 1908 article "The Geographic Landscape and Its Investigation". In the paper, he defined two types of landscape: *Urlandschaft* (original landscape) – that is, what exists before any human intervention; and *Kulturlandschaft* (cultural landscape), or the landscape after modification by human hand. This might take many different forms – from urban settlements and transportation networks, to agricultural and industrial activities.

Challenging Ratzel's environmental determinism, Schlüter proposed that human agency was an active force in transforming the natural landscape into a cultural landscape. He also believed that the main focus of geography should be the examination of this evolution.

Kulturlandschaft, as defined by Otto Schlüter, is the landscape shaped by human culture. It includes the agricultural system of dividing land into fields to manage different crops.

Human intervention

Carl Sauer built on Schlüter's theory in his 1925 paper "The Morphology of Landscape", in »

Carl Sauer

The most influential proponent of the theory of cultural landscape, Carl Sauer was born in 1889 in Missouri, US. As a child, he was sent to study in his father's homeland of Germany for five years. After returning to the US, he attended Central Wesleyan College in his home town of Warrenton, where his father was the school botanist.

After receiving a doctorate degree in geography from the University of Chicago in 1915, Sauer spent a few years teaching in Michigan; then, in 1923 he became founding chairman of the geography department at the University of California, Berkeley. He went on to serve in that post for more than 30 years, establishing Berkeley as a leading centre for geographic research, especially in the fields of cultural landscapes and historical geography. Sauer died at Berkeley in 1975.

Key work

1925 "The Morphology of Landscape"

which he defined a cultural landscape as a natural landscape that had been modified by a cultural group over time. Stressing the importance of culture as a force in shaping the visible features of Earth's surface, Sauer went on to declare that the physical environment is not merely a backdrop against which human culture unfolds, but the medium with and through which human cultures are shaped.

Sauer also underlined the need to understand a landscape's historical development, arguing that cultural landscapes could be seen as living records of human evolution. If we know how to interpret the landscape we can identify the forces that have acted to produce what we see today.

Landscape preservation

The UNESCO World Heritage Convention was established in 1972 to recognize and protect natural sites and cultural properties. Since 1992, it has also acknowledged the importance of cultural landscapes – from sacred sites, to settlements and agricultural systems – which, the Convention states, represent the "combined works of nature and man" and illustrate how human societies have evolved within their physical environments, harnessing the opportunities they offered, or circumventing their constraints.

Because few places in the world have not experienced some impact from human activity, it could be argued that most landscapes are cultural landscapes.

Types of cultural landscape

The UNESCO Convention identifies three main forms of cultural landscape. The first is a landscape that has been intentionally designed and created by humans, including parklands and gardens constructed for aesthetic reasons.

> Any landscape is composed not only of what lies before our eyes but what lies within our heads.
> **Donald W. Meinig**
> "The Beholding Eye: Ten Versions of the Same Scene", 1979

Examples of this include the Palace and Park of Versailles, France. From 1682, when King Louis XIV moved the court and government there, until the French Revolution of 1789 and the end of the monarchy, Versailles was considered a template for royal residences in Europe, and its influence is visible in the architecture of the monumental Royal Palace of Caserta, Italy, and Nymphenburg Palace, Germany; and in the formal gardens at Hampton Court Palace, UK.

The second kind of cultural landscape is one that has evolved organically over time to meet the social, economic, administrative, or religious needs of a community. Its current form has been influenced by its connection with – and in response to – the surrounding natural environment. These landscapes fall into two subcategories: relict (or fossil) and continuing landscape.

A relict cultural landscape is one that still bears the physical traces of a development that has come to an end, either abruptly or over a period of time. An example of an organically grown relict cultural landscape is the ancient city of Bam, in Iran, a fortified settlement that developed

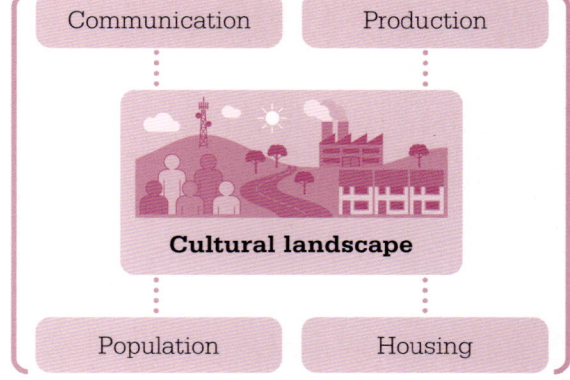

Sauer's theory of cultural landscape states that, over time, a culture group (the agent) acts on the natural landscape (the medium), shaping it to suit its needs. The resulting forms are the visible expressions of human activity on the cultural landscape.

HUMAN GEOGRAPHY

along the main trading routes of Central Asia in a desert landscape. In particular, the Convention acknowledges Bam's mud-brick construction techniques and water-management system as representations of the interaction between humans and landscape.

A continuing landscape is closely associated with a society's traditional way of life, and it still plays an active role in it as it continues to change and develop. An example of this is Cuba's Viñales Valley, an area with long-standing agricultural traditions, especially with regards to the cultivation of tobacco.

The final category is the associative cultural landscape. This refers to a geographical area that may be largely untouched by human activity but is recognized for its cultural, spiritual, or artistic significance for a particular group of people. Tongariro National Park, in New Zealand, belongs in this category; indeed, it was the first cultural landscape ever to be recognized by UNESCO (in 1993), which acknowledged the profound spiritual significance of this environment to the Māori people, who believe the mountains represent their ancestors.

Indigenous cultures

Often shaped by thousands of years of traditional use, Indigenous landscapes hold profound cultural, spiritual, and historical value for Indigenous peoples. Former United Nations secretary general Javier Pérez de Cuéllar often spoke of the need to preserve these physical landscapes as a fundamental step towards understanding the historical development of Indigenous cultures, as well as recognizing their presence within the environment and their contribution to shaping it.

The International Institute for Environment and Development has estimated that the lands and territories of Indigenous peoples

Mount Ngauruhoe, in New Zealand's Tongariro National Park, is considered *tapu*, or sacred, by the Māori people, who respect it as a living entity.

Rio de Janeiro in Brazil is classed as a UNESCO cultural landscape for its unique interaction between urban development and natural beauty.

are home to some 80 per cent of the world's biodiversity. Preserving them can contribute to developing sustainable land use practices, maintaining or enhancing natural values in the landscape, and protecting biological diversity, as well as safeguarding the heritage of Indigenous peoples. If we are to achieve the aim of protecting the environment, we have to understand the people who shaped it, and who were shaped by it. ∎

WE ARE BECOMING MORE AND MORE INTERESTED IN THE RELATION OF OUR NUMBERS TO OUR WELFARE

DEMOGRAPHIC TRANSITION MODEL

DEMOGRAPHIC TRANSITION MODEL

IN CONTEXT

KEY FIGURES
Warren Thompson (1887–1973), **Frank W. Notestein** (1902–83)

BEFORE
1798 Thomas Malthus argues that population will outpace resources, leading to famine.

1907 An equation to forecast population growth is published by US chemist Alfred Lotka.

1922 US newspaper proprietor E.W. Scripps establishes the Foundation for Research in Population Problems.

AFTER
1934 Adolphe Landry, a French economist, publishes *The Demographic Transition*.

1965 Danish economist Ester Boserup argues that population growth can lead to innovation.

2022 Ukraine's birth rate plummets as a result of the Russian military invasion.

During the course of the 19th century, many commentators – from British philanthropist Robert Owen, to German philosopher Karl Marx – criticized British economist Thomas Malthus for predicting that population growth would inevitably lead to food shortages and increased mortality. They argued that he had been overly pessimistic, failing to take account of technological innovation and social change. However, there was no serious investigation of the link between economic growth and population until the 1920s.

In 1929, American demographer Warren Thompson, having analysed population trends in some countries for the years 1908–27, published a paper called "Population" in the *American Journal of Sociology*. He proposed that all countries are at one of three stages of development. The first stage is characterized by high birth and death rates, the second by high birth rates and declining death rates, and the third by low birth and death rates.

Thompson further explained his thinking by adding that as a country industrializes, it undergoes a demographic transition, moving from the first stage to the second, and then on to the third.

The Industrial Revolution began in the UK in the 18th century. Advances in technology introduced new ways of working and living, changing society and triggering demographic transition.

The need to plan

Government agencies and planners took little notice of Thompson's idea at the time, but after World War II there was a greater understanding of the need to plan for the future, and an important part of this lay in forecasting population change. The United Nations Commission

> It is only when rising levels of living… and rising hope for the future give new value and dignity… [that] fertility comes under control.
> **Frank W. Notestein**
> "Population – The Long View", 1945

Warren Thompson

Born in Nebraska, US, in 1887, Warren Thompson received a doctorate in sociology from New York's Columbia University in 1915. He later headed the Scripps Foundation population think-tank. He and his assistant Pascal Whelpton analysed demographic trends, becoming the pre-eminent forecasters of US population change.

Thompson's views were sometimes controversial. In 1929, he forecast that Japan, which was undergoing rapid population growth at the time, would be forced to "expand by the acquisition of more territory", a theory that proved to be correct.

After World War II, Thompson went to Japan as an adviser to US General Douglas MacArthur, who led the Allied occupation of the country. Thompson stepped down at Scripps in 1953. He died in 1973.

Key work

1929 *Danger Spots in World Population*

HUMAN GEOGRAPHY 179

See also: Cultural landscape theory 172–75 ▪ Central place theory 182–83 ▪ The concentric zone model 184–87 ▪ Censuses and population geography 188–89 ▪ Theories of development 190–91

on Population and Development, established in 1946, began to make long-term projections, and US demographer Frank W. Notestein reprised and developed Thompson's ideas, which he called the demographic transition. His 1945 paper "Population – The Long View" provided a conceptual framework for population projections.

Notestein highlighted some general trends, notably that a decline in the mortality rate takes place before a decline in fertility. He stressed the important role of access to contraception and social and economic advance, and he forecast global declines in birth and death rates. Summarizing, he explained that a combination of higher living standards, improved health, better education, and rising expectations for the future would reduce the birth rate.

A contemporary view

Contemporary demographic transition models interpret the ideas advanced by Thompson, Notestein, and others, and usually include five stages. The first typifies most of pre-Industrial Revolution human existence, featuring very high birth and death rates, resulting in a stable or slowly increasing population. Birth rates were high because there was no contraception, and children were a crucial source of labour and family support. Many died young, however.

After the Industrial Revolution, major improvements in sanitation and water quality, food production, and healthcare reduced the rate of child mortality dramatically. Yet the expectation that many children would still die young meant that birth rates remained high. As a result, the population increased markedly in stage-two "early expanding" societies. Britain was in this stage around 1760–1880, India 1950–2000, and several less economically developed countries,

These children in Niamey – the capital of Niger, west Africa – represent the country's young population. In 2025, the median age in Niger was estimated to be 15.6 years.

especially in sub-Saharan Africa, reached that stage in the first quarter of the 21st century. For example, in 2023, the birth rate in Niger was 42 per 1,000 people, and the death rate was 8.9, which indicates a rapidly growing population. Other nations at that stage of development in the mid-2020s included Afghanistan, the Democratic Republic of Congo, Mozambique, and Sudan.

At stage three (called the "late expanding" phase) of the demographic transition, several factors contribute to a declining »

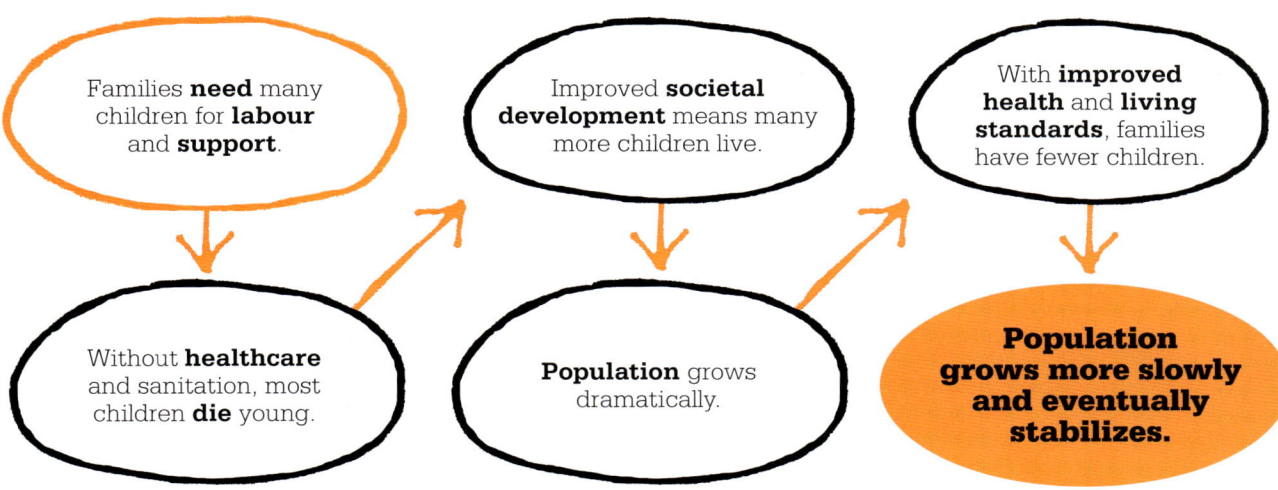

180 DEMOGRAPHIC TRANSITION MODEL

birth rate. Contraception is more widely available, and many families make a conscious decision to have fewer children because there is a greater chance their children will survive to adulthood.

Fewer people are needed to work the land because of mechanization; and with the legal requirement in most countries for children to go to school, child labour is impractical. Additionally, changing societal norms mean that more women can work outside of the home.

Britain was at this stage between about 1880 and 1940, while India was generally considered to be at stage three in the mid-2020s. Its death rate declined sharply from 28.2 per 1,000 in 1950, to 7.5 in 2024, but its birth rate has fallen even more dramatically: from 44.2 to 16.7. Consequently, although India's population is still increasing and it is now the most populous country on Earth, its rate of growth has slowed greatly. Other countries generally considered to be in this stage-three group include Colombia, Mexico, and South Africa.

Elderly healthcare in countries like India, with falling death rates, often focuses on conditions such as diabetes, heart disease, and dementia, which are more prevalent in older age groups.

The impact of affluence

As the process of economic development and growing affluence continues, stage four means both birth and death rates will be very low, at up to around 15 births per 1,000, and 12 deaths per 1,000 or fewer. Excluding immigration and emigration, a nation's overall population will be stable or increasing slowly. The UK and the US are both at this "low fluctuating" stage. The birth rate in the UK was 11.2 per 1,000 in 2024, and its death rate was 9.5. However, both trends have been inconsistent. Unlike in

India, there was a steady increase in the UK birth rate between 1950 and 1964, followed by a series of decreases and increases, with subsequent peaks in 1990 and 2010, with each one smaller than the previous. Meanwhile, the death rate steadily declined until 2011, when it started to increase slightly, this being blamed on government austerity policies.

The US birth rate has halved since 1960, from 23.7 per 1,000, to 12.0. Its death rate fell fairly

The five stages of demographic transition

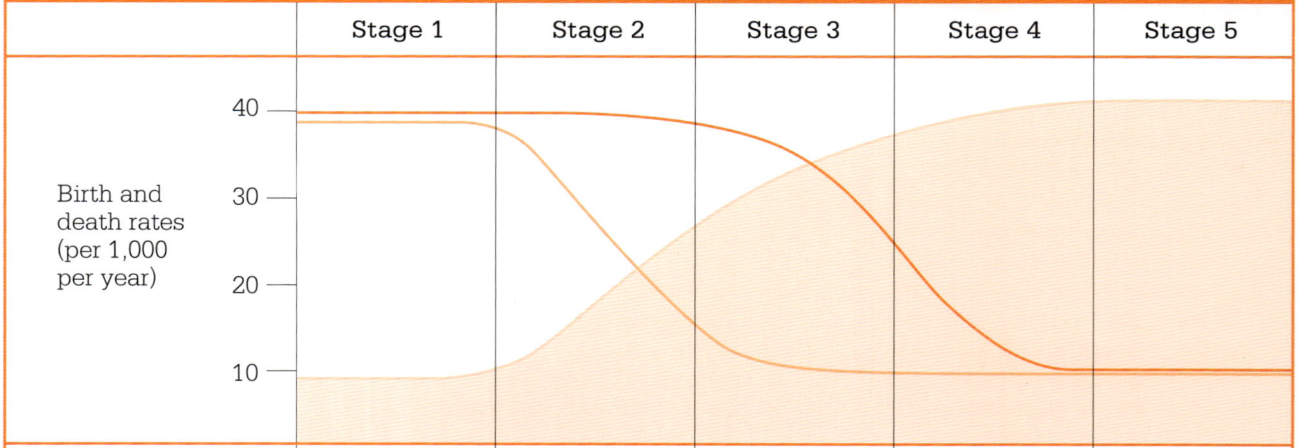

	Stage 1	Stage 2	Stage 3	Stage 4	Stage 5
— Birth rate	High	High	Falling	Low	As yet unknown
— Death rate	High	Rapid fall	Falls slowly	Low	As yet unknown
Total population	Stable or slow increase	Rapid increase	Increase slows	Slows then stabilizes	As yet unknown

HUMAN GEOGRAPHY 181

Japanese family sizes have been falling since the 1970s. In 2024, the average number of children per family was at a record-low 1.15, contributing to a overall shrinking population.

steadily up to 2009, when it was 7.9 per 1,000, but has since increased to 9.2. The overall population is increasing slowly. The median age of the UK and US populations is 40 and 38 years respectively, and with a smaller proportion of women of childbearing age, the low birthrate trend is likely to continue – leading in turn to an increased population of older people.

The population profile in Western Europe highlights some of the limitations of the demographic transition model: while death rates remain stable, reflecting ageing demographics, birth rates stay low, influenced by social and cultural factors, rather than economic trends. Other factors that affect nations' demography are immigration and emigration, which are independent of birth and death rates, and fertility.

A shrinking population?

Many demographers now believe that some countries have entered a fifth stage – one of declining population, where fertility is below the replacement level of 2.1 children per woman. Despite the advances of medical science, if the average age of a country increases, the death rate will also go up.

Japan has experienced a population decline of about 6 million since 2010. In 2024, it registered a birth rate of 7.0 per 1,000 and a death rate of 11.7, meaning the median age has risen to more than 49 years old. Other factors that limit the number of children born in more economically developed countries include the often prohibitive costs involved with starting or enlarging a family.

A complex reality

Critics argue that the reality of demographic change is much more complex than the five-stage model. They point to government interventions to boost or cut birth rates. In Estonia, for example, where women's fertility rate is 1.6 children – well below the rate required to maintain the population – the government

In the long run... the control of mortality without the control of fertility is impossible.
Frank W. Notestein
"Population – The Long View", 1945

provides extended parental leave, improved one-off payments for each birth, and financial allowances for each child. Its aim is to increase the population and reduce the average age of its citizens. In contrast, China's state-imposed one-child policy, enforced between 1979 and 2015, contributed to a fall in the birth rate to 10.5 per 1,000 in 2024.

Unforeseen events can have a dramatic effect on population. The HIV/AIDS pandemic massively increased the death rate of many sub-Saharan African countries, and the Covid-19 pandemic of 2020 also resulted in a high number of excess deaths. Meanwhile, Israel's 2025 food blockade in Gaza disproportionately affected Palestinian children under five, with devastating demographic consequences in the long term.

With a global population of more than 8 billion, there has never been a greater need for demographic modelling. It can help to inform governmental policy decisions, ranging from future maternity, education, employment, and social care needs, to a country's international aid requirements. ∎

THE RANGE OF A GOOD IS GREATER WHEN IT IS OFFERED IN A LARGER CENTRAL PLACE
CENTRAL PLACE THEORY

IN CONTEXT

KEY FIGURE
Walter Christaller
(1893–1969)

BEFORE
1915 Charles Galpin, a US rural sociologist, recognizes that towns offering the same level of services should be equally spaced.

1923 US geographer John Harrison Kolb produces a model in which small settlements are divided into five categories, with populations ranging from less than 200 to more than 5,000.

AFTER
1954 August Lösch's theory of profit maximization proposes that industry will locate where it can maximize its profit.

1982 French historian Fernand Braudel explains in *The Wheels of Commerce* that the rise of cities is inextricably linked to the development of capitalism.

Geographers in the early 20th century attempted to explain the location, number, and size of settlements – and also their relationships with one another. They were keen to establish why there are more villages than small towns, and more of the latter than large towns and cities. They also queried why smaller settlements are situated closer together than larger ones.

German geographer Walter Christaller set out to determine if there were patterns governing the size, number, and distribution of settlements, which he called central places – that is, those that provide goods and services to the surrounding area. In his 1933 work *The Central Places in Southern Germany*, he proposed that such patterns do exist, in a concept that became known as central place theory.

A hierarchy of places

Christaller argued that the main purpose of a central place is to provide goods and services for the surrounding market area. Those providing a wider range of goods and services have larger market areas. He called these higher-order central places. Those offering a smaller range of goods and services – lower-order central places – have smaller market regions, are more numerous, and are situated closer together.

To explain the different levels of settlement, Christaller examined the role of consumers and retailers and introduced the concepts of "range" and "threshold". In this context, range is the furthest distance a member of a population will travel to buy a particular item. For example, the range for the purchase of basic groceries is only

Our existence in time is determined for us, but we are largely free to select our location.
August Lösch
The Economics of Location, 1940

HUMAN GEOGRAPHY 183

See also: Location theory 150–51 ▪ Rank-size rule 160 ▪ Primate city rule 161 ▪ Spatial interaction theory 192–93 ▪ Behavioural geography 194–95

a short distance, but the range for new cars is much further. Because everyone needs basic groceries daily, all central places will sell these, but only the largest will have vehicle retailers.

Christaller's threshold is the minimum level of demand needed to ensure that sales of a particular item are profitable. If car showrooms appeared in all lower-order central places, most would soon close for failing to meet the required sales threshold.

Model assumptions

When building his central place model, Christaller made several assumptions about the nature of the area on which it was based. It had a uniform population density; all its consumers had the same income and the same demand for goods and services; and both consumers and retailers acted rationally to minimize costs and maximize profits. He postulated that one higher-order central place will be surrounded by a hexagonal market area and six centres of the next order, and each of these will be surrounded by six smaller market areas.

German economist August Lösch expanded on central place theory in his 1940 book *The Spatial Organization of the Economy*, focusing more on specialized goods and manufacturing and allowing for a greater variety of market areas. Neither his nor Christaller's ideas reflect the great complexity of settlement patterns, but they do help to explain the hierarchy of central places, market areas, and the spread of retail outlets. ▪

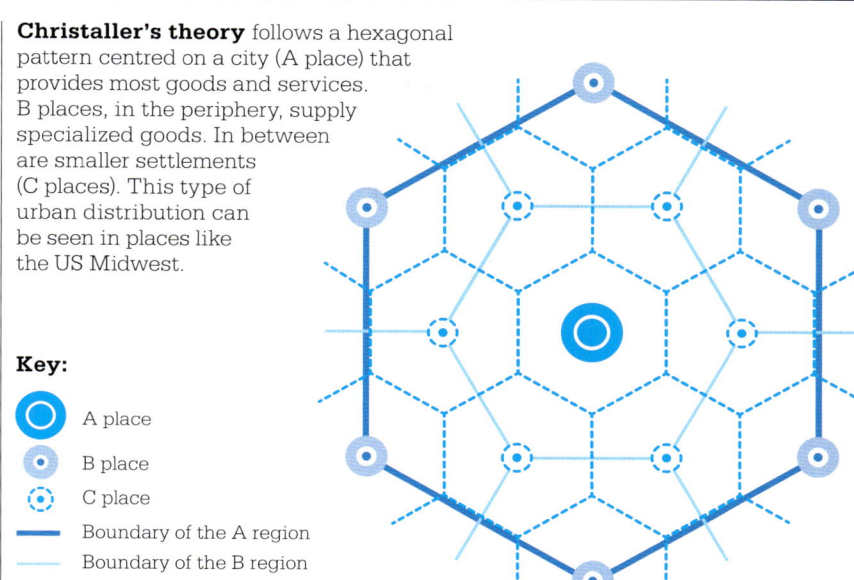

Christaller's theory follows a hexagonal pattern centred on a city (A place) that provides most goods and services. B places, in the periphery, supply specialized goods. In between are smaller settlements (C places). This type of urban distribution can be seen in places like the US Midwest.

Key:
- A place
- B place
- C place
- Boundary of the A region
- Boundary of the B region
- Boundary of the C region

The hamlet of Pischlach and the surrounding villages in Bavaria, southern Germany, are a real-world example of the relatively close proximity of smaller settlements.

Walter Christaller

The son of a children's novelist and a pastor, Walter Christaller was born in Berneck, southern Germany, in 1893. His studies in economics and philosophy at the University of Freiburg were cut short by World War I. He was wounded during the war and returned home, before joining the University of Erlangen in 1929. It was there that he published his central place theory.

By the outbreak of World War II, Christaller's work had become known to the Nazi government, and in 1940 he was given a job in Heinrich Himmler's Planning and Soil Office, with responsibility for establishing the economic geography of German-occupied Poland. Post-war, he focused on tourism geography and, in 1964, received the Association of American Geographers' Outstanding Achievement award. He died five years later.

Key work

1933 *Central Places in Southern Germany*

THE CITY IS THE NUCLEUS

THE CONCENTRIC ZONE MODEL

IN CONTEXT

KEY FIGURE
Ernest Burgess (1886–1966)

BEFORE
1826 German economist Johann Heinrich von Thünen presents a model of land use, with activities arranged in concentric circles around market towns.

AFTER
1945 In "The Nature of Cities", American geographers Chauncy Harris and Edward Ullman propose a model for a city with multiple centres.

1964 Argentine-American planner William Alonso explains bid rent theory, the rationale for rents being highest in central business districts.

1982 Leo van den Berg publishes *Urban Europe: A Study of Growth and Decline*, which outlines a spatial-model cycle of urban development.

Planners, architects, and sociologists in the early 20th century studied the way in which urban spaces were organized spatially, with the aim of developing planning guidelines and making cities better places to live and work. In 1925, Ernest Burgess constructed an urban model that highlighted the relationship between urban space, social class, and social mobility.

Haphazard development
City planning was not a new concept: as early as the 3rd millennium BCE, for example, Harappa, a major settlement of the Indus Valley civilization in

HUMAN GEOGRAPHY 185

See also: Rank-size rule 160 ▪ Primate city rule 161 ▪ Central place theory 182–83 ▪ Spatial justice 206–07

> According to **traditional land use models**, cities tend to **expand outward** from the central business district (**CBD**).

> Some models show that **industrial sites** are located **near the CBD**; in others, these sectors are **along transport corridors**.

> The location of **low- and high-income housing** may depend on their **distance from industry** or transport links.

Land use models developed by the likes of Burgess and Hoyt explain the various zoning patterns in urban centres.

Ernest Burgess

Born in Ontario, Canada, in 1886, Ernest Burgess studied in Oklahoma, US, and was awarded a doctorate in sociology from the University of Chicago in 1913. He taught at other colleges before becoming a professor at his alma mater in 1916.

During his long career, Burgess founded the Chicago School of Sociology and made significant contributions to the sociology of marriage, the parole system, and the impact of retirement on the elderly.

Burgess countered the eugenicist view that crime in poor neighbourhoods was caused by genetics; with fellow academic Robert E. Park, he wrote the influential *Introduction to the Science of Sociology* in 1921. Four years later, the pair published their theory of concentric zone city development. Burgess stayed at the University of Chicago until his death in 1966.

Key work

1925 *The City* (with Robert E. Park)

South Asia, had been meticulously planned. However, most modern cities had grown in a haphazard fashion, and during the 19th century, especially in Europe and North America, they had expanded at a speed never previously seen. In 1800, just 4 per cent of the world's population lived in urban areas, but by 1925 this figure had risen to 18 per cent.

The biggest cities in the world – New York, London, Chicago, Berlin, and Philadelphia – were blighted by multiple problems rooted in deep societal issues related to unregulated capitalism. One of these was a complete absence of planning for the needs of any but the wealthiest citizens. Industrial capitalism required large numbers of workers for its ever-growing factories. The housing for these labourers and their families was erected quickly and cheaply, within urban developments that were unplanned, even chaotic.

Cities such as Manchester in northern England – where the population grew from 90,000 in 1800 to 700,000 a century later – were smoky, dirty, smelly, and noisy places, with poor transport, squalid housing for most people, wretched sanitation and health, and a high rate of child mortality. People such as Burgess sought not only to study the contemporary conditions but also to shape future possibilities. »

Urban housing for 19th-century industrial workers was built quickly and cheaply around the factories, and it consisted of long rows of terraced dwellings with shared walls.

186   THE CONCENTRIC ZONE MODEL

After studying Chicago – the fourth-largest city in the world in 1925 – he developed the concentric zone model to explain how urban settlements functioned.

Outward expansion

Burgess described how cities expand outwards from the central business district (CBD, zone I), which is occupied by offices, large stores, hotels, restaurants, banks, and theatres. Often referred to as the financial district, the CBD is where the urban transport infrastructure converges, making it the most accessible area of a city. Because of its central position, its shops and other businesses have the highest footfall and sales, so rents are more expensive than anywhere else in the city. However, the CBD is not only an area for people to visit to purchase goods, entertainment, and services – it also provides employment, with jobs ranging from bank executives, to chefs, cleaners, and caterers.

Adjacent to the CBD in the Burgess model is a transition zone (zone II). This is where many industrial activities base themselves in order to take advantage of the markets in zone I

> The main fact of expansion [is] the tendency of each inner zone to extend its area by the invasion of the next outer zone.
> **Ernest Burgess**
> *The City*, 1925

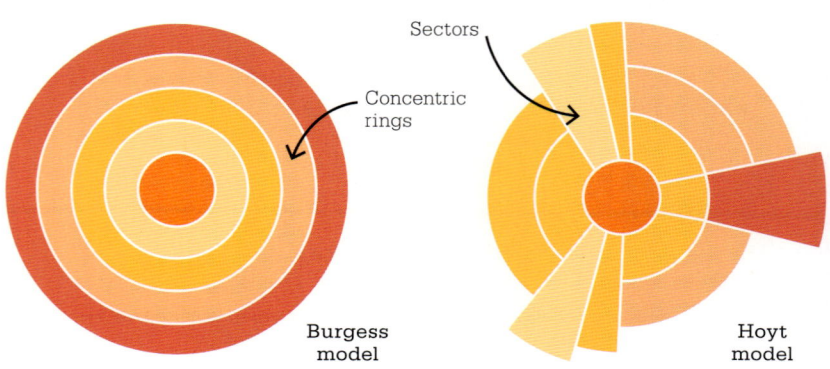

Burgess's and Hoyt's theories emphasize different aspects of urbanization: while the former's concentric model focuses on social class distribution in circular zones spreading out from the centre, Hoyt's sector model highlights transport links and economic corridors, resulting in wedge-shaped growth.

Key:
- I: Central business district (CBD)
- II: Transition zone
- III: Working-class housing
- IV: Middle-class housing
- V: Commuter zone

and the labour in zone III. Another feature of this area is the presence of older, decaying, low-rent housing, which is often populated by first-generation immigrants and poor families. Its houses and tenements may once have been occupied by more affluent people when the city was much smaller. Most of the main railway terminals are also located here.

The next concentric ring (zone III) comprises working-class housing – not affluent, but not as run-down as the previous zone. From here, people can easily travel to work in transition-zone factories or perform menial tasks in the CBD without having to pay the higher rents of the outermost parts of the city or the higher fares to commute long distances to work.

Surrounding this working-class residential zone is a ring of middle-class housing (zone IV), with more spacious, newer properties. Property prices are higher, and it costs more to commute to the CBD, so only people with a relatively large disposable income can afford to live here.

Finally, the outermost commuter zone (zone V) houses an even more affluent community of people with large homes who can afford to commute to the city centre for work and travel there for pleasure. Before the massive growth of car ownership – in the 1930s in the US, and the 1950s in much of Europe – most of these suburbs were well served by railway stations providing them access, at a cost, to the CBD. Burgess also described how the zones were not static but tended to grow outwards from the centre and expand into the adjacent zone.

Hoyt's sector model

In 1939, American economist Homer Hoyt countered the Burgess model, arguing that cities are primarily laid out in wedge-shaped or spoked zones of different land use leading out from the CBD, rather than concentric rings. For Hoyt, these sectors developed primarily along transportation lines, so concentrations of factories follow railway lines, rivers, and canals, with some very close to the city centre and others on its outskirts. The noise and pollution of these industrial areas means that people who can afford to move away will do so, leaving behind poor families in run-down residential areas.

In Hoyt's model, workers in better-paid jobs live some distance from the grime of industry, and the middle-class residential areas are still further afield. Finally, the most affluent people live in quiet, leafy residential suburbs, far removed from the excesses of industry.

Too simple?

Critics of the concentric zone and wedge models argue that they oversimplify the complexities of city structure. Since the mid- to late 20th century, in particular, these theories have been faulted for failing to take into account greatly increased rates of car ownership and the expansion of relatively cheap public transport options, both of which have enabled easier commutes from low-cost housing at city margins or even outside city boundaries.

The models are also not applicable to many cities outside the US. For example, Paris, France, has a ring of expensive housing around its CBD, with most of its very large working-class districts on the outskirts of the city. This is also true in parts of London, Britain, such as Notting Hill and Islington, where developers have refurbished Victorian dwellings, attracting affluent residents. However, despite their weaknesses, the Burgess and Hoyt models are helpful tools in explaining urban land use.

Urban regeneration

In 1982, Dutch urban geographer Leo van den Berg and others put forward a spatial-cycle model, involving four stages of city development. This has proved valuable for planners trying to revitalize urban centres.

In the first stage, population growth happens in the city centre. Next, in a process of suburbanization, the city expands outwards, creating new residential areas as people move away from the centre. As a result, the city's population at its core grows slowly or even declines. In the third stage, counter-urbanization, the central districts and the suburbs decline as people abandon the city and only commute in to work, leaving the city centre "dead" at night. Lastly, re-urbanization occurs when there is regeneration, and a new influx of people into the centre. ■

Regenerating the Motor City

Detroit is an example of Leo van den Berg's spatial-cycle model at work. Nicknamed Motor City, the city went into a long decline as its car industry slumped. With fewer jobs, people moved away from the centre, buildings were abandoned, and the urban landscape fragmented. Detroit's population fell 60 per cent between 1950 and 2010; in 2013 the city filed for bankruptcy.

In the late 2010s, planners kick-started the fourth stage of city development: re-urbanization. Abandoned office blocks in the CBD were renovated and reoccupied; in addition, the city oversaw the creation of safe new pedestrian spaces in formerly dangerous parts of the city, and embarked on a campaign to attract new residents and businesses.

Paris, France, contradicts the Burgess and Hoyt models with working class areas, or *banlieues*, such as Saint-Denis, Saint-Ouen, and L'Île-Saint-Denis, some distance from the city's heart.

HUMAN CAPITAL IS THE MOST CRITICAL CAPITAL FOR CONTEMPORARY SOCIETIES
CENSUSES AND POPULATION GEOGRAPHY

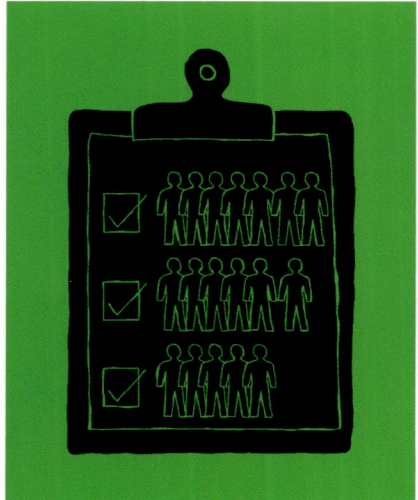

IN CONTEXT

KEY ORGANIZATION
United Nations (1945–)

BEFORE
2 CE China's Han dynasty census records a population of 57.7 million people in 12.4 million households; Chengdu is the largest city, with 282,000 inhabitants.

1790 According to the first US census, which does not include Indigenous Americans, there are just under 4 million people in the country.

1841 The British census is the first to record individuals by name.

AFTER
2010 In China, 10 million census workers collect data on 1.3 billion people.

2020 The 24th decennial US census is the first to allow people to respond online or by phone, as well as via paper forms.

In the aftermath of World War II, there was an urgent need across the world for reliable, comparable data to assess population trends and analyse socio-economic developments. Designing poverty-reduction strategies, monitoring the effectiveness of government policies, and tracking the progress of national and international development goals depended on the availability of this information. In 1947, the newly established United Nations (UN) set up its Statistics Division (UNSD) to assist with this task.

The traditional census is among the most complex and massive peacetime exercises a nation undertakes.
Principles and Recommendations for Population and Housing Censuses, 2015

The UNSD assimilates statistical information collected by member nations, develops standards for census enumeration, and provides experts and advisers on enumeration methodology, geographic information systems (GIS), and data processing.

A history of censuses

The first census known to historians was conducted around 3800 BCE in the Babylonian empire; officials counted livestock and agricultural produce. Later, the rulers of ancient Egypt, Greece, Rome, and Han-dynasty China also ran censuses of their domains. Each of these served a valuable role, but they could not be directly compared with each other because their methodologies, accuracy, and scopes varied enormously.

The first "scientific" census – that is, one that was replicable – was the one organized by the British government in 1841. Respondents had to answer questions relating to their age, occupation, and place of birth, whereas previous censuses had been essentially just a head count. Since 1801, a national census has been carried out in the UK every ten years.

HUMAN GEOGRAPHY 189

See also: Malthusian theory 148–49 ▪ Rank-size rule 160 ▪ Demographic transition model 176–81 ▪ Theories of development 190–91 ▪ Gender and space theory 210–13 ▪ Sustainable development 244–45 ▪ The environmental justice movement 246–47

Tabulators at the Census Bureau in Washington, DC, record the information gathered for the 1940 US census, which is said to have missed just 5.4 per cent of the population.

In the second half of the 20th century, there was a big increase in the number of nations carrying out censuses and in the scope of information being collected.

The largest census ever

The launch of the World Population and Housing Census Programme (WPHCP) in 2015 marked a step-change in the scope of censuses. This programme was designed to assist the implementation of the UN's Agenda for Sustainable Development 2030 by ensuring that nations ran population and housing censuses at least once in the period 2015–24 (the "2020 round") and at least once in 2025–34 (the 2030 round). Despite the problems linked with the Covid-19 pandemic, which caused many countries to delay their census, the 2020 round was extraordinarily successful, collecting data for 7.7 billion men, women, and children – about 94 per cent of the global population. It was the largest coordinated census ever conducted. As well as tracking population growth and distribution, the census investigated healthcare, housing, education, sanitation, and infrastructure, helping to monitor progress towards the Sustainable Development Goals (SDGs) for 2030.

Challenges remain – not least, getting complete coverage in remote areas and among marginalized communities. Even with advanced digital tools, achieving this goal will be expensive. The paradox is that those countries most likely to benefit from SDGs will find it most difficult to finance the census work that guides solutions, so they will need support from wealthier nations. ▪

Digital methodology

The UN took the opportunity presented by the launch of the WPHCP to survey census methodology around the world. This has changed dramatically in the 225 years since the first US census was conducted over a nine-month period by federal marshals on horseback. Now, digital tools are commonplace in the collection, processing, and analysis of respondents' information. Most commonly, enumerators conduct face-to-face interviews using electronic questionnaires on mobile phones, or respondents fill out online questionnaires. Face-to-face interviews, with paper questionnaires, are also used, but this traditional technique is likely to become less common as digital data collection grows.

A data collector from the Pakistan Bureau of Statistics (PBS) carries out an interview for the country's first digital census, in 2023.

THE STORY OF EACH NATIONAL ECONOMY
THEORIES OF DEVELOPMENT

IN CONTEXT

KEY FIGURE
Walt Rostow (1916–2003)

BEFORE
1948 The US government's Marshall Plan, which is aimed at revitalizing the European economy after World War II, is a major influence on economic growth models.

1953 Ragnar Nurske, an Estonian-American economist, argues that "underdeveloped" nations need to invest heavily in several different industries simultaneously.

AFTER
1970 Immanuel Wallerstein, an American sociologist, argues that Rostow's stages of growth model is too simplistic.

1995 Colombian-American anthropologist Arturo Escobar suggests that the concept of international development is a mechanism of control similar to colonialism.

It is common practice in geography and economics to categorize countries in terms of their development, with the world's nations divided into the "developed" and the "developing". However, who judges a country's place on the developmental scale? One of the key thinkers in the field of development studies was Walt Rostow, an economist and government official in President John F. Kennedy's administration.

Stages of economic growth

In 1960, Rostow developed a model for growth that proposed five stages through which all countries must pass to become fully developed. Each country starts as a traditional society, characterized by an agriculture-based, labour-intensive subsistence economy. In the next stage, the preconditions for economic take-off are in place, with the country starting to develop a manufacturing base. Industrialization begins to gain in importance during stage three, or take-off, followed by a drive to maturity. This stage takes place over a long period, during which the national economy grows and diversifies. The final stage is that of high mass consumption, with the country's economy flourishing in a capitalist system.

A theory for the Cold War

Prior to Rostow, ideas of development were generally based on the assumption that the "Western world" of wealthy, powerful countries was the

Walt Rostow, shown here with US president Lyndon Johnson in 1968, believed that economic development and industrialization could help to prevent the spread of communism.

HUMAN GEOGRAPHY 191

See also: Organic theory 154–55 ▪ Heartland theory 158 ▪ Rimland theory 159 ▪ World-systems theory 198–99 ▪ Territoriality theory 202–03

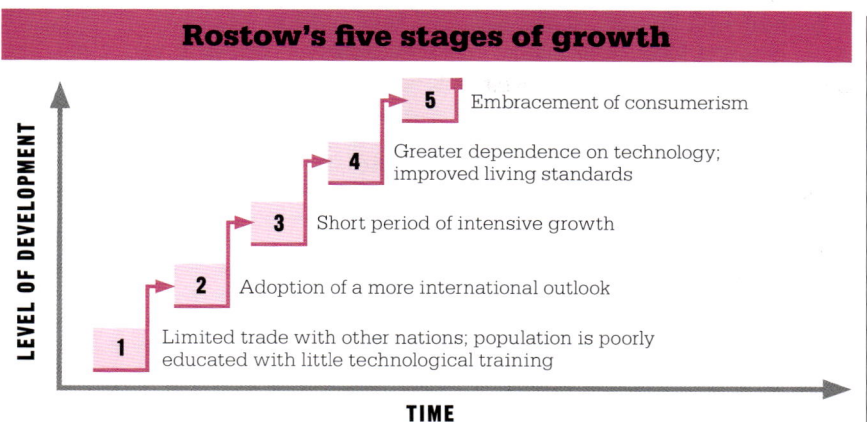

Rostow's five stages of growth

- 1: Limited trade with other nations; population is poorly educated with little technological training
- 2: Adoption of a more international outlook
- 3: Short period of intensive growth
- 4: Greater dependence on technology; improved living standards
- 5: Embracement of consumerism

Political and social institutions are reshaped… to permit the pursuit of growth to take root.
Michael Pacione
Urban Geography, 2005

model that other "less developed" countries should aspire to. An avowed anti-communist, Rostow also skewed his model towards Western industrial, urbanized, and capitalist countries. However, his "stages of growth" model for economic development was aimed not simply at aiding lower-income countries in the development process, but doing so in a way that would advance the global influence of the US over that of the communist USSR. Developed at the height of the Cold War, Rostow's model became an important tool of US foreign policy.

Questionable assumptions

Rostow argued that development results from changes in traditional values, exposure to Western society and norms, and increased investment. However, his view does not address the problems faced by previously colonized countries, nor the limits placed on universal mass consumption by environmental concerns. It also rests on the presumption that the US is the model to which other countries should aspire.

An alternative development strategy based on export-led growth aims at growing productivity by focusing on foreign markets. This "comparative advantage" model, favoured by institutions such as the IMF, focuses on goods and services produced by one country and sold profitably to another. ■

Walt Rostow

Named after poet Walt Whitman, Walt Rostow was born in New York in 1916 to a Russian-Jewish immigrant family. He graduated from high school at the age of 15, entering Yale University on a full scholarship. After later studying at Balliol College, Oxford, UK, he returned to Yale, where he completed his PhD in 1940.

Rostow served as an intelligence analyst during World War II, and in 1961 he became director of policy planning in the Department of State. As a staunch anti-communist and firm advocate for US intervention in Vietnam, he first suggested invasion in 1963. After Kennedy's assassination, he continued as an advisor to new president Lyndon Johnson.

In 1969, Rostow was appointed professor of economics at the University of Texas at Austin, where he remained until his death in 2003.

Key work

1960 *The Stages of Economic Growth*

NEAR THINGS ARE MORE RELATED THAN DISTANT THINGS
SPATIAL INTERACTION THEORY

IN CONTEXT

KEY FIGURE
Waldo Tobler (1930–2018)

BEFORE
1945 US geographer Edward Ackerman advocates for a quantitative shift in geography to create a discipline able to address the needs of the future.

1962 The Canadian government commissions British geographer Roger Tomlinson to develop the world's first geographic information system (GIS).

AFTER
1970s Geodemographic segmentation, developed by British demographer Richard Webber, classifies households and neighbourhoods by socioeconomic activity.

2000s British architect Ruth Conroy Dalton uses virtual reality and visual simulation to explore wayfinding, navigation behaviour, and pedestrian movement patterns.

Spatial interaction theory describes the flow of people, goods, and information between different locations – from commuting, to freight delivery – and the decision-making processes involved. US geographer Waldo Tobler played a key role in shaping this theory. In 1969, he presented the First Law of Geography, which states: "Everything is related to everything else, but near things are more related than distant things." This apparently simple statement encapsulates all the complexities of connections and movement that spatial interaction theory aims to explain.

Distance decay

Applying his theory to predict the volume and patterns of spatial interactions, Tobler showed that distance and connectivity shape relationships between locations, with certain locations exerting a

The **movement** of a person or material between locations can be **quantified**.

An analysis of these **spatial interactions** helps with planning and modelling.

Reducing the distance or costs of travel between two areas increases the interactions between them.

> What is where, why there, and why care?
> **Charles Gritzner**
> *Journal of Geography*, 2002

stronger influence because of their centrality or proximity. According to Tobler's law, the likelihood of interaction between two locations generally decreases as the distance between the locations increases. This inverse relationship between distance and interaction is referred to as distance decay. People are more likely to visit the shop that is nearest to their home than one further away, for example; likewise, the frequency of phone calls between two cities diminishes with increasing separation.

Distance decay is not a simple linear relationship but is weighted differently case by case. The advent of fast and private communication networks in the first quarter of the 21st century means that the nearest shop is now competing with online stores. To the shopper, the online store is in fact nearer, despite the goods being further away. The change in distance to point of sale means that goods now move in a completely new pattern: instead of large deliveries being taken weekly to a few local retailers, goods are being distributed over a wider area, with small deliveries made every few minutes.

Friction of distance

Understanding the rate of distance decay is crucial for building accurate spatial interaction models. A useful tool is the concept of friction of distance, which attempts to quantify the deterrent effect that distance has over human activity. Covering a distance uses energy, time, and money; the greater the distance, the greater this outlay, and the less likely that an interaction takes place. Friction in this context is not solely a function of physical distance; it also takes into account terrain (or ocean), transport infrastructure, and human barriers such as foreign borders.

The impacts of distance decay and friction of distance are key elements in cost-distance analysis. This technique evaluates each potential spatial interaction not by its physical distance but by its cost, which takes into account time, resources, other expenses, and the perceived effort or friction involved. Online retailers use this analysis to identify the routes with the lowest costs for delivering to their customers, while real-world retailers seek the store locations that are easiest for customers to reach based on the cost of their journeys in time as well as distance.

Multiple applications

Spatial interaction theory allows geographers and planners to understand a region and the activities within it in a more nuanced and effective way than by simply using proximity. The insights from spatial interaction theory find uses in urban planning, modelling commuter transport networks, and evaluating locations for new shops and homes. They can also play a part in tracking the spread of epidemic diseases. ■

Freight-forwarding trucks wait in line to have their paperwork and goods checked at a border crossing. Open borders, such as in the Schengen area, reduce the friction of distance.

EVERY CITIZEN HAS HAD LONG ASSOCIATIONS WITH SOME PART OF HIS CITY
BEHAVIOURAL GEOGRAPHY

IN CONTEXT

KEY FIGURE
Kevin Lynch (1918–84)

BEFORE
1945 The ways in which people respond to hazards such as flooding is the focus of a study by American geographer Gilbert White.

1948 Edward Tolman, a US psychologist, explores the idea of cognitive maps.

1953 Swiss psychologist Jean Piaget investigates schemas, mental frameworks that help people organize and interpret their knowledge of the world.

AFTER
1981 US criminologists Paul and Patricia Brantingham suggest that crime occurs in places where offenders, targets, and opportunities intersect.

2000 Patricia Gilmartin, an American geographer, theorizes that perceived distances are affected by how safe a route is believed to be.

Behavioural geography attempts to understand human activity within a space, place, and environment. It analyses data on the behaviour of individual people – looking at what they actually do and why they do it, rather than what they might be expected to do – recognizing that no two individuals will behave in the same way.

An important aspect of behavioural geography is the idea of the cognitive, or mental, maps that people create – the focus of a study by American urban planner Kevin Lynch in the late 1950s. He argued that models of human activity and interaction could be improved by incorporating more realistic assumptions about human behaviour. Studied activities include where people live, how they get to work and spend their leisure time, and how they exchange ideas and information.

The geography you carry in your mind, your mental map… is… as important as recording the facts of human existence on the surface of the Earth.
Reginald Golledge
Nature, 2009

Cognitive maps

For a mental map to be functional, it must have four elements. First, a person has to recognize where they are and be able to identify common objects such as particular types of buildings or pathways. Second, they need to understand how to get from one place to another. Third, they must be able to modify and update their mental maps according to changes experienced within the environment. Finally, they need to be willing to take new or alternative courses of action according to their updated mental map.

Behavioural geographers maintain that the time and trouble it takes to get from one place to another is an important determinant of human activity. Subjective, rather than objective, distance is what matters – not how

See also: Environmental determinism 156 ▪ Spatial interaction theory 192–93 ▪ Gentrification 196–97 ▪ Humanistic geography 200–01 ▪ Time-geography 266–67

The US city of Boston was the focus of much of Lynch's fieldwork. His 1960 community map shows what he called the five elements of the city, including landmarks such as Boston Common.

far away something actually is, but how far away the subject thinks it is. What someone believes about the world determines what they will or will not do. For example, is it an easy trip to the cinema or is it just too much trouble?

Lynch published his findings in *The Image of the City* (1960). He used simple sketches of maps created from memory by a number of participants to reveal five elements of the city: paths, edges, districts, nodes, and landmarks. Paths are the routes that people take repeatedly. Edges are natural or artificial breaks that impede the flow of travel, such as a river or a wall. Districts are areas with common identifying characteristics that can be entered and left, such as a shopping precinct. Nodes are junctions, where a decision needs to be made about where to go next. Finally, landmarks are identifiable points of reference that are relied upon for navigation.

Anchor-point theory

Australian-American geographer Reg Golledge looked at how people's spatial behaviour is driven by their subjective perceptions as much as, if not more than, objective reality. Specifically, he developed his anchor-point theory, which postulates that when people form mental maps they do not memorize every detail but select a few landmarks, or nodes, such as home or workplace, and use these as anchor points from which to orient themselves and estimate distances and routes to other places. ▪

Kevin Lynch

The youngest child of an Irish-American family, Kevin Lynch was born in Chicago, Illinois, in 1918. He enrolled at Yale University in 1935 but left to study architecture under Frank Lloyd Wright at the latter's Taliesin studio, Wisconsin. Lynch then enrolled at Rensselaer Polytechnic, New York City, to study engineering, but he went to work for Chicago architect Paul Schweikher before graduating.

In 1941, when the US entered World War II, Lynch was drafted into the Army Corps of Engineers. After the war, he was awarded a degree in city planning at the Massachusetts Institute of Technology, where he began teaching in 1948. He was appointed professor in 1963 and professor emeritus in 1978. Throughout his academic career, Lynch continued to practice architecture and to write about planning. He died in 1984.

Key work

1960 *The Image of the City*

THE WHOLE SOCIAL CHARACTER OF THE DISTRICT IS CHANGED
GENTRIFICATION

IN CONTEXT

KEY FIGURES
Ruth Glass (1912–1990),
Neil Smith (1954–2012)

BEFORE
1953 British magazine *Good Housekeeping* describes how run-down Georgian properties in London can be restored.

1961 US writer Jane Jacobs defends the positive qualities of New York's low-income neighbourhoods in the face of large-scale regeneration.

AFTER
1971 The Lafayette Square Restoration Plan is drawn up for a neglected district of St Louis, US.

1989 After the reunification of Germany, Prenzlauer Berg and other low-income neighbourhoods in Berlin undergo rapid gentrification.

2025 A redevelopment plan gets under way to rehabilitate the Dharavi slum district in Mumbai, India.

In the 1950s and 60s, Ruth Glass described how London neighbourhoods previously starved of investment were being "uplifted" as wealthier residents moved in. At the time, such run-down areas included parts of Notting Hill, Camden, and Islington. Housing and business rents were low, so low-income families could afford to live there; some were able to open small shops and cafés to cater for the local population. Many were white low-income families who had lived in these communities for generations; others, especially in Notting Hill, were first-generation immigrants from the Caribbean, who often lived in poorly maintained rented accommodation.

A double-edged sword

In 1964, Glass coined the term "gentrification" to describe how working-class neighbourhoods were changing, with wealthier people taking advantage of the low prices to buy and restore properties, open their own businesses, and demand infrastructure improvements. Outwardly, this seemed like a positive development; however,

Ruth Glass

Born Ruth Lazarus in Berlin, Germany, in 1912, Ruth Glass was forced to flee to the UK in 1932 because of her Jewish heritage. After studying at the London School of Economics and New York's Columbia University, she returned to the UK in 1943 to work at the Association for Planning and Regional Reconstruction.

In 1950, Glass moved back into academic life at University College London, where she later established the Centre for Urban Studies. She had a particular interest in London's housing problems, which she researched and wrote about extensively, giving evidence to several government enquiries.

A lifelong Marxist, Glass maintained that sociological research must have a purpose, namely "to influence government policy and bring about social change". She died in London in 1990.

Key work

1964 *London: Aspects of Change*

HUMAN GEOGRAPHY

See also: Migration theory 152–53 ▪ Geography as human ecology 162–63 ▪ Cultural landscape theory 172–75 ▪ Theories of development 190–91 ▪ The urbanization of capital 204–05 ▪ Spatial justice 206–07

> Rent gap grows: **capital flows out** of an area which, although well placed close to the city centre, becomes **run down**.

> Capital in-flow: houses and businesses are **restored**, infrastructure is **improved**, and more **affluent people move in**.

> **Low-income displacement: property prices and rents rise, more middle-class people move in, and low-income residents move out to the suburbs.**

Glass warned that rising rents were forcing the existing low-income residents out of these areas.

Capital flows

In his 1979 paper "Toward a Theory of Gentrification", Scottish-born geographer Neil Smith looked at gentrification through the lens of capital flows in and out of neighbourhoods. He theorized that the reason why some parts of cities are gentrified and others are not is due to the so-called "rent gap". This is the difference between the current rent a landlord charges on housing or business units and the potential rent that would be charged if the units were redeveloped. The wider the gap, he argued, the greater the likelihood of the area being prioritized for regeneration.

Gentrification accelerated in the early 21st century, spreading from Europe and North America to every continent as a by-product of globalization. Smith cited the transformation of Beijing, China, in the run-up to the 2008 Olympics as one of the most extreme examples.

Due to its impact on low-income inhabitants, gentrification has since acquired a bad reputation. In what Smith called a rebranding exercise, similar projects are now promoted as "regeneration" to emphasize urban improvement and downplay the displacement of residents.

Not everyone agrees with Smith. British-born Canadian geographer David Ley argues that gentrification is driven not by the market, but by the consumption patterns of individuals, who spend money to improve the aesthetics of any area, attracting others who wish to do the same. The reality is probably a combination of both processes.

The need to engage

Local governments also play a key role. Zoning policies can worsen or mitigate the negative impacts of gentrification. A 2025 report by the National Community Reinvestment Coalition in the US pointed out that genuine community engagement is essential where neighbourhood revitalization schemes are being planned. Otherwise, there is a risk of vulnerable communities being displaced, inequality deepening, and the historic and cultural fabric of neighbourhoods being erased. ■

The Bird's Nest was built in Beijing for the 2008 Olympic Games. As many as 600,000 low-income people were displaced to make way for stadia, shopping malls, and other structures.

Once this process of "gentrification" starts… it goes on rapidly until all or most of the working-class occupiers are displaced.
Ruth Glass
London: Aspects of Change, 1964

THE ONLY KIND OF SOCIAL SYSTEM IS A WORLD-SYSTEM
WORLD-SYSTEMS THEORY

IN CONTEXT

KEY FIGURE
Immanuel Wallerstein
(1930–2019)

BEFORE
1949 Argentine economist Raúl Prebisch highlights the worsening terms of trade faced by developing countries.

1967 According to German sociologist Andre Gunder Frank, the development of core nations depends on the underdevelopment of peripheral countries.

AFTER
1978 Samir Amin, an Egyptian economist, proposes a global law of value that takes into account the exploitation of the "periphery" by core nations.

1994 US-based Italian economist Giovanni Arrighi writes *The Long Twentieth Century*, the first in a trilogy of books exploring changes in global capitalism.

In the 1970s, in an attempt to account for the continuing exploitation of developing countries by developed countries, Immanuel Wallerstein set out his world-systems theory. He posited that a country's position within the capitalist world-system determines its economic development.

Development theories
In the years after World War II, modernization theory aimed to shape the development of former European colonies along capitalist lines to achieve standards of living characteristic of Western Europe and the US. Underdevelopment – believed to be the consequence of weak levels of production – was seen as a stage common to all developing countries. The theory forecast that greater international trade would result in higher earnings, and increased investment would stimulate competition, increase productivity, and drive growth in developing countries.

By the late 1950s, however, it had become apparent that underdeveloped countries remained resolutely underdeveloped. American economist Paul Baran advanced a counter theory, which suggested that developing countries were structurally different from developed countries and could not progress along the same lines. This notion formed the basis of what came to be known as dependency theory, which emphasizes the role played by imperialism in shaping post-colonial states.

The central plank of dependency theory is that core nations at the centre of the world economy exploit those at the periphery, creating fundamental discrepancies that hold back the latter's development. The unequal relationship of

In the 19th and 20th centuries there has been only one world-system in existence, the capitalist world-economy.
Immanuel Wallerstein
The Capitalist World-Economy, 1979

HUMAN GEOGRAPHY 199

See also: Heartland theory 158 ▪ Rimland theory 159 ▪ Theories of development 190–91 ▪ Globalization 250–53

Immanuel Wallerstein

The son of Polish-Jewish immigrants to the US, Immanuel Wallerstein was born in New York City in 1930. He was awarded his bachelor's degree in sociology from his home city's Columbia University in 1951, before serving in the US Army from 1951 to 1953.

Following his discharge from the army, Wallerstein wrote his MA thesis on McCarthyism (the campaign against communists, real or imagined, in US society in the early 1950s). From that point, he considered himself to be a "political sociologist". He completed his doctorate at Columbia University in 1959.

Wallerstein taught sociology at Columbia until 1971, taking part in campus protests against the Vietnam War. In 1976, he was appointed head of the Fernand Braudel Center for the Study of Economies at Binghamton University, New York. Wallerstein died in 2019.

Key work

1974 *The Modern World-System*

dominance and exploitation between centre and periphery is reinforced by market forces.

Economic hierarchy

In the 1970s, a theory emerged that sought to account for the exploitation of the periphery from the perspective of the core. Wallerstein suggested that the world economic system is divided into a hierarchy of three types of countries: core, semi-peripheral, and peripheral. The division of labour between these regions determined their relationships with each other.

Wallerstein explained that the dominant core countries, such as the US, Germany, and Japan, are characterized by high levels of technology, industrialization, and urbanization, and relatively low levels of workforce exploitation. In contrast, peripheral countries, including most African nations, are usually agrarian and less industrialized than the core countries on which they rely for capital investment. Semi-peripheral countries such as Brazil, India, Mexico, and South Africa have a level of development intermediate between the core and periphery.

While Wallerstein acknowledged that nations can move between the different categories over time, he maintained that the basic principles always apply. The core owns and controls most of the world's capital and technology and has a huge influence over world trade and economic agreements. Peripheral countries are generally a source of low-cost labour, raw materials, and agricultural produce for the core nations, which set the prices for the products the periphery relies on for export income. Semi-peripheral countries tend to exploit the periphery, just as the core exploits both semi-peripheral and peripheral countries. ▪

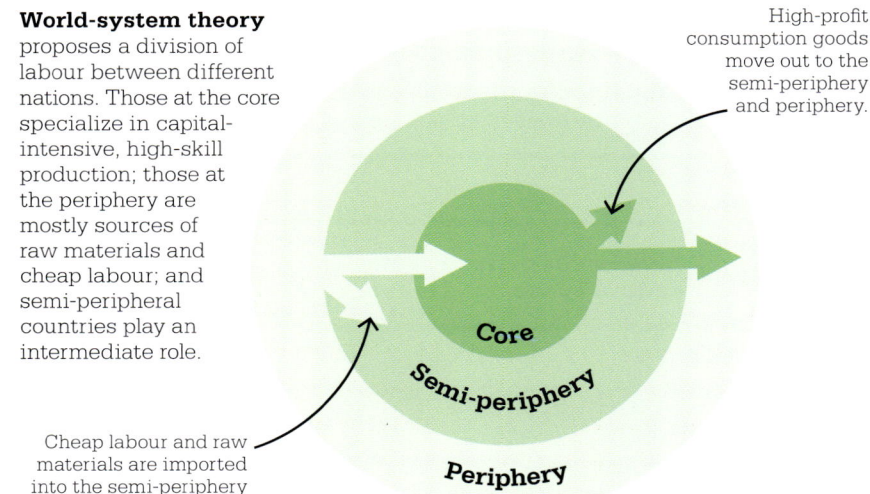

World-system theory proposes a division of labour between different nations. Those at the core specialize in capital-intensive, high-skill production; those at the periphery are mostly sources of raw materials and cheap labour; and semi-peripheral countries play an intermediate role.

Cheap labour and raw materials are imported into the semi-periphery and core nations

High-profit consumption goods move out to the semi-periphery and periphery.

SPATIAL FEELINGS AND IDEAS IN THE STREAM OF EXPERIENCE
HUMANISTIC GEOGRAPHY

IN CONTEXT

KEY FIGURE
Yi-Fu Tuan (1930–2022)

BEFORE
1927 In *Being and Time*, German philosopher Martin Heidegger makes a distinction between neutral "space" and significant "place".

1960 US urban theorist Kevin Lynch's *The Image of the City* considers how people construct mental maps of their surroundings.

AFTER
1986 In *Human Territoriality*, US geographer Robert D. Sack examines the various reasons why individuals, communities, and nation states define and enforce territories.

2005 Nigerian researcher Oladele A. Ogunseitan identifies four elements of topophilia: familiarity, mental stimulation, sensory richness, and ecodiversity – all of which shape people's happiness.

Humanistic geography defines locations and areas – and the human activities that take place there – according to their emotional and social importance. The field emerged in the 1970s as a backlash to the quantitative revolution that had taken place over the preceding decades. That revolution had transformed geography into a science that required systematic data collection to investigate any topic, even those that involved people. Chinese-American geographer Yi-Fu Tuan argued that such an approach missed an entire level of understanding. Instead of treating people solely as economic or spatial data points, geography

---

Space is an **open, quantifiable, and neutral** concept, with **no special significance** attached to it.

As **people move through a space**, they project onto it **qualities** related to their **experience**.

Lived experience and memory can turn **neutral space** into a **meaningful place**.

Emotional investment in a place results in a deep bond with it, or topophilia.

HUMAN GEOGRAPHY

See also: Possibilism 157 ▪ Spatial interaction theory 192–93 ▪ Territoriality theory 202–03 ▪ Local and global 208–09 ▪ Gender and space theory 210–13

Place is security, space is freedom: we are attached to the one and long for the other.
Yi-Fu Tuan
Space and Place: The Perspective of Experience, 1977

should also explore the emotional attachments people have to their physical surroundings. Tuan refined the matter-of-fact term "topophilia" – first introduced by British author W.H. Auden in 1948 to mean "love of place" – into a more emotionally rich concept.

Space and place

Tuan's 1977 book *Space and Place: The Perspective of Experience* was pivotal in the establishment of humanistic geography, creating a framework for examining the nuanced relationship between the two titular concepts. "Space" and "place" might seem synonymous, but Tuan suggested that they represent distinct ways of understanding how people construct a "sense of place".

Traditionally, geographers have regarded a space as also a place – and vice versa. The important information could be the set of coordinates that define the boundary of the space and place, the numbers of people living inside it, and the activities that take place there. However, Tuan gives almost opposite definitions to the terms. For him, space is an undifferentiated area with no clear classification or definition of boundaries. The important thing is how we feel when we are in a space – or, more commonly, moving through it.

Tuan links space to feelings of openness, hope, and freedom, but also to the anxiety and untethered bleakness instilled by an undefined vastness. He offers a deserted beach and a bustling city as examples of space. Either may offer a break from one's past and roots, providing the chance to think anew and explore new ideas, both objective and subjective.

By contrast, a place is defined by a person's experiences, their memories, and their emotions. It is a geographical representation (in part, at least) of a person's identity. This is where a person lives, where they have ties to other people and objects.

Space can become place as a person bestows these values onto it – chief among them, the biological needs for food, shelter, and companionship. Such a place must be protected; therefore, unlike space, it is given a boundary and classified by who and what it contains. Just as with space, place can also be imbued with negative feelings, such as boredom, frustration, and oppression.

Subjective dimensions

Tuan showed that geographical work was incomplete without acknowledging the subjective, emotional, and cultural dimensions that make coordinates and polygons on a map into meaningful locations. By accounting for the lived experience and other qualitative aspects of a location, humanistic geography has become a cornerstone of geography and has found applications in wider fields. ■

An ocean-facing beach is identified as space – vast and open; however, to a surfer, who might visit often and see it through the lens of their positive experience, it becomes a place.

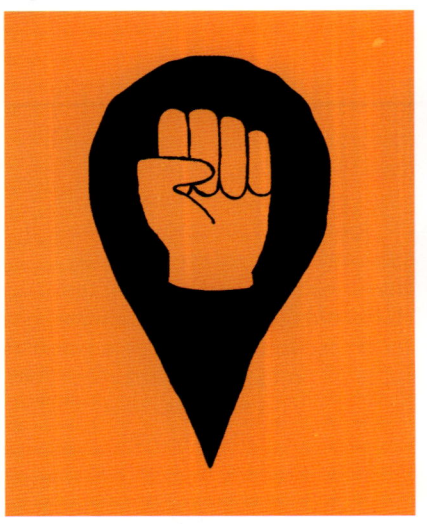

PLACE HAS POWER
TERRITORIALITY THEORY

IN CONTEXT

KEY FIGURE
Robert D. Sack (1944–)

BEFORE
1897 In *Political Geography*, German geographer Friedrich Ratzel suggests that nations can ensure their survival only by expanding their territories.

1967 Israel starts occupying Palestinian land in the West Bank, asserting control through the establishment of illegal settlements and checkpoints.

AFTER
2018 Emmanuelle Boulineau, a French geographer, says that cross-border cooperation zones in the European Union (like the Danube Region) contrast with the notion of hard territoriality.

2022 *Territory*, by British-Canadian geographer Nicholas Blomley, examines how territoriality reinforces social and economic inequalities, limiting opportunities for marginalized populations.

Territories – or, rather, their borders – are a fundamental part of geography. Unlike natural landscape features, though, these areas are created entirely by humans, and they often change. Territoriality is the way people control an area and communicate their ownership to others.

Nuanced and intricate, the territoriality of human communities goes beyond a purely instinctive or biological behaviour. In his 1986 book *Human Territoriality: Its Theory and History*, American geographer Robert D. Sack proposed that the adoption of territory is a deliberate use of power that aims to influence and control the people living both inside the territory and beyond it. His theory hinges on the fact that claiming and defending territories are fundamental aspects of human society, and they serve

People need an **organized space** to achieve their **social, political, and economic objectives**.

This is achieved by **regulating access into** and **use of a defined area**.

Territorial control reinforces **social hierarchies** and authority within the area.

Territorial strategies affect social identity, settlement patterns, and resource distribution.

See also: Organic theory 154–55 ▪ Heartland theory 158 ▪ Rimland theory 159 ▪ Spatial interaction theory 192–93 ▪ World-systems theory 198–99

many purposes, such as creating identity, providing protection, and facilitating social interactions.

Territoriality tendencies

Sack gave several reasons – or, as he called them, tendencies – for human territoriality. The first tendency is classification, which allows the grouping of people, land, and resources inside a boundary. It is simpler for a person or community to say that everything within a certain area belongs to them than it is to make a detailed record of their possessions. A clearly defined territory signals ownership, making any intrusion or use of resources by outsiders immediately obvious.

A second tendency is communication. The territory needs only one sign that it exists: the boundary. Past this point, the nature of the space a person is in changes, but the only sign of that change is the relative position of the boundary. This links to a third tendency: the boundary is the only feature of the territory that has to be controlled and regulated.

Further to this, the activities and resources protected by a boundary must have a higher value than the resources required to defend that boundary. This "value versus cost" relationship acts as a strong control mechanism on the size of a territory and the length of its boundary.

A base for power

Territoriality also forms a concrete base for power. The power conferred on a group, a nation, or an individual is proportionate to the size, nature, and location of the territory, and it can be used as leverage when dealing with the people and rulers of other territories. A leader can also use the territory to shift blame and deflect criticism.

In democratic countries, the authority of a leader derives from the laws of the land, which apply equally to all. As a result, power is about the system, not personal relationships; and a leader controls a territory, not the people in it.

Sack's territoriality theory looks into the ways in which people use the physical spaces they inhabit –

Territoriality for humans is a powerful geographic strategy to control people and things by controlling area.
Robert D. Sack
Human Territoriality: Its Theory and History, 1986

whether offices, cities, or countries. The seemingly natural divisions of the world are, in fact, the result of ongoing strategic efforts to define, control, and use areas for social, economic, and political purposes. ■

East German soldiers supervise the construction of the Berlin Wall in 1961. The Communist government of the city's eastern sector ran strict controls on entry and exit of people and goods.

THE CAPITALIST PRODUCTION AND RECONSTRUCTION OF SPACE
URBANIZATION OF CAPITAL

IN CONTEXT

KEY FIGURE
David Harvey (1935–)

BEFORE
1925 American sociologists Ernest Burgess and Robert E. Park collaborate on developing the concentric zone model of city development.

1970 In *The Urban Revolution*, French sociologist Henri Lefebvre argues that cities are not passive backdrops but are actively shaped by human interactions.

AFTER
1991 Dutch-American sociologist Saskia Sassen's *The Global City* identifies a small group of cities that serve as hubs for globalization.

1996 According to Scottish geographer Neil Smith, intentional flows of capital out of and into city districts are responsible for processes of dereliction and gentrification.

Until the late 20th century, geographers had mostly considered urbanization through a functionalist lens. In 1915, Scottish sociologist Patrick Geddes had explained how environment, culture, and economic activities were key factors in the development of urban areas, while German geographer Walter Christaller's 1933 central place theory had offered a market-based explanation for the spatial relationship between cities and towns. However, neither had taken into account the politics of power in shaping urban spaces. It was David Harvey who explained the role of cities in the capitalist economic system in his 1985 book *The Urbanization of Capital*.

Investment and surplus
In the 1860s, German philosopher Karl Marx stated that the capitalist system relied on manufacturing. According to Marx's theory of a "circuit of capital", capitalists invest in factories, machinery, raw

---

Capitalists invest in manufacturing until their rate of **profit falls** below a certain level. → With manufacturing **no longer worthwhile**, they **switch** to **real estate**.

↓

The **building boom runs out of steam** when there is an **oversupply** of office and apartment spaces. ← A **construction boom** follows, with new office and apartment buildings **transforming cities**.

↓

Demand and prices fall, and capitalists look elsewhere for a return on their investments.

HUMAN GEOGRAPHY

See also: The concentric zone model 184–87 ▪ Theories of development 190–91 ▪ Gentrification 196–97 ▪ Spatial justice 206–07 ▪ Globalization 250–53

> [Urban transformation] played a crucial role in the stabilization of global capitalism after World War II.
> **David Harvey**
> "The Right to the City", 2008

materials, and workers to produce goods and surplus value, with the latter being reinvested to make more of both. Greater investment increases production to the point that the market becomes saturated, and capitalists then stop investing because they are not getting a good return. Harvey argued that one way of getting around this obstacle was to invest away from manufacturing, and into real estate, transport systems, and other infrastructure – the building blocks of cities.

Harvey cited as an example the reconstruction of Paris in the 1850s and '60s, when French emperor Napoleon III appointed Georges-Eugène Haussmann to transform the city with grand buildings, new parks, and broad boulevards. Relying on the state to provide the initial finance, a large "army" of surplus labour carried out the work, which created huge opportunities for capital accumulation through construction, rents, retail, and transport provision.

Changing times

In his 2008 article "The Right to the City", Harvey returned to the theme of the urbanization of capital. He described how the worldwide growth of cities accelerated in the late 20th and early 21st centuries in response to a reduced rate of profit from manufacturing. For Harvey, urbanization is not simply a by-product of capitalism, but a central process to its very survival.

Harvey's theory shifted the way geographers study cities. Whereas urban geography was once primarily concerned with land use, history, transport patterns, and demographics, Harvey saw cities as intrinsic to the evolution of capitalism. He modified anti-capitalist theory by shifting the primary emphasis from workplace struggles to a fight over who controls city space, what is built, and where. His concepts also influenced many other geographers, helping Neil Smith to propose rent gap theory and Edward Soja to theorize on spatial justice. ▪

Shanghai's financial district shows how surplus capital flooding into an urban space fuels capitalist growth while also leading to social inequality.

David Harvey

Born in Gillingham, Kent, UK, in 1935, David Harvey studied geography at St John's College, Cambridge, earning his doctorate there in 1961 with a thesis on the history of hop production in his native county.

After teaching for several years at the University of Bristol, in 1971 Harvey moved to Johns Hopkins University, Baltimore, US. There, he witnessed great poverty, racism, and social upheaval, which led him to a radical, Marxist geographical outlook, as expressed in books such as *Social Justice and the City* (1973).

Harvey remains an activist as well as an academic, backing social movements opposing racism, sexism, exploitation, and capitalism. He is professor of anthropology and geography at the City University of New York.

Key works

1985 *The Urbanization of Capital*
2008 "The Right to the City"

THE ACCUMULATION OF CAPITAL HAS ALWAYS BEEN A PROFOUNDLY GEOGRAPHICAL AFFAIR
SPATIAL JUSTICE

IN CONTEXT

KEY FIGURE
Edward Soja (1940–2015)

BEFORE
1968 US radical geographer Bill Bunge founds the Detroit Geographical Expedition to map the concentration of social problems in the inner city.

1970 Henri Lefebvre, a French sociologist, argues in *The Right to the City* that cities should be democratically shaped by their residents.

AFTER
2005 The World Social Forum in Porto Alegre, Brazil, defines the right to the city as the collective right of all residents to access and share its resources equitably.

2010 In *The Just City*, American urban geographer Susan Fainstein argues that all urban planning should incorporate the central concepts of diversity, democracy, and equity.

In *The Condition of the Working Class in England* (1845), German philosopher Friedrich Engels examined uneven social conditions in Britain's industrial cities. He noted how workers were segregated into overcrowded and unsanitary neighbourhoods, which resulted in higher rates of mortality from disease compared to the surrounding countryside.

Many working-class, inner-city areas have long had to deal with issues such as more pollution, poorer public transport, an absence of green spaces and playgrounds, schools that are less well funded, and inferior infrastructure. In 1989, American urban geographer Edward Soja articulated a theory of spatial justice – without using those exact words – in his book *Postmodern Geographies*.

Locational injustice factors

Soja's theory emphasized that where people live and work determines their ability to

---

Full engagement with communities means that **the needs of no group are ignored**.

Inclusive **zoning** and progressive pricing policies ensure that a range of **housing types** and community facilities is **available for all**.

Accessible, **subsidized public transport** results in no one being prevented by cost from reaching **employment, education, and retail hubs**.

Public spaces are designed to be **accessible, safe, and inclusive**, catering for the needs of **all residents**.

There is a fair distribution of resources and opportunities across different areas of the city.

HUMAN GEOGRAPHY 207

See also: Gentrification 196–97 ▪ The urbanization of capital 204–05 ▪ The tragedy of the commons 220–21 ▪ The environmental justice movement 246–47 ▪ Globalization 250–53

Combining the terms spatial and justice opens up a range of new possibilities for social and political action.
Edward Soja
"The City and Spatial Justice", 2009

Edward Soja

The son of Polish immigrants, Edward Soja was born in 1940 in New York City's Bronx borough, where he grew up "nurtured in its dense diversities". While studying at Syracuse University, New York, he visited Kenya to study urban planning under the tutelage of Mozambican anthropologist and revolutionary Eduardo Mondlane.

In 1972, Soja began a long academic career at the University of California, Los Angeles, where his special interests were regional development, planning and governance, and the spatiality of social life. He became one of the key players in the "spatial turn" in geography, and in 1996 he advanced the concept of "thirdspace". A month before he died, he received the 2015 Vautrin-Lud Prize, one of the highest awards in geography.

Key works

1989 *Postmodern Geographies*
1996 *Thirdspace*
2010 *Seeking Spatial Justice*

access resources, services, and opportunities, because power shapes the organization of space. While it is easy to see where spatial injustice occurs by examining of a range of datasets, Soja explained that it is far more difficult to understand the underlying processes that have produced it – or to find solutions.

According to Soja, the three broad-brush drivers of locational injustice are class, race, and gender. Among the many other factors that may be superimposed on these are gerrymandering, where poor neighbourhoods get less political representation than affluent ones; and residential segregation, both official and unofficial.

A bridge to action

Soja's ideas bridged theory and practice. Since space is not a fixed entity but is actively shaped by human activities and power relations, spatial injustices can be challenged and reversed through social and political interventions. These can come from "above" – that is, from government or planners; or from "below", through grassroots campaigns run by those at the receiving end of injustice.

In a 2009 lecture, Soja used the example of members of the Bus Riders Union (BRU) in Los Angeles, US. In 1996, the BRU had successfully challenged the city's Metropolitan Transit Authority when the latter proposed to invest disproportionately more in mass transit from relatively wealthy suburbs at the expense of inner-city bus services used by the working poor. After it cited race discrimination and the need for spatial justice, the BRU eventually won a court order that prioritized new buses, better routes, shorter waiting times, and cheaper bus passes. Soja heralded the hugely improved bus service as, unusually, one that "benefits the poor more than the rich".

While accepting that perfectly "complete socio-spatial, pure distributional justice" is never achievable, Soja argued that when social unevenness crystallizes into more lasting structures of privilege, such interventions become necessary. His theory is important because it demands that urban planning, design, and policy should function in such a way that everyone has access to resources, opportunities, and power. ■

The Bus Riders Union's protest against a proposal to cut inner-city bus routes in Los Angeles is an example of marginalized communities demanding equitable access to urban resources.

PLACES ARE NOT CLOSED OR BOUNDED
LOCAL AND GLOBAL

IN CONTEXT

KEY FIGURE
Doreen Massey (1944–2016)

BEFORE
1857–58 In the *Grundrisse*, German philosopher Karl Marx explains that capital is by its very nature expansive and always pushes outwards, looking for new markets and shaping local geographies.

1987 In an article in the *Journal for Critical Social Science*, German political scientist Elmar Altvater describes how globalization is "shrinking" the world.

AFTER
2009 Israeli geographers Nurit Alfasi and Tovi Fenster explore how cities integrate global and local dynamics, and identify them as "local globalities" or "global localities".

2022 The life and work of Doreen Massey are celebrated over 10 episodes in the podcast "Spatial Delight".

In his 1990 book *The Condition of Modernity*, British-American geographer David Harvey described how late 20th-century developments in communications and transport, accelerated by capitalist globalization, had made the world a smaller place. These days, for example, a publishing work meeting can simultaneously include an editorial team in the UK, a design team in India, and a marketing team in the US. Harvey believed that the speed of human interactions caused people to feel a sense of placelessness.

The following year, Doreen Massey put forward a different view in her article "A Global Sense of Place", which appeared in *Marxism Today*. She proposed that places should not be defined solely by their physical location or the flows of capital through them. Instead, they should be imagined as "networks of social relations and understandings".

Multiple identities

For Massey, places have multiple identities, reflecting the complex web of social relations and experiences within them. Places are neither static nor isolated, but constantly changing, and they are connected to the wider world, as well as to their near neighbours.

Massey wrote of two senses of "place". One is reactionary, resisting change and wanting to keep a location in its present form, or even take it back to a historic state. The other is progressive, viewing places

"Smaller world" communication advances like internet-hosted virtual meetings are now commonplace, with many people taking part in video conferencing every day.

HUMAN GEOGRAPHY

See also: Geography as human ecology 162–63 ▪ Cultural diffusion 164–71 ▪ Spatial interaction theory 192–93 ▪ Behavioural geography 194–95 ▪ Humanistic geography 200–01 ▪ Globalization 250–53

Each place is the focus of a distinct mixture of wider and more local social relations.
Doreen Massey
"A Global Sense of Place", 1991

Local markets can be places where global heritage thrives. Britain's capital, London, has residents from more than 250 countries, making it one of the most diverse cities in the world.

as dynamic, interconnected, and containing multiple histories and varied identities.

Using as an example the area where she lived – the district of Kilburn, northwest London – Massey described its history as encompassing many communities, cultures, and traditions brought from all parts of the world. She described talking to people in shops and on the street about a huge range of global and local topics, reflecting residents' truly global heritage and kaleidoscopic lived experience. Massey was arguing not that Kilburn was unique, but that its mixture of thousands of different "trajectories" could be replicated anywhere and showed that the "global" and "local" are not separate but intertwined.

A far-reaching impact

South African sociologist Laurine Platzky gave an example of how global and local interact when she described the impact of gold price changes, which are determined by international events, on miners and their families. If the price of gold increases, mining companies employ more men, and they are then able to send money home to their families in rural areas such as the Transkei. The families are then more secure financially and will spend more money on food, clothes, and other goods from local stores, boosting the local economy. However, should the price of gold fall, miners are laid off and return to their families. With dramatically reduced income, domestic finances – and relationships also – will be strained, local shops will lose business and may have to close. One decision taken thousands of kilometres away may have a positive or devastating impact on a multitude of local communities. ∎

Doreen Massey

Experiencing the disadvantages of growing up in a working-class community in Manchester, Britain, strongly influenced Doreen Massey's world view. Born in 1944, she won a scholarship to St Hugh's College, Oxford, where she obtained a geography degree in 1966. The following year, Massey began to work at the Centre for Environmental Studies, a UK government think-tank. In 1982, she was appointed chair of geography at the Open University.

Among many contributions in her field, Massey proposed the spatial divisions of labour theory in 1984. She suggested that capitalism creates an uneven distribution of economic activities, leading to inequalities between regions and social classes. She also wrote about gender and space – for example, how women were affected by views that they should stay at home. Massey died in 2016.

Key work

1991 "A Global Sense of Place"

SPACES AND PLACES ARE GENDERED
GENDER AND SPACE THEORY

IN CONTEXT

KEY FIGURES
Linda McDowell (1949–),
Gillian Rose (1962–)

BEFORE
1845 In *The Condition of the Working Class in England*, German philosopher Friedrich Engels describes how working-class women also perform unpaid labour at home.

1987 Scottish academic Liz Bondi teaches the first gender-focused geography course at Edinburgh University.

AFTER
1995 US geographers Susan Hanson and Geraldine Pratt explore the gendering of the labour market in Worcester, Massachusetts.

2021 In Sweden, the Global Utmaning think tank launches a digital toolbox to encourage female participation in urban planning processes.

Gendered spaces are environments that are socially constructed to be associated with specific genders. An idealized division between the woman's domain of household and family and the man's sphere of paid labour and civic life had become the cultural norm in 19th-century industrial societies. Starting in the 1970s, influential feminists and social campaigners such as British geographers Doreen Massey and Linda McDowell – and later Gillian Rose – argued that this division meant restricting women's role in society to that of "invisible" unpaid labour looking after men and rearing children.

HUMAN GEOGRAPHY 211

See also: Geography as human ecology 162–63 ▪ Cultural diffusion 164–71 ▪ Cultural landscape theory 172–75 ▪ Spatial interaction theory 192–93 ▪ Humanistic geography 200–01

Young women in a Chinese factory make clothes destined for foreign markets. In the highly gendered garment industry, women represent 80 per cent of the global workforce.

In her 1993 book *Feminism and Geography*, Rose suggested that the discipline was too structured around male perspectives and interests, and that women's voices and experiences needed to take a more central role in research.

Gendered workplaces

"A Woman's Place?", a 1984 article co-authored by McDowell and Massey, examined how the explosive growth of industry in 19th-century Britain had disrupted previous relations between men and women, with traditional forms of domestic production torn apart.

The article explored gendered workplaces in different regions and industries. In the coalfields of Wales, for example, the paid work was all done by men, who became their households' breadwinners. Men worked long hours in dangerous, dirty conditions, and the burden of domestic labour fell to their wives or mothers, leading to a definite separation of labour along gender lines. Conversely, in the cotton towns of Lancashire, England, about 60 per cent of the workforce was female.

Gender divisions in the labour market continue to exist, with men largely employed in construction, industry, and transport; and women in health and education. Some male-dominated work environments might not be welcoming to women. In recent years, several measures have been adopted to make workplaces more inclusive. Among them are flexible working arrangements to help women with childcare responsibilities; paid parental leave; equal pay policies; and initiatives to open traditionally male workplaces – such as the »

Linda McDowell

British geographer Linda McDowell was born in Stockport, near Manchester, in 1949. She studied for her first degree at the University of Cambridge, then researched housing change in London for her doctorate at the Bartlett School of Planning, part of University College London. She subsequently lectured at the University of California, Los Angeles, and at her alma mater.

McDowell's research interests focus on the connections between economic restructuring, labour market change, and age, class, ethnicity, and gender. Between 2007 and 2009, she co-directed a project on the political involvement of South Asian women in the UK, and in 2008 she became a fellow of the British Academy. McDowell is now emeritus professor at St John's College, Oxford.

Key works

1999 *Gender, Identity and Place*
2003 *Redundant Masculinities?*

212 GENDER AND SPACE THEORY

police, armed forces, and fire service – to women. However, even in countries where such legislation exists, barriers remain.

A feminist geography

In 1994, McDowell's *Gender, Identity and Place* advanced the theory that people and places are gendered; as a consequence, "social and spatial relationships are mutually constituted". Gender is not just an individual attribute, it is actively constructed: the way that people act, what they do, and how they are perceived are all influenced by gendered norms and expectations.

Since places are also gendered, people's behaviours are shaped by where they live, work, spend free time, and pass through. McDowell explored five gendered spheres: the human body, the home, the workplace, public spaces, and the nation state. Like Rose, she believed that while urban design is often assumed to be gender-neutral, in reality cities have mainly been designed under the assumption that men commute to and from work while women stay at home. In most parts of the world, that is no longer the case, but the legacy design is at best not welcoming, and at worst it is positively dangerous for some. The experience of women, girls, and gender-diverse people has not been taken into consideration.

Challenges in public spaces

The way streets, buildings, parks, and other urban features are planned and built determines who feels safe, welcome, visible, and empowered in them. For example, poorly lit station platforms, long underpasses with no escape routes, and parking garages with limited visibility can make women in particular feel vulnerable.

Transport network planners tend to prioritize train- and car-centric traffic, rather than pedestrian infrastructure – such as wide, unobstructed pavements and step-free routes – that would improve access for women with prams, the disabled, and the elderly.

A 2021 UK study found that women were twice as likely as men to list personal safety as a barrier to walking, cycling, or using public transport. The fact that women feel deterred from using these forms of transport not only restricts their freedom of movement, which may in turn limit their educational and professional opportunities; it also has negative implications for the environment, as Earth-friendly transportation is ditched in favour of cars, with the consequent rise in rates of carbon emissions and pollution. In 2023, this study was echoed by a survey conducted in New South Wales, Australia, which revealed that women were nearly twice as likely as men to feel unsafe walking after dark.

Public space design also reflects gender bias: recreational spaces often include football pitches and other areas geared towards historically "male-dominated" activities, but comparatively few spaces catering to the perceived needs and interests of women and the elderly. Likewise, the provision of public toilets is typically grossly inadequate for women and people with accessibility needs.

Additionally, because of their disproportionate role in childcare, adult care, and other domestic responsibilities, women are more

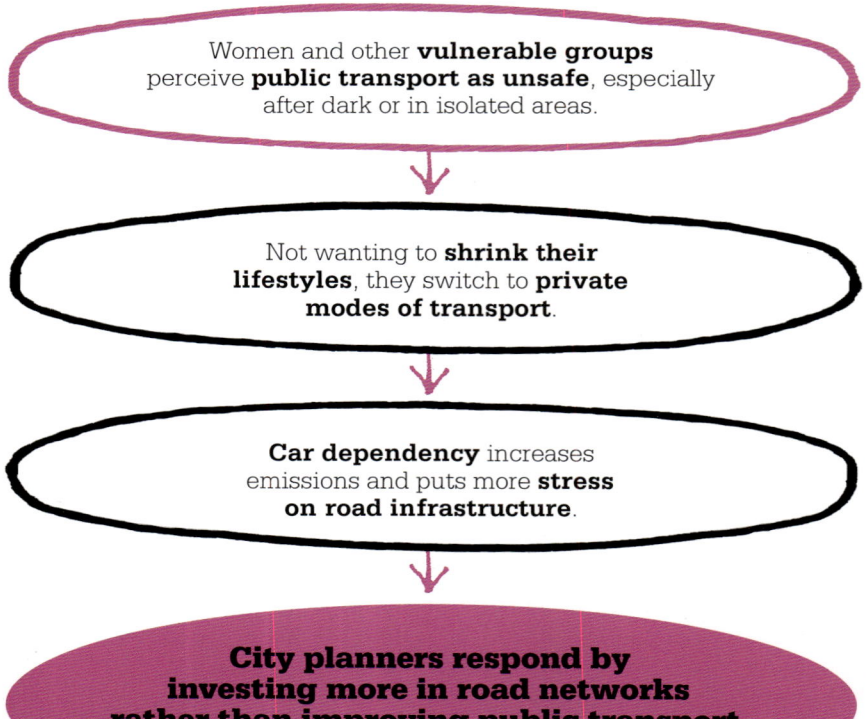

At a "Reclaim the Night" protest in 2024, women demanded justice for a doctor murdered at a Kolkata hospital. She had been attacked while sleeping in a seminar room after a 36-hour shift.

likely than men to "trip chain" – or make a succession of short trips – rather than undertaking a single commute to and from work. This adds to cost and increases safety risks, underlining the need for public transportation connection hubs to be safe and secure.

Unsafe streets

Massey argued that women are often confined to private places such as households, whereas men can generally move unhindered through public spaces. Women are effectively imprisoned by "differentiated mobility" due to obstacles that prevent them from initiating "flows and movement".

Violence against women in public spaces is still a fact of life. According to the World Population Review, in 2023, only 25 per cent of South African women said they felt safe walking alone at night. In many parts of the world, the same is true for members of LGBTQI+ communities, more than half of whom in southern and eastern Africa have experienced violence. The authorities cannot always be relied on to assist: in the Caribbean, 59 per cent of trans and gender-diverse people have faced violence at the hands of the police.

There is also a long history of pushing back against violence. In the US, the Take Back the Night campaign, launched after Sue Smeeth was stabbed to death while walking home from work in Philadelphia in 1975, demanded that women should be able to go where they wanted to, without fear. In late 1970s Britain, when the police advised women to stay out of public spaces after dark following the Yorkshire Ripper murders in northern England, women took to the streets chanting "No curfew on women; curfew on men".

Gender-sensitive planning

A 2021 gender-equity study on public transport in the Los Angeles area resulted in immediate reforms, such as the introduction of on-demand bus stops in the hours of darkness. Three years later, a survey of women's attitudes to public spaces in London found that safety and comfort depend on thoughtful design. Women feel safer in well-lit areas with clear sightlines, multiple entry and exit points, and either CCTV or trusted, diverse human security. They also value inclusive amenities such as accessible toilets, baby-changing and breastfeeding facilities, children's play areas, clean seating, and phone-charging points.

Gender inclusivity

There is a growing realization that buildings, infrastructure, and transportation have to be seen primarily as spaces for human connection. That means they have to be designed for the needs of everyone, regardless of gender or any other characteristic – with input from women and girls and members of LGBTQI+ communities guaranteed. The new ethos should be that gender inclusivity is a necessity, not a luxury. ∎

Increasingly, urban planning is, or should be, concerned to establish the diversity of interests in contemporary cities.
Linda McDowell
"Women, Men, Cities", 2001

THE BORDER BETWEEN REAL AND DIGITAL VIRTUAL WORLDS IS ALREADY POROUS
DIGITAL SPACES AS CULTURAL SPACES

IN CONTEXT

KEY FIGURE
Rob Kitchin (1970–)

BEFORE
1984 In his novel *Neuromancer*, US sci-fi author William Gibson coins the term "cyberspace" to mean a vast digital network.

1991 *Neverwinter Nights* is the first massively multiplayer online role-playing game (MMORPG), allowing players to engage with each other as they explore a world of characters and environments.

AFTER
1999 Martin Dodge, a British geographer, calls for a new field of geographic enquiry, cybergeography.

2025 British artificial intelligence (AI) researcher and Google DeepMind CEO Demis Hassabis states that the impact of AI will be 10 times bigger than the Industrial Revolution.

Digital space, which has grown exponentially since the 1990s, comprises a vast range of online platforms, websites, applications, and virtual realities where people can interact with digital content and with each other. It encompasses the Internet, social media channels such as Facebook and TikTok, online marketplaces including eBay and Amazon, meeting platforms, and all kinds of games.

In 1998, Irish geographer Rob Kitchin published "Towards Geographies of Cyberspace",

Cyberspaces coexist with geographic spaces providing a new layer of virtual sites superimposed over geographic spaces.
Rob Kitchin
"Towards Geographies of Cyberspace", 1998

a paper in which he summarized the contemporary state of cyberspace – that is, the digital space created by the Internet's interconnected networks. According to Kitchin, cyberspace has transformed space-time relations, and created "new social spaces" with qualities different to traditional geographic spaces.

A world of cyberspace

By February 2025, estimates suggested there were 5.56 billion Internet users across the world. This new digital reality has transformed geographers' work, with digital mapping and software packages now indispensable tools; however, the changes in the way the wider population communicates and interacts have been even more revolutionary.

Digitization of information provides instant access to vast amounts of content, changing the way people learn about the world and moulding opinions. Culture, politics, music, fashion, and business now all have an immediate online global reach – and people can produce and share their own content. Some people have striven to reach as large an

HUMAN GEOGRAPHY 215

See also: Digital twins of Earth 60–61 ▪ Geography as human ecology 162–63 ▪ Cultural diffusion 164–71 ▪ Spatial interaction theory 192–93 ▪ Behavioural geography 194–95 ▪ Humanistic geography 200–01

audience as possible. In 2025, the world's top influencers – Internet users who post on a variety of digital channels to shape opinions and influence buying behaviour – have millions of followers.

The code/space concept

In 2011, Kitchin advanced the concept of "code/space", which he defined as any space dependent on software-driven technology. In his book *Code/Space and Everyday Life*, which he co-wrote with British geographer Martin Dodge, Kitchin illustrated this idea by using the example of a supermarket where all checkout pricing is reliant on software. The checkout assistant, or the customer, passes barcodes across a scanner, and in this way each item is priced. If the software crashes, there is no other way to price the goods, so shoppers can no longer purchase anything, and the code/space supermarket becomes nothing more than a non-code/space warehouse. The supermarket code/space is also global because a shopper buying the same items in a supermarket on the other side of the world will be reliant on the same software.

From kinship to toxicity

Free apps allow people to communicate instantly and globally, help families and friends to cement bonds (so-called "digital kinship"), enable co-thinkers to launch campaigns, and contribute to the formation of new global communities in digital space. Taking digital kinship a stage further, dating apps have reshaped people's romantic encounters with strangers, changing them from chance to choice.

The global population has embraced the social, creative, and commercial opportunities offered by digital spaces. Concerns about disinformation and toxic discourse, however, are leading to growing calls for the regulation of these environments. Online disinformation can act as cyberwarfare, manipulating narratives, destabilizing democracies, and creating a post-truth world where objective facts are dismissed as fake news. Children are increasingly exposed to harmful content, cyberbullying, and unrealistic standards, leading to self-esteem and mental health issues. Meanwhile, server farms sustaining all this data need vast amounts of electricity and water, which results in a massive environmental cost. ▪

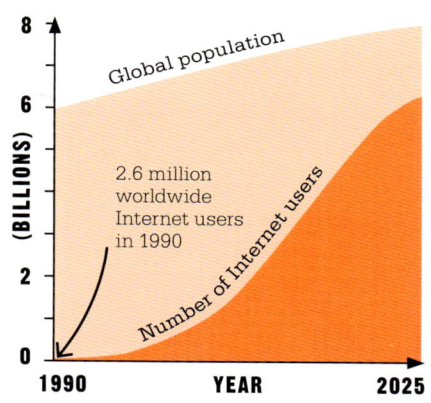

Internet use over time has shifted from being the tool of a few academics and researchers to an everyday platform for billions of people.

The world of gaming

Virtual realities abound in the world of gaming, where massively multiplayer online role-playing games (MMORPGs) such as *World of Warcraft* and *Final Fantasy XIV* are "always on".

These shared virtual worlds enable players to create their own characters (or avatars) and interact with several thousand other players from around the world at any one time. Featuring immersive environment, complex in-game economies, and collaborative quests, MMORPGs provide players with spaces in which they can explore creativity and cooperation, leading to the formation of communities with shared cultures, landscapes, and social norms.

However, these gaming communities come with the risks associated with all online activity. These include addiction, social isolation offline, economic exploitation, and exposure to toxic culture, which can harm the mental health of the most vulnerable players.

Virtual reality headsets allow people to feel as though they are inside the game, giving them an immersive experience.

ENVIRON
GEOGRA

MENTAL
PHY

INTRODUCTION

William Forster Lloyd warns of the "Tragedy of the Commons" and stresses the need to **restrict human population growth**.

1833

The National Forest Commission is created in the US to advise on the sustainable management of the nation's forests.

1896

Gilbert F. White's paper "Human Adjustment to Floods" advocates working with the natural environment to **mitigate flood damage**.

1945

Charles David Keeling begins to measure the levels of CO_2 **in air and water**.

1958

1859

John Tyndall describes how carbon dioxide (CO_2) and water absorb infrared radiation, creating a **greenhouse effect**.

1896

Svante Arrhenius predicts that increasing industrial activity could lead to **higher levels of atmospheric CO_2**.

1949

André Aubréville introduces the concept of **desertification** in his *Climates, Forests, and Desertification in Tropical Africa*.

1962

Rachel Carson's ***Silent Spring*** describes how DDT enters the food chain and can cause long-term damage to animals and humans.

Although it incorporates elements of both physical and human geography, the principal focus of environmental geography is how humans and Earth's systems interact and affect each other. It is an area that has attracted increasing attention over the past century as the negative impact of modern industrial society has become increasingly evident.

The discipline has its roots in the 19th century, when British economists Thomas Malthus and William Forster Lloyd began to challenge the idea that the natural world is an inexhaustible resource for humans to exploit. Both warned of the effects on nature of unchecked population growth.

Environmental degradation as a result of industrial practices, such as logging, mining, and railroad building, prompted the first formal attempts to manage and conserve natural resources. In the US at the turn of the 20th century, forestry pioneer Gifford Pinchot and naturalist John Muir campaigned for forest management and the establishment of national parks.

Managing resources
By the 1940s, the need to mitigate human harm to the natural world had become increasingly apparent. French ecologist André Aubréville pointed to intensive farming as a cause of soil erosion. In the US, Gilbert F. White recognized that structural modifications to rivers and floodplains, such as dams and levees, could worsen flood losses, and recommended that human behaviour adapt to the landscape instead. Bioregional theory took White's concept of living in harmony with nature a step further, calling for communities to fulfil their basic needs from local resources. Finding ways to manage resources sustainably has since become an integral part of environmental geography.

Climate change
The natural greenhouse effect of atmospheric carbon dioxide (CO_2) was identified during the 19th century. Although later theorists suggested that industrial emissions could increase CO_2 levels, thus exacerbating its warming effect, it was American geochemist Charles David Keeling's careful readings that made a clear link between the burning of fossil fuels and the continuing rise in the level of atmospheric CO_2. Scientists began

ENVIRONMENTAL GEOGRAPHY

1972 — James Lovelock's **Gaia hypothesis** proposes that Earth's living organisms and their inorganic environment interact to form a self-regulating system.

1980s — Robert D. Bullard argues for **environmental justice** in local planning to protect marginalized communities from living with increased environmental risks.

1993 — Norman Myers highlights the problem of **climate migration**, warning that climate change could prompt millions of environmental refugees

1996 — David Harvey describes the **detrimental impact of globalization** on the environment in *Justice, Nature and the Geography of Difference*.

1970s — Allen Van Newkirk and Peter Berg argue that communities should live more **sustainably** and in harmony with nature, **using local resources**.

1987 — The United Nations publishes the Brundtland Report on **sustainable development** – "Our Common Future".

1990s — William Rees and Mathis Wackernagel develop the idea of **"ecological footprint"** to limit overuse of resources.

2000 — Paul Jozef Crutzen and Eugene F. Stoermer propose a **new geological epoch** – the Anthropocene.

---

to track the effects of climate change – from the rise in global temperatures year on year to the bleaching of corals in warming seas and advancing glacier melt.

A growing movement

Environmental politics and activism gained momentum in the 1960s. Figures such as American scientist and writer Rachel Carson helped raise public awareness by highlighting the destructive effect of the pesticide DDT on ecosystems. In the early 1970s, British scientist James Lovelock's Gaia hypothesis offered some hope that Earth's organisms and inorganic elements, working together, could forever maintain conditions for life. However, some 30 years later, Lovelock conceded that climate change could upset Earth's synergy.

In the 1980s, the UN's Brundtland Report aimed to catalyse global support for sustainable development. The following decade, Swiss academic Mathis Wackernagel developed the ecological footprint concept along with his doctoral supervisor William Rees at the University of British Columbia, Canada. This provided a framework for quantifying ecological impact.

Global impact

Since the 1980s, when scientists noted an accelerating rate of damage to Earth's systems, German sociologist Ulrich Beck argued that risk is an inevitable byproduct of industrial and technological societies. He characterized these risks as indiscriminate, global in scope, and hard to perceive without scientific expertise.

During the same period, American sociologist Robert D. Bullard highlighted how marginalized groups are more exposed to the adverse effects of environmental degradation, such as industry and pollution. On a wider scale, British-American geographer David Harvey argued that globalization has led to uneven development, while accelerating capitalist growth and resultant harms to the natural world.

Such is the weight of evidence that, in 2000, some scientists have called for the naming of a new geological epoch – the Anthropocene – to reflect the impact of human activities such as industrialization and nuclear testing. In light of these challenges, environmental geography has a key role to play in offering a framework for a more sustainable future. ∎

WHY IS THE COMMON ITSELF SO BARE-WORN?
THE TRAGEDY OF THE COMMONS

IN CONTEXT

KEY FIGURE
William Forster Lloyd
(1794–1852)

BEFORE
350 BCE Aristotle states: "That which is common to the greatest number has the least care bestowed upon it."

1798 CE British cleric and economist Thomas Malthus warns that increases in human population would eventually reduce the ability of the world to feed itself.

AFTER
1968 American ecologist Garrett Hardin publishes an article in *Science*, "The Tragedy of the Commons", which claims the inevitability of resource overexploitation.

2009 American political economist Elinor Ostrom wins the Nobel Prize in Economic Sciences for research that showed how people organized to manage shared resources.

A community has a **desirable common resource**.

⬇

A number of **people share it**.

⬇

People overuse it for personal gain.

⬇

As there is **no incentive to conserve it**, overuse leads to **depletion**.

⬇

The common resource of the community is lost.

In the late 18th century, the work of the British cleric and economist Thomas Malthus stated that unchecked population growth would inevitably lead to a scarcity of resources. This controversially disputed the contemporary belief in social progress, yet his ideas gained traction and, in 1833, the English economist William Forster Lloyd delivered "Two Lectures on the Checks to Population" at Oxford University. These lectures, subsequently published in pamphlet form, argued that individual self-interest undermined the long-term sustainability of shared resources.

A common property

To illustrate the idea, Lloyd chose as a hypothetical example the situation of cattle herders, who share an area of common land on which they are entitled to let their cows graze. If one herder exceeded their agreed number of cows, they would personally profit but it would be to the detriment of the others as the quality of the land would diminish due to overgrazing. If other herders also exceeded their share, the pasture would become a wasteland, to the detriment of all.

ENVIRONMENTAL GEOGRAPHY 221

See also: Malthusian theory 148–49 ▪ Natural resource management and conservation 222–27 ▪ Desertification 228–29

Lloyd's lectures were an indirect riposte to the Scottish economist Adam Smith. In his work *The Wealth of Nations* (1776), Smith argued that, with the supply and demand mechanisms of a free market, the self-interest of individuals would promote the general good of society, as if guided by "an invisible hand". In contrast, Lloyd's work asserted that, in the context of a common shared property, self-interest contradicted the collective good.

Cattle graze in this 1776 painting, showing how English common land was an open field where local people had the right to share its resources.

The theory was largely ignored for over a century, until 1954, when Canadian economist H. Scott Gordon published "The Economic Theory of a Common-Property Resource: The Fishery", which proposed the idea of fishing quotas as a policy solution to overfishing. The issue of "the commons" had become a foundational idea in environmental sustainability and resource management.

Managing the commons

In 1968, American ecologist Garrett Hardin published an essay – "The Tragedy of the Commons" – which applied Lloyd's cattle-herding parable to deforestation, overfishing, and overgrazing, as well as to the pollution of air and water. Arguing that the self-interest of individuals would inevitably strain resources to their limit, Hardin claimed this "tragedy" had no "technical" solution and would require a collective change in human morality.

Therein is the tragedy. Each man is locked into a system that compels him to increase his herd without limit – in a world that is limited.
Garrett Hardin
"The Tragedy of the Commons", 1968

William Forster Lloyd

Born in 1794 in Bradenham, Buckinghamshire, UK, William Forster Lloyd was educated at Westminster School and studied mathematics and classics at Christ Church College, Oxford. He went on to be ordained as an Anglican cleric, although he was never appointed to a position. In 1832, he became Drummond Professor of Political Economy at Oxford and, over the next five years, delivered a series of lectures that challenged the intellectual canon. As well as his theories on population control, Lloyd also delivered a series of lectures on the Notion of Value, in which he introduced the concept of diminishing marginal utility. Lloyd was elected a fellow of the Royal Society in 1834 and died in 1852.

Key works

1832, 1833 *Two Lectures on the Checks to Population*
1833 *Lecture on The Notion of Value*

Hardin's article proved influential, yet it was criticized for ignoring the possibility of cooperation and management. In the 1990s, Hardin himself conceded that a better title would have been "The Tragedy of the Unmanaged Commons".

In 2009, American political scientist Elinor Ostrom was awarded the Nobel Prize for Economics for her work that demonstrated how, in many circumstances throughout history, human societies have valued and protected the resources they held in common, such as forests, fisheries, and irrigation systems. ∎

ALL OUR LIVING TREES WILL CLAP THEIR HANDS

NATURAL RESOURCE MANAGEMENT AND CONSERVATION

224 NATURAL RESOURCE MANAGEMENT AND CONSERVATION

IN CONTEXT

KEY FIGURES
John Muir (1838–1914),
Gifford Pinchot (1865–1946)

BEFORE
1st century CE Roman Stoic philosopher Seneca the Younger calls for humans to live in harmony with nature.

1864 In his book *Man and Nature*, American diplomat George Perkins Marsh argues that humans have a duty to preserve the natural world.

AFTER
1916 US president Woodrow Wilson signs the act that establishes the US National Parks Service (NPS).

1961 The World Wildlife Fund (WWF), dedicated to the conservation of the natural environment, is founded.

1992 More than 150 nations attend the Earth Summit in Rio de Janeiro, producing five international agreements on environmental protection.

Climb the mountains and get their good tidings. Nature's peace will flow into you as sunshine into trees.
John Muir
Our National Parks, 1901

Natural resources, such as water, sunlight, minerals, and forests, occur through natural processes without the need for human intervention, and benefit people. Natural resources that can renew themselves – known as renewables – require management. Forests, for example, are renewable with natural resource management, as they can be replanted and allowed to regenerate.

Indigenous peoples have long been aware of the need to manage natural resources, practising traditional methods for millennia, such as performing controlled burns of the land, which limits wildfires, terracing slopes to create arable land and prevent erosion, or placing taboos on hunting animals during their reproductive seasons.

In the 19th century, following the unprecedented depletion and spoiling of nature caused by rapid industrialization, the management and conservation of resources became a more pressing concern for industrializing nations. This was partly a cultural reaction – endorsed by writers and artists – against scientific and technological advances that seemed to subjugate nature to humans. It was also driven by fear of the destruction of nature, particularly the loss of forests.

The mountainous terrain of Hot Springs, Arkansas, was the first federally protected area in the US. It became a national park in 1921.

Those calling for the protection of nature and its resources began to talk of both "conservation" and "preservation". The terms were largely interchangeable at first, but from the late 19th century divisions began to emerge, particularly in the US, between "conservationists", such as forestry pioneer Gifford Pinchot, who believed in the sustainable and regulated use and management of natural resources, and "preservationists", such as naturalist and writer John Muir, who wanted to protect nature from any human intervention. Their views helped to inspire the environmental movements that emerged in the 20th century but also created an ongoing debate over which approach best serves the survival of the natural world.

The path to national parks
One early advocate of both resource management and conservation was British writer and polymath John Evelyn. In his book *Sylva, or A Discourse of Forest-Trees and the Propagation of Timber in His Majesties Dominions*, published in

ENVIRONMENTAL GEOGRAPHY

See also: Geography as human ecology 162–63 ▪ The tragedy of the commons 220–21 ▪ Climate change 232–39 ▪ Silent Spring 240 ▪ Monitoring environmental change 284–89 ▪ Monitoring biodiversity 306–07

1664, he called for the conserving of forests by ensuring that trees were replanted as quickly as they were cut down. His advocacy was driven partly by concerns within the British Navy about depletion of the forests that supplied wood for shipbuilding.

More than a century later, naval expansion in the US led to a series of laws, beginning in 1799, aimed at conserving oak forest. By the middle of the 19th century, however, the country had reached a deforestation crisis as vast amounts of wood were demanded for fuel, home-building, and railroad construction, and large swathes of forest were cleared for the agriculture needed to feed a booming population.

In 1847, Vermont congressman George Perkins Marsh warned of the potential negative impacts of human activity on the land, including deforestation, and called for a more conservation-minded approach to managing natural resources. His concerns for animal and land conservation were shared by US president Abraham Lincoln, who in June 1864 signed the Yosemite Valley Grant Act, which ceded Yosemite Valley and Mariposa Grove (later parts of Yosemite National Park) to the state of California. In 1872, despite fierce debate surrounding the US government's right to take land out of private hands and into public ownership, Yellowstone National Park was established as America's first national park. Following this precedent, others soon followed across the world. Royal National Park was created near Sydney, Australia, in 1879, followed by Banff in Canada (1885), Tongariro in New Zealand (1887), Hluhluwe–Imfolozi Park in South Africa (1895), and Abisko, the first of nine national parks set up in Sweden in 1909.

Private organizations dedicated to conservation also emerged during the same period. In Australia, the Field Naturalists Club of Victoria was set up in 1880, and in 1895 Britain's National Trust was founded. In France, the Society for the Protection of Landscapes was established in 1901, and 1909 saw the formation of the Swiss League for the Protection of Nature, and, in Germany, the founding of the Verein Naturschutzpark (Nature Conservation Park Association).

A champion of the outdoors

The founding of further American national parks was driven largely by the passionate support of John Muir. Inspired by his love for glaciers and forests, he became a central figure in the debate over »

Conservation is the application of common sense to the common problems for the common good.
Gifford Pinchot
"The ABC of Conservation", 1909

- Making sure **resources** last from **one generation to the next**.
- Distributing resources in an **equitable way**.
- Addressing **environmental impacts** of **the extraction of** natural resources.
- Using resources **with care** and within **impartially agreed limits**.
- **Effective and sustainable natural resource management.**
- **Disposing of waste safely** when resources are extracted or used.

land use. He first visited Yosemite Valley in 1868 and returned the following year, settling in the area to study its plants, animals, and landscapes. In 1876, Muir urged the US government to adopt a forest conservation policy, and in the 1880s he focused his writing on the destruction of natural resources around Yosemite. He was alarmed at the extensive damage caused by livestock animals to the delicate ecosystems of the surrounding High Sierra, referring to sheep as "hoofed locusts".

In 1889, Muir took Robert Johnson, editor of the popular *Century Illustrated Monthly Magazine*, to see for himself how sheep were damaging the land. Following Johnson's publication of Muir's disclosures, the US Congress passed a bill to create a new federally administered park surrounding the original Yosemite Grant area, and in 1890 Yosemite National Park was established,

The flooding of the Hetch Hetchy river valley to create a reservoir was opposed by Muir, who wanted to preserve the landscape. Pinchot saw its potential as a sustainable natural resource.

followed in the same year by Sequoia National Park. Two years later, Muir co-founded the Sierra Club, an organization devoted to protecting the environment, and then played a key role in the establishment of two further national parks: Mount Rainier (1899) and Grand Canyon (1908).

To preserve or conserve?

At first, Muir found common ground with the forest protection ideas espoused by Gifford Pinchot, the first professional forester in the US and an early adopter of the word "conservation" to apply to the policy of sustainable land management. Both men became involved in the National Forest Commission, created in 1896 by the National Academy of Sciences to provide policy recommendations for the administration of America's forests. Muir and Pinchot formed a firm friendship and took part in a four-month survey of the American West, which resulted in a proposal that the US government assume protection of more than 8.5 million ha (21 million acres) of forest. When the commission delivered its report, in 1897, it met

Clean air, clean water, open spaces – these should once again be the birthright of every American.
President Richard Nixon
State of the Union address, 1970

stiff opposition from mining and logging interests. Pinchot lobbied congressmen in Washington, DC, argued with the logging and mining companies, and pressed for a forest system that would satisfy the needs of everyone.

Pinchot's influence over forestry policy continued, and in 1905 conservation-minded US president Theodore Roosevelt made him the first chief of the US Forest Service, with the two continuing to work closely together on government regulation of natural resources. In 1908, James R. Garfield, US Secretary of the Interior, granted San Francisco rights to create reservoirs in Yosemite from Lake Eleanor and by damming the Tuolumne River in the Hetch Hetchy Valley. Pinchot agreed with the decision, leading to fierce clashes with Muir, who described the project as "this outrageous scheme".

Muir wanted to *preserve* the wilderness, setting it aside and leaving it alone, free from any exploitation, whereas Pinchot believed in *conserving* the land, making sustainable use of available resources. The issue was decided in 1913, when construction of the dam was authorized.

ENVIRONMENTAL GEOGRAPHY

Gifford Pinchot talks to a top-hatted Theodore Roosevelt on a river steamer in 1907. Both men shared a commitment to the conservation and sustainable management of natural resources.

Rise of environmentalism

From the 1950s, awareness rose about the broader, damaging human impact on natural systems, such as pollution, deforestation, and fossil fuel emissions. Recognition of the need to protect our oceans came more slowly. The International Whaling Commission, one of the first major international marine conservation organizations, was founded in 1946, but further significant national marine protections generally arrived later, in the 1960s and 1970s.

In 1956, the UK government introduced a Clean Air Act to combat air pollution, and the US followed with similar legislation in 1963. A year earlier, American marine biologist Rachel Carson had published *Silent Spring*, a book that brought to a wide audience an understanding of the ecological problems caused by industrial and commercial activity. From this increased recognition of threats to

the natural world an environmental movement grew that looked beyond the conservation of natural habitats and management of resources to embrace a global picture of human-induced threat to the planet.

The movement's first major victory came at the end of 1970, when US president Richard Nixon established the Environmental Protection Agency (EPA), with powers to ensure the protection of habitats in the US as well as act on issues of public health. Other conservation triumphs followed. In 1973, after the Indian government granted permission for the felling of trees in the Uttarakhand region to make sports goods, environmental campaigner Sundarlal Bahuguna began a peaceful protest against the deforestation. This became the Chipko movement, dedicated to preserving the local ecosystem and a traditional way of life. Their campaigns led to a 15-year logging ban in the region and the introduction of community-based, sustainable forestry management. Similarly, in 1977, environmental activist Wangari Maathai founded the Green Belt Movement in Kenya to combat the degradation of habitats and promote conservation through tree-planting initiatives.

A continuing debate

The tensions between preservation and conservation exemplified by John Muir and Gifford Pinchot continue. While many believe in a need for action to ensure biodiversity and environmental protection, others argue that this has to be achieved in a way that will not deny access to resources for future generations. ∎

John Muir

Born in 1838 in Dunbar, Scotland, John Muir emigrated to the US with his family in 1849. He entered the University of Wisconsin in 1860 but left after three years to travel through the northern US and Canada. After nearly losing his sight in a workshop accident, he walked from Indianapolis to the Gulf of Mexico, sailed to Cuba and Panama, and arrived in California in 1868, settling in Yosemite Valley. From 1874, a series of articles launched his success as a writer. He married in 1880 and moved to Martinez, California, in 1890, where in later life he wrote more than 300 books and articles about his travels. He died in a Los Angeles hospital in 1914.

Muir has been criticized for derogatory comments he made about Indigenous people he met on his early American travels. He later revised his views, especially after visiting Alaska, where he gained a deeper appreciation of Indigenous peoples and their custodianship of the natural environment.

Key work

1901 *Our National Parks*

MAN-INDUCED SOIL EROSION
DESERTIFICATION

IN CONTEXT

KEY FIGURE
André Aubréville
(1897–1982)

BEFORE
1814 Alexander von Humboldt deplores the deforestation and soil degradation caused by European plantations in Latin America.

1930s The American Dust Bowl is caused by drought and also farming methods that destroy the topsoil.

AFTER
1977 The United Nations Conference on Desertification is prompted by the Sahelian drought and famine (1972–75).

1990 British geographer Michael Mortimore documents the adaptive strategies of farmers in arid north Nigeria.

2024 China completes its Great Green Wall, a 3,000-km (1,865-mile) vegetated belt around the Taklamakan Desert, increasing forest coverage by 15 per cent.

Desertification was first introduced as a concept by French plant ecologist André Aubréville in his 1949 work *Climates, Forests, and Desertification in Tropical Africa*. He wrote that human activities could transform tropical rainforest, first into savannah, and then into desert. In his view, desertification was not an extension of existing desert but an ecological process by which humans caused productive land to become arid. Although Aubréville's definition applied to the degradation of tropical forests, today the term desertification is usually associated with the degradation of drylands (arid and semi-arid zones), which change from savannah to desert-like conditions.

Drylands cover 41 per cent of Earth's land surface, and include grassland, savannah, shrubland, and

Drylands are divided into four groups according to their aridity levels, which are determined by the ratio of annual precipitation to potential evapotranspiration.

Type	Average annual rainfall	
Dry, sub-humid	Average annual rainfall of 500–750 mm (20–30 in) and can support crops	
Semi-arid	Average annual rainfall of 250–500 mm (10–20 in) and is prone to drought	
Arid	Average annual rainfall of 100–250 mm (4–10 in); supports sparse, drought-tolerant vegetation	
Hyper-arid	Average annual rainfall less than 100 mm (4 in); true deserts with little or no vegetation	

ENVIRONMENTAL GEOGRAPHY 229

See also: Pedology 128–31 ▪ Geography as human ecology 162–63 ▪ Natural resource management and conservation 222–27 ▪ Climate change 232–39

Trees are planted in a suburb of Rosso, Mauritania in 2022 – part of the Great Green Wall project that aims to help re-green areas of the Sahara desert.

woodland. Between 1982 and 2015, 6 per cent of the world's drylands underwent desertification, largely as a result of unsustainable land-use practices, such as deforestation and overgrazing resulting in soil erosion, loss of nutrients, and changes to natural cycles of carbon, nitrogen, and water. Over a quarter of drylands are already degraded, putting about a billion people at risk.

Aubréville's findings

From 1925, while working in the colonial forestry service, Aubréville studied the tropical forests of Equatorial Africa. He noted an anomaly surrounding the equatorial forest: a belt of land, hundreds of kilometres wide, covered with only sparse vegetation, instead of dense forest as elsewhere.

Aubréville observed how local populations were using fire to clear land for crops. Where this was practised at the edge of forests, it created gaps into which fires could penetrate. Over time, these fires reduced the tree canopy and eroded the soil. He noted how the increase in population, and the adoption of intensive farming of industrial (non-food) crops for export were depleting the tropical forest. Aubréville predicted that this would ultimately result in ecological transformation of the landscape.

A complex picture

From the mid-1960s, when the West African Sahel, a semi-arid dryland region, experienced numerous droughts, Aubréville's term came into broad use to describe the seemingly irreversible process taking place in large parts of Africa. In the 1980s, satellite technology enabled the long-term observation of Sahelian vegetation, revealing the droughts to be an early manifestation of global climate change, caused by the emission of greenhouse gases.

In 2004, German geographer Helmut Geist and Belgian geographer Eric Lambin listed climatic, economic, political, and population factors in their paper "Dynamic Causal Patterns of Desertification". In 2016, research by British geographer Michael Mortimore and British anthropologist Roy Behnke illustrated how the patterns are further complicated by the natural variability of drylands over long timescales, and how the process is not irreversible.

In 2021, at the United Nations Convention to Combat Desertification, the international community pledged to restore 1 billion hectares (3,861,000 sq miles) of degraded land by 2030. ∎

André Aubréville

Born in Pont-Saint-Vincent in 1897, André Aubréville served in World War I and, after gaining an engineering degree in 1922 and graduating from the Nancy School of Forestry in 1924, he joined the colonial forestry service. In 1939, he was appointed Inspector-General of Waters and Forests in French West Africa, while simultaneously serving as President of the Botanical Society of France. In 1955, he became a Professor at the Natural History Museum in Paris, and in 1968, he was elected a member of the French Academy of Sciences. During his career, he published more than 300 papers and 20 books on studies he conducted throughout Africa, southeast Asia, South America, and New Caledonia. Aubréville died in Paris in 1982.

Key work

1949 *Climates, Forests, and Desertification in Tropical Africa*

By restoring land, we restore life, restore our economies, restore our communities, and so much more.
Ibrahim Thiaw
Executive Secretary, UNCCD, 2022

FLOODS ARE "ACTS OF GOD", BUT FLOOD LOSSES ARE LARGELY ACTS OF MAN
FLOODPLAIN MANAGEMENT

IN CONTEXT

KEY FIGURE
Gilbert F. White (1911–2006)

BEFORE
340 BCE In *Meteorologica*, Greek philosopher Aristotle correctly attributes the Nile floods to spring and summer rain in the Ethiopian highlands.

1887 In China's worst-ever floods, up to two million people die when rainfall swells the Huang-Ho (Yellow) River, its dams burst, and it overflows, despite flood barriers.

AFTER
2020 Japanese scientists claim that flood-control basins may be as rich or richer than other waters in some insect, fish, bird, and plant species.

2022 The UN Environment Program reports that, since 2000, flood-related disasters have increased by 134 per cent, with the current and future likelihood of flooding increased by climate change.

Over the centuries and worldwide, structures such as dams, floodwalls, and levees have been built to control flooding – but with limited success. In 1945, American geographer Gilbert F. White proposed a different way. In his PhD dissertation at the University of Chicago, he argued for a more non-structural, holistic approach, where humans worked with – instead of against – the natural environment. His paper, "Human Adjustment to Floods", became one of the most influential texts in natural hazards research.

Among other issues, White highlighted the dangers of building on floodplains – riverside areas that are naturally prone to flooding.

For centuries, people **build settlements on the floodplains** of major rivers that are used for transport.

To control flooding, they **construct dams and levees** (raised banks) that **sometimes fail**.

Instead of trying to control the floods, people should **adjust to the floodplain environment**.

Rather than building on floodplains, these should be **designated agricultural and natural areas** that can absorb floodwater.

ENVIRONMENTAL GEOGRAPHY 231

See also: The hydrological cycle 66–69 ▪ Climate change 232–39 ▪ Environmental impact assessment 272–73 ▪ Early warning systems 296–303 ▪ Monitoring environmental change 284–89

Mississippi River flooding in June 2011 broke previous records after earlier, exceptionally heavy rainfall. As White predicted, its levees have not proved enough to contain high water levels.

Instead, he suggested that zoning laws could designate floodplains for agricultural or recreational use to minimize potential flood damage and preserve the land's natural function. The 1968 US National Flood Insurance Act reinforced White's crusade and also stipulated that only those who complied with the regulations would be eligible for subsidized flood insurance.

Unnatural disasters

The Mississippi River in the US was the subject of White's studies. Historically, locks, dams, and levees were built along its length to curb floods and facilitate trade. White observed that diverting the river and building on its floodplains resulted in costly and extensive flood damage. New Orleans lies on a Mississippi floodplain, and in 2005, Hurricane Katrina produced a storm surge that swamped the city after its levee system failed. In 2011, flooding produced the highest Mississippi flood levels in a century.

A holistic approach

White's natural approach has had a significant impact. The governments of many countries around the world now promote natural flood management by, for example, reinstating the natural course of some rivers, creating new floodplain wetlands, and restoring vegetation along rivers and on coasts to mitigate flooding and storm damage. The Associated Programme on Flood Management, founded in 2001, and the UN's Environment Program both support flood management projects worldwide that integrate land and water resources to reduce the impact of flooding. However, many regions remain vulnerable to catastrophic flooding, particularly in Southeast Asia, where more than 1 billion people are at risk. ▪

Gilbert Fowler White

Born in Chicago, US, in 1911, Gilbert Fowler White studied at the University of Chicago and then prepared his doctorate while working for the Mississippi Valley Committee of the Public Works Association. After completing his studies, he performed relief work in France during World War II and was interned in Germany.

From 1955, White was professor of geography at the University of Chicago and a key figure in its programme of natural hazard research. In 1970, he moved to the University of Colorado. A year later, he and meteorologist Thomas Malone proposed a global programme of ongoing environmental monitoring, discussed at the UN Conference of Human Environment in 1972.

White's work also involved water management in several developing countries and international cooperation on shared river basins such as the Jordan and Nile. He died in Boulder, Colorado, in 2006.

Key work

1945 "Human Adjustment to Floods"

AN UNCONTROLLED INCREASE IN ATMOSPHERIC CO$_2$
CLIMATE CHANGE

234 CLIMATE CHANGE

IN CONTEXT

KEY FIGURES
Guy Stewart Callendar (1898–1964), **Charles David Keeling** (1928–2005)

BEFORE
1824 French mathematician Joseph Fourier theorizes that Earth's atmosphere insulates the planet in some way.

1856 Experiments by American scientist Eunice Foote suggest that CO_2 and water vapour absorb heat and might affect temperature.

1859 Irish physicist John Tyndall publishes his findings that CO_2 and water absorb infrared radiation.

AFTER
1985 Scientists discover a hole in the ozone layer of Earth's atmosphere, prompting a ban on most ozone-depleting gases.

2015 The Paris Agreement legally binds 196 nations to cut greenhouse gases.

By fuel combustion, man has added about 150,000 million tons of carbon dioxide to the air during the past half century.
Guy Stewart Callendar

Naturally occurring gases, such as carbon dioxide (CO_2), methane, nitrous oxide, water vapour, and traces of other gases trap reflected heat from the Sun. Although only a fraction of atmospheric gases, they warm the planet in what has become known as the greenhouse effect.

The effect has allowed life to thrive on Earth for millennia, and atmospheric CO_2 is central in the carbon cycle, in which carbon is cycled between organisms and the environment via photosynthesis, respiration, and decomposition. However, human activities since the Industrial Revolution have generated vast amounts of CO_2, upsetting this cycle and the planet's natural balance, and therefore its climate.

In 1938, engineer Guy Stewart Callendar first demonstrated the connection between Earth's rising temperature and the generation of heat-trapping CO_2. Twenty years later, Charles David Keeling's more precise measurements indicated the rapid rate of the warming.

Early warnings

From the mid-19th century onwards, industrial and population growth has been driven by the burning of fossil fuels, including natural gas for heating, petroleum and oil for

Smoke pours from the chimneys of a coal-fired power station, sending CO_2 and other greenhouse gases into Earth's atmosphere. Coal is the most polluting fossil fuel.

transport, and coal for generating electricity. Such activity generates huge amounts of CO_2, which can remain in the atmosphere for centuries, trapping more and more heat, leading to global warming.

One of the first scientists to link industrial activity with the production of CO_2 was Swede Svante Arrhenius, who received the Nobel Prize in Chemistry in 1903. While he did not explicitly state that burning fossil fuels would cause global warming, he noted that industrial advances could affect atmospheric CO_2 levels. Arrhenius predicted that levels might rise by close to 50 per cent, with a global temperature increase of 5–6°C (9–11°F). He was, however, thinking in terms of thousands of years and saw the changes as beneficial, particularly for the coldest of regions.

The scientific consensus that CO_2 played little part in climatic changes remained unchallenged for 30 years, until the unexpected involvement of Callendar, a British steam and power generation engineer, who was also an amateur meteorologist. Putting aside his

ENVIRONMENTAL GEOGRAPHY

See also: The Greenhouse effect 80–83 ▪ Extreme weather 142–43 ▪ Climate migration 254–55 ▪ Climate modelling 268–71 ▪ Monitoring environmental change 284–89 ▪ Global Earth observation programmes 308–09

engineering work, he carried out painstaking surveys of data, including CO_2 levels and readings of temperature, from around 150 weather stations across the world.

In his paper "The artificial production of carbon dioxide and its influence on temperature", published in 1938, Callendar stated that there was an "increase of mean temperature, due to the artificial production of carbon dioxide". He calculated that the average global temperature had risen by 0.3°C (0.54°F) in the preceding 50 years, or 0.06°C (0.1°F) per decade, a figure since confirmed by the Intergovernmental Panel on Climate Change (IPCC). His calculations also suggested that it would rise a further 0.39°C (0.7°F) to 0.69°C (1.24°F) by the year 2000 – an underestimation, as recent IPCC figures suggest it warmed by as much as 0.9°C (1.62°F) between 1850 and 2000.

Before Callendar published his deliberations, it was thought that as the atmosphere held so much

Greenhouse gases, of which CO_2 is the most abundant, have greatly increased as a result of human activity. Methane is generated mainly from fossil fuels, agriculture, and landfill waste. Nitrous oxide is also produced by agriculture (largely by fertilizer use) and the manufacture of synthetic goods. Fluorinated gases (F-gases), used in air-conditioning and refrigeration, are entirely human-made and, although less abundant, are extremely potent in terms of their heat-trapping properties.

more water vapour (another greenhouse gas) than CO_2, the climatic effect of CO_2 was minimal. Callendar reported that, because CO_2 and water vapour absorb different wavelengths of heat radiation, more heat is trapped than was first believed. He also noted that water vapour disperses quickly, while CO_2 accumulates at a much higher level in the atmosphere, where it remains

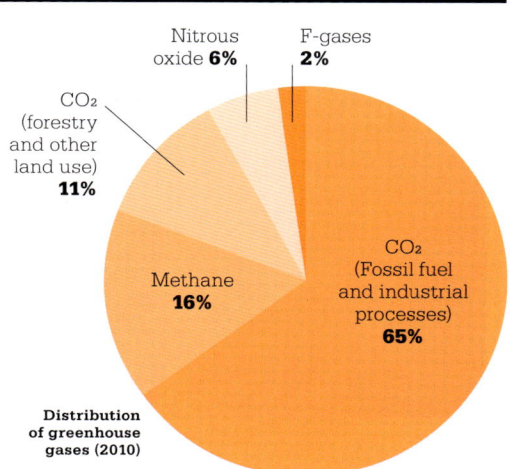

Distribution of greenhouse gases (2010)
- CO_2 (Fossil fuel and industrial processes) **65%**
- Methane **16%**
- CO_2 (forestry and other land use) **11%**
- Nitrous oxide **6%**
- F-gases **2%**

as a potent, heat-trapping layer. The scientific community did not take Callendar's speculations seriously at first, but in time he was proved correct. He had shown that air temperatures had increased significantly and proposed that people were unknowingly raising the temperature by burning fossil fuels. His important findings became known as the "Callendar Effect", the foundation of the concept of global warming.

Accurate figures

Callendar's discoveries were enough to provoke further research into Earth's climate. After World War II, this was spurred on by significant increases in scientific funding, particularly from American military agencies keen to keep at the forefront of meteorological knowledge during the Cold War era. Research was also boosted by the arrival in the 1950s of powerful digital computers able to make complex calculations.

In 1958, Keeling, an American geochemist, began to measure and compare the amounts of CO_2 in air and water. This had never been »

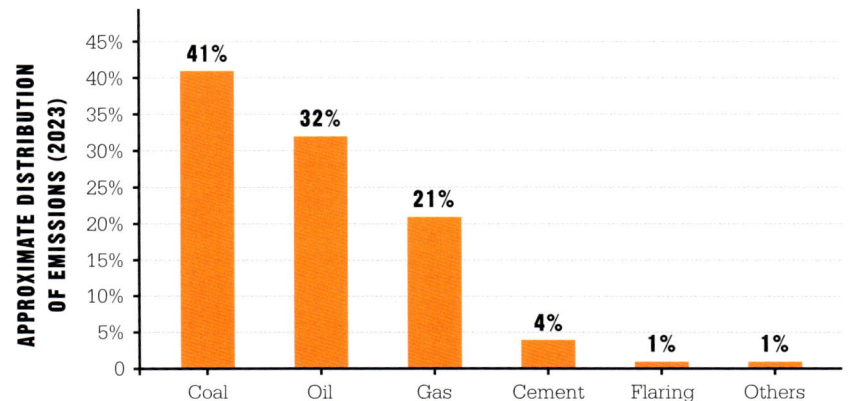

Coal is the fossil fuel that emits the most CO_2 into the atmosphere. Cement production, which involves huge amounts of energy, is also a major contributor to CO_2 emissions, as is flaring – the burning of natural gas during oil and gas extraction.

APPROXIMATE DISTRIBUTION OF EMISSIONS (2023):
- Coal 41%
- Oil 32%
- Gas 21%
- Cement 4%
- Flaring 1%
- Others 1%

236 CLIMATE CHANGE

Charles David Keeling

Born in 1928 in Scranton, Pennsylvania, Charles David Keeling was raised in Chicago, where his early interest in science was sparked by his astronomy-loving father. Keeling "drifted into chemistry", and after gaining a degree in 1948 and a PhD in 1953 he joined the new department of geochemistry at the California Institute of Technology. There, he developed a device for measuring atmospheric CO_2 and in 1956 joined the Scripps Institution of Oceanography, in San Diego, where he began to devote his studies to CO_2.

By 1958, Keeling had enough funding to take CO_2 readings in Antarctica and Hawaii, and in 1960 he published the first of his ground-breaking results, proving the rising levels of CO_2 in the environment. He was appointed professor of oceanography at the Scripps Institution in 1968 and continued his research there until his death in 2005.

Key work

1960 *"The concentration and isotopic abundances of carbon dioxide in the atmosphere"*

done before in a precise or reliable way, so Keeling designed his own apparatus to record ultra-accurate readings of CO_2 levels. He took air samples from weather observatories in Antarctica and from the Mauna Loa volcano in Hawaii – sites chosen for their pure and consistent airflow.

By 1960, Keeling had started to identify patterns. He noticed that samples taken at night contained more CO_2 than those recorded in the day. This, he realized, was because photosynthesis – the sunlight-driven process by which green plants absorb CO_2 to create their own food energy – happens by day and not at night. Keeling also noted a seasonal variation in the carbon cycle, most prominent in the northern hemisphere where more vegetation is located: in autumn, when photosynthesis slows and rotting plant material releases CO_2, levels of the gas are high, while in the spring, when plants use up more CO_2, levels are lower. More significantly, Keeling discovered that global atmospheric CO_2 levels were rising year on year, and by analysing the contents of his samples he identified the origin of the raised CO_2 as fossil fuels.

As well as confirming what Arrhenius and Callendar had suggested – that human-produced emissions of CO_2 were contributing to the greenhouse effect – Keeling also revealed the alarming rate at which Earth's atmosphere was warming. Although forests, oceans, and other systems play a key role in the carbon cycle by absorbing and storing around half of human-produced CO_2, the remainder stays in the atmosphere, growing year by year. From the steady rise in CO_2 levels, Keeling devised a graph, later called the Keeling Curve, which monitored this progressive build-up and remains a key long-term record of atmospheric CO_2 concentration and measure of the rate and extent of climate trends.

Modelling the climate

In 1967, data that included Keeling's research was fed into the world's first accurate climate change

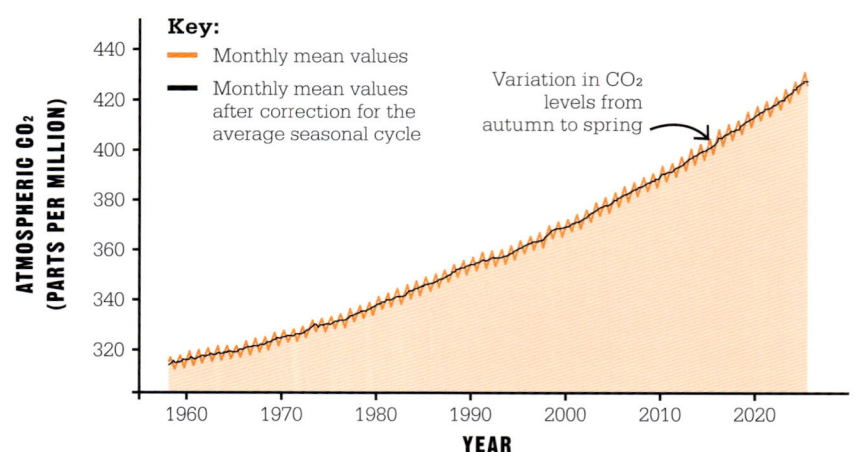

The jagged nature of the Keeling Curve – showing the year-on-year increase in atmospheric CO_2 – reflects the seasonal fluctuations in gas measurements.

computer model. The software program was designed and set up by American Richard Wetherald and Japanese-American Syukuro Manabe, both physicists and climatologists, working at the Geophysical Fluid Dynamics Laboratory, part of the US Weather Bureau. Their model looked at the relationships between climate components, such as atmosphere, oceans, and clouds, allowing them to predict future trends. They estimated that a doubling of CO_2 in the atmosphere (with relative fixed humidity) would raise global temperature by about 2°C (3.6°F). Although not set as a target by Manabe and Wetherald, from the 1990s, the figure began to be used in international climate agreements as an accepted limit for global temperature rise above pre-industrial levels. On 17 November 2023, for the first time, that limit was exceeded.

When NASA's Nimbus III satellite was launched into Earth orbit in 1967, its software and instruments began transmitting back the first accurate readings of global atmospheric temperatures. The data provided independent confirmation that Earth's lower atmosphere was warming.

Far from being self-stabilizing, the Earth's climate system is an ornery beast which overreacts even to small nudges.
Wallace S. Broecker
Nature magazine, 1995

By the 1980s, as new climate observation technologies continued to evolve, so did the language of climate awareness. The term "greenhouse effect" – previously the preserve of scientists – entered political and public discourse, as did "global warming", first used in the title of a 1975 paper by US scientist Wallace S. Broecker. Climatologists increasingly turned away from both terms, viewing them as descriptions of only one simplified aspect of the complex Earth systems that human actions have distorted. They referred instead to "climate change", a term that encompasses not just temperature fluctuations, but also changing climatic patterns and features, such as rainfall and wind.

The effects of warming
The effects of a persistently warming world are many, most of them destructive. Melting glaciers and ice sheets, together with the expansion of ocean water as it warms, cause sea levels to rise, posing threats to coastal areas, including the potential inundation of major cities such as Bangkok, Amsterdam, and New Orleans. As

Ice melts into the sea around the Arctic Svalbard islands, the fastest-warming place on Earth. Retreating glaciers leave exposed ground that leaks methane into the atmosphere.

sea levels rise, saltwater can also seep inland, salinizing freshwater resources, making them unusable for drinking or agriculture. Readings from tidal gauges and, since the 1990s, from satellites have revealed that the average global sea level has increased by about 21 cm (8¼ in) since 1900. The rate of increase has varied: from around 1.3 mm (¹⁄₂₀ in) per year between 1901 and 1971 to 3.7 mm (³⁄₂₀ in) per year between 2006 and 2018.

Global warming may also intensify extreme weather, including floods, heatwaves, hurricanes, tornadoes, droughts, and wildfires. Fluctuations in temperature and precipitation patterns can disrupt agriculture and increase desertification, turning productive land into wasteland, resulting in food shortages and even famine.

Climate change can also alter natural ecosystems and habitats, impacting biodiversity and the »

238 CLIMATE CHANGE

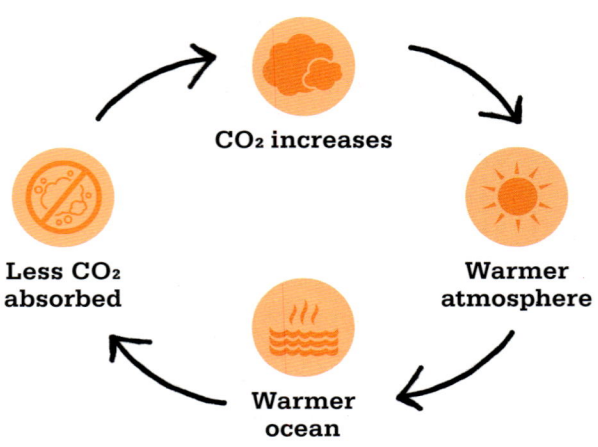

Positive feedback loop

Oceans absorb and release CO_2 as part of the carbon cycle. However, warm water cannot store as much dissolved CO_2 as cold water, so more of the gas remains in the atmosphere, further warming it, which, in turn, raises sea and ocean temperatures.

distribution of species, including shifts in migration patterns. Extreme weather can displace human populations, and higher temperatures can adversely affect physical and mental health, leading to increased hospital admissions and higher mortality rates.

Shifting the blame
Until the turn of the 21st century, it was still difficult to link extreme weather events conclusively to climate change caused by global warming, and counter claims that natural catastrophes, such as floods, wildfires, or severe drought, were simply a normal part of Earth's climatic adjustment. A shift began to happen, however, in 2004, when the journal *Nature* published a paper by three British climate scientists – Peter Stott (from the UK's Meteorological Office) and Daíthí Stone and Myles Allen (from the University of Oxford). It set out to demonstrate that human activity had contributed decisively to a European heatwave in 2003 – an event that had killed tens of thousands of people in probably the hottest summer since 1500.

Using computer programs, the team of scientists ran a range of climate simulations across Europe, both with global-warming emissions generated by humans, such as CO_2, and without the emissions. Their modelling predicted average temperatures across the 1990s that were 0.5°C (0.9°F) higher in the simulations with the human element than those without. They concluded that "human-induced increases in atmospheric concentrations of greenhouse gases and other pollutants" had at least doubled the risk of a heatwave in 2003. Their paper went on to suggest that by the 2040s more than half of the summers in Europe would be hotter than the summer of 2003.

The article in the *Nature* journal helped to generate a new branch of climatology – "extreme event attribution" – in which researchers take an extreme weather event, such as a drought or flood, and apply modelling to it to show whether the incident could only have happened because of human-induced climate change. One principal proponent, German climatologist Friederike Otto, of the Environmental Change Institute at the University of Oxford, helped to found World Weather Attribution (WWA) in 2014, working with climate scientists worldwide to quantify the influence of human-induced climate change on extreme weather events. WWA has carried out more than 100 attribution studies of such events, including a heatwave in July 2021, in the Pacific northwest of Canada and the US, where the temperature rose as high as 49.6°C (121.3°F), breaking records by as much as 5°C (9°F). The study found that climate change made extreme heat 150 times more likely and increased its intensity by 2°C (3.6°F).

Accelerating the change
The Arctic region is, according to climate models, warming at between twice and four times the rate of the rest of the planet, with up to 60 per cent of the difference caused by a phenomenon known as ice-albedo feedback. Albedo is a measure of the reflectiveness of a surface – the more light it reflects, the higher the albedo. Arctic ice has a high albedo, and as the ice melts, less sunlight is reflected back into space. This warms the atmosphere even more – as well as warming the dark, low-albedo sea.

What turns weather into a disaster is not how much it rained but how vulnerable people are and how well prepared.
Friederike Otto
Interview in *The Guardian*, 2025

ENVIRONMENTAL GEOGRAPHY

Climate change is also causing the permanently frozen ground in the Arctic – the permafrost – to thaw, releasing methane into the atmosphere. Although much less abundant than CO_2, methane is 84 times more potent and is also released from the Arctic sea floor as ice melts and the sea warms.

The more additional pressures there are on the planet's climate (known as positive feedback loops), the greater the chances of a "runaway greenhouse event". For that to happen, Earth would have to pass tipping points, such as the collapse of the Greenland and West Antarctic ice sheets. Although human-induced climate change is extremely unlikely to cause such an event, there have been warning signs, including the collapse of two ice shelves – Larsen A, attached to the Antarctic Peninsula and covering 2,000 sq km (722 sq miles), in 1995, and Larsen B in 2002. Larsen C ice shelf remains but suffered a major rift in 2017.

If the entire Antarctic ice cap melted, sea levels could rise by around 58 m (190 ft). According to studies by researchers at NASA's Jet Propulsion Laboratory (JPL) in 2020, just a 1 m (3^1/$_3$ ft) rise in sea levels by the end of the 21st century would inundate the land occupied by up to 640 million people – around 10 per cent of the world population.

Mitigating climate change

Since the late 1980s, a series of international agreements have called for actions to combat global warming and climate change, such as insulating houses, planting trees,

In a CCS project in Reykjavik, Iceland, hot water mixed with CO_2 is carried in pipes from a direct-air capture plant, then pumped deep into the island's bedrock.

increasing flood defences and cultivating drought-resistant crops. The first large-scale underground carbon capture and storage (CCS) project was launched in 1996 at the Sleipner gas field in the Norwegian sector of the North Sea, where about 1 million tonnes (1.1 million tons) of CO_2 are captured each year and stored in a saline formation around 1,000 m (3,280 ft) underground. In 2020, the Alberta Carbon Trunk Line (ACTL), in Canada, became the world's largest CCS project, capable of storing up to 14.6 million tonnes (16.1 million tons) of CO_2 a year.

Geoengineering scientists in a new area of study known as solar radiation management (SRM) have also begun to consider potential means of modifying the Sun's heating effect by either reducing the amount of solar radiation that enters Earth's atmosphere or reflecting it back. Experiments, for instance, have been carried out to mimic the cooling effect of volcanic eruptions by injecting reflective sulphur particles into the stratosphere. As yet theoretical ideas include large sun shades or mirrors in space, high-albedo buildings and crops, and more controversial measures such as marine cloud brightening (MCB) – creating larger, whiter high-albedo clouds by shooting saltwater particles into them to encourage water condensation.

Some scientists are concerned about SRM's high costs and its potential side-effects on Earth's climate and ecosystems. Others fear it might be used as a pretext to avoid cutting human-produced emissions. Although SRM might reduce temperatures, its effects would vary regionally. CO_2 would remain in the atmosphere, its levels continuing to increase until humans resolve to reduce, and finally end, their dependence on fossil fuels. ∎

> We cannot allow ourselves to choose between climate change mitigation and adaptation. We simply need both.
> **Kofi Annan**
> Former secretary-general of the UN, 2015

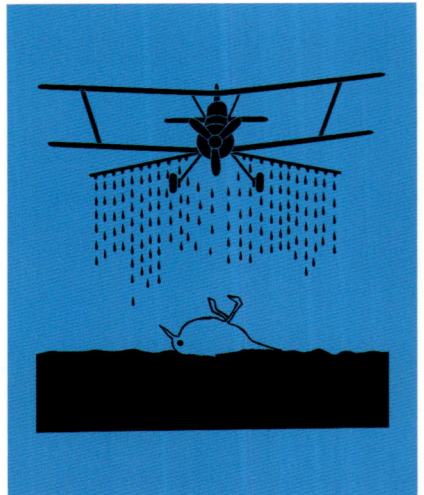

THIS SUDDEN SILENCING OF THE SONG OF THE BIRDS…
SILENT SPRING

IN CONTEXT

KEY FIGURE
Rachel Carson (1907–64)

BEFORE
1874 Austrian PhD student Othmar Zeidler condenses chlorobenzene with chloral hydrate to produce DDT.

1945 "DDT: The insect-killer that can be either boon or menace", by American naturalist Edwin Way Teale, is published in *Nature Magazine*.

AFTER
1967 British naturalist Derek Ratcliffe explains that when birds of prey ingest DDT, it can fatally weaken eggshells.

2006 The drug diclofenac is banned for veterinary use across South Asia after it decimates vulture populations feeding on carcasses.

2023 America's Environmental Working Group reports that potentially toxic PFAS ("forever chemicals") contaminate wildlife across the globe.

In 1939, Swiss chemist Paul Müller examined a chemical synthesized 65 years earlier – dichlorodiphenyltrichloroethane (DDT) – and discovered its potential for controlling disease-vector insects, such as mosquitoes. DDT was used during World War II to control insect-borne diseases such as malaria and typhus and, after the war, was widely adopted as an agricultural pesticide.

However, growing concerns emerged regarding the detrimental effects of DDT on human health and the environment. In 1958, American biologist and nature author Rachel Carson was alerted to bird deaths on Cape Cod after DDT spraying. She tried to highlight the issue in the press, and when no newspaper was interested, decided to write about it herself. In 1962, she published the book *Silent Spring*.

Carson's book explained how DDT entered the food chain and was stored in the fatty tissues of animals, including humans, where it could cause both cancer and genetic damage. She wrote how a single

Spraying crops with DDT helped control disease-spreading pests but leached poisons into the food chain.

application could kill not only pest insects but also those essential for crop pollination, and concluded that DDT and other long-lived insecticides harmed wildlife and contaminated food supplies.

The chemical industry responded to *Silent Spring* by launching a smear campaign, attacking Carson's credibility. However, a scientific enquiry, ordered by US president John F. Kennedy, endorsed her findings. Agricultural use of DDT was banned in many countries by the late 1980s, and worldwide in 2004. ■

See also: Biomes and ecological zones 136–39 ▪ Natural resource management and conservation 222–27 ▪ Sustainable development 244–45

ENVIRONMENTAL GEOGRAPHY

THE GUARDIAN OF LIFE FOR ALL OF ITS EXISTENCE
THE GAIA HYPOTHESIS

IN CONTEXT

KEY FIGURE
James Lovelock (1919–2022)

BEFORE
1785 Scottish geologist James Hutton suggests that geological and biological processes are interconnected.

1845–62 *Cosmos,* by German polymath Alexander von Humboldt in five volumes, outlines his theory of the coevolution of Earth's systems.

1926 Ukrainian geochemist Vladimir Vernadsky explains in *The Biosphere* that living matter is Earth's transforming geological force and creates its atmosphere. His book remains largely unknown until it is translated into English in 1998.

AFTER
2004 In a scientific paper, German earth systems scientist Axel Kleidon describes how biota (living things) shape energy flows in Earth's dynamic system.

British scientist James Lovelock worked for NASA during the 1960s on life-detection technologies for space missions. The atmosphere around Mars, he noted, was chemically largely stable, unlike Earth's, which has highly reactive gases. Life on Mars was therefore unlikely, he concluded, but Earth's atmosphere was the result and proof of its living organisms, constantly renewing the gases. The realization was key to his Gaia hypothesis, proposed in 1972.

Developed with American microbiologist Lynn Margulis, the Gaia hypothesis maintains that all Earth's living organisms are capable of homeostasis, that is, they interact with their inorganic surroundings to form a self-regulating, synergistic system that maintains conditions for life. Via the carbon cycle, for instance, organisms regulate Earth's climate and chemical composition. Their biological processes influence the environment, which in turn influences the organisms' evolution, creating a series of feedback loops that maintain the system's stability.

Some critics of Lovelock's theory pointed to scientific evidence of mass-extinction events, arguing these showed biological processes to have the potential for destruction as well as self-regulation. In 2006, Lovelock published *Revenge of Gaia,* stating that global warming, largely caused by burning fossil fuels, is now destroying Earth's synergistic system, potentially leading to mass extinction and a dead planet. ■

The entire range of living matter on Earth… could be regarded as… a single living entity capable of maintaining the Earth's atmosphere to suit its overall needs…
James Lovelock
Gaia: a new look at life on Earth, 1979

See also: Uniformitarianism 75 ■ The Greenhouse effect 80–83 ■ Biomes and ecological zones 136–39 ■ Monitoring environmental change 284–89

HOW DO WE REDISCOVER WHERE WE ACTUALLY LIVE?
BIOREGIONALISM

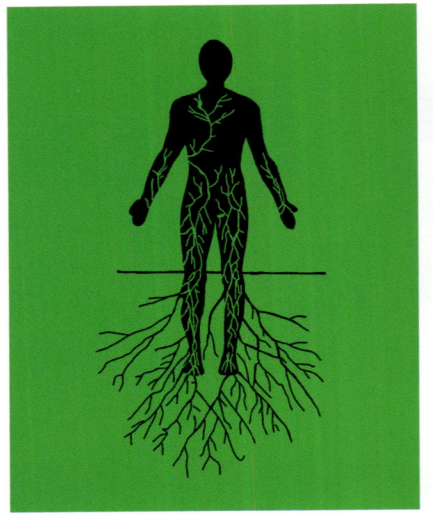

IN CONTEXT

KEY FIGURES
Raymond Dasmann
(1919–2002), **Peter Berg**
(1937–2011)

BEFORE
1898 British town planner Ebenezer Howard proposes "garden cities" to combine the urban and natural world.

1915 In his book *Cities in Evolution*, Scottish biologist and sociologist Patrick Geddes calls for urban planning that connects cities to their surrounding regions.

1960s A "back-to-the-land" movement emerges in the US, based on self-sufficiency and human harmony with nature.

AFTER
1984 The first North American Bioregional Congress meets in the Ozark Mountains.

2006 Totnes, UK, becomes the first officially-declared "Transition Town", focused on community self-sufficiency.

In the early 1970s, alarmed by the signs of environmental degradation they saw around them, a number of North American academics proposed that people should live in much closer harmony with the natural world. In his book *Environment, Power, and Society* (1971), ecologist Howard T. Odum emphasized the importance of integrating human activity with ecosystems, while in *A Landscape for Humans* (1972), ecologist Peter van Dresser argued for greater local self-sufficiency in the production of goods, services, and amenities.

Environmental writer Peter Berg and Allen Van Newkirk, an American-Canadian poet, whose 1975 article "Bioregions" helped to popularize the term, developed the ideas of Odum and van Dresser. In 1977, Berg and American conservation biologist Raymond

- **Restore** and **maintain natural features** wherever possible.
- **Develop sustainable means** to satisfy human needs.

These are guidelines for strengthening a bioregion.

- Use **education** and **media** to heighten awareness of **issues relevant to the bioregion**.
- **Protest** against **ecological damage** and **social injustice**.

ENVIRONMENTAL GEOGRAPHY 243

See also: Geography as human ecology 162–63 ▪ Local and global 208–209 ▪ The Gaia hypothesis 241 ▪ Sustainable development 244–45 ▪ The environmental justice movement 246–47 ▪ Globalization 250–53

The defence of bioregions from globalist intrusions is a persistent issue.
Peter Berg
"Bioregionalism", 2002

Dasmann presented their exposition of bioregional theories in their article "Reinhabiting California".

Harmony with nature

The central tenet of Berg and Dasmann's article was that human communities must live sustainably with nature, meeting their basic needs from local or regional sources. Communities should aim to halt and reverse the multiple problems created by unsustainable growth, such as biodiversity loss, climate change, soil degradation, and polluted seas.

Bioregions were not to be confused with scientifically classified ecoregions, focusing solely on flora and fauna, but rather defined by their natural characteristics, such as watersheds, geology, soils, and native species. Berg and other campaigners argued that bioregions are the world's natural "countries" and should also emphasize humans' place within them, as cultural traditions and practices are often interconnected with ecosystems. A bioregion's shared qualities, they argued, are more cohesive than the often arbitrary political boundaries of a nation state. Democratic community and environmental values also provide core principles, such as energy independence, a net carbon footprint, and shared land ownership with Indigenous peoples.

North America's bioregion

Among the many proposals for bioregions, the best-known is Cascadia, with land drawn from British Columbia, Canada, and the US states of Oregon, Idaho, and Washington, and neighbouring states. As mapped, it has abundant natural resources in the form of hydroelectric and wind power, and rich cultural traditions rooted in its many Indigenous peoples.

While critics argue that the Cascadia project is utopian, the Transition Network programme it has spawned is successful and growing globally. There are now over 2,000 Transition communities in more than 50 countries, sharing three common goals: cutting fossil fuel use, sourcing of food and energy locally, and strengthening community connections. ▪

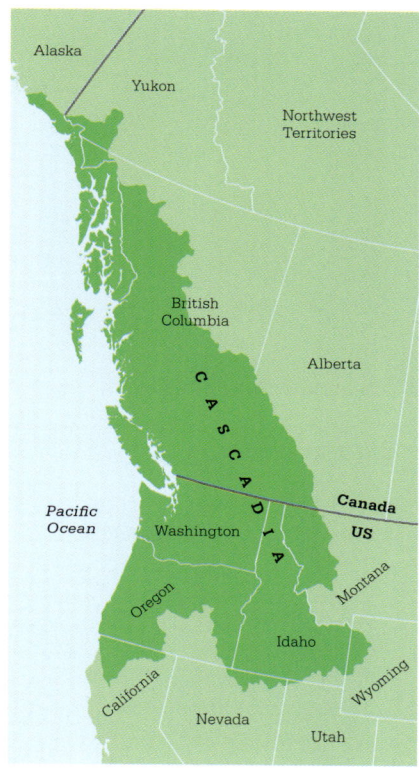

The Cascadia bioregion, named after its primary mountain range the Cascades, is defined by shared ecological and cultural connections.

Peter Berg

Berg was born in New York City in 1937, later moving to Florida, where he developed an interest in the natural world. Aged 17, he dropped out of university, hitch-hiked across the US, and joined the Beat movement subculture in California. After spells in the army and activism in the Civil Rights Movement in New York, he returned to the West Coast in 1964 and joined a mime troupe, later co-founding the Diggers, a street anarchist group. After they disbanded in 1969, Berg took a road trip and noted environmental problems, from deforestation in the Ozark Mountains to a radioactive waste dump in New Mexico. In 1973, he launched the Planet Drum Foundation to encourage harmonious living with the natural environment. Berg promoted bioregionalism until his death in 2011.

Key work

2014 *The Biosphere and the Bioregion* (collected writings)

244

FOR THE ENTIRE PLANET INTO THE DISTANT FUTURE
SUSTAINABLE DEVELOPMENT

IN CONTEXT

KEY ORGANIZATION
Brundtland Commission
(1983–87)

BEFORE
1713 German scientist Hans Carl von Carlowitz proposes that the relationship between logging and reforestation must remain *nachhaltig* or "sustainable".

1905 Gifford Pinchot, head of the US Forest Service, regulates the US logging industry for sustainability.

AFTER
1992 The Rio de Janeiro Earth Summit agrees to strategies for sustainable development in the 21st century.

2000 UN member states adopt eight Millennium Development Goals (MDGs) to accelerate sustainable development.

2015 All UN member states sign up to 17 Sustainable Development Goals (SDGs) to be achieved by 2030.

Early examples of sustainable development arose to deal with specific crises. In 1933, for example, US President Franklin D. Roosevelt established the Soil Erosion Service with the aim of restoring the American Midwest "Dust Bowl" to productivity. However, from the 1960s, there was a growing awareness of the need to tackle the issues of environmental degradation in a more holistic way with long-term strategies.

Dust Bowl formation over Stratford, Texas, in 1935 was the result of soil erosion, unsustainable wheat-farming practices, and drought.

In 1983, the UN-formed Brundtland Commission was tasked with addressing how the world could integrate environmental concerns into political, economic, and social decision-making. In 1987, it published *Our Common Future*, which defined sustainability as "meeting the needs of the present without compromising the ability of future generations to meet their own needs".

The Brundtland Report

Our Common Future became known as "The Brundtland Report", named for its chair, Norwegian politician Gro Harlem Brundtland.

ENVIRONMENTAL GEOGRAPHY 245

See also: The tragedy of the commons 220–21 ▪ Natural resource management and conservation 222–25 ▪ The environmental justice movement 246–47

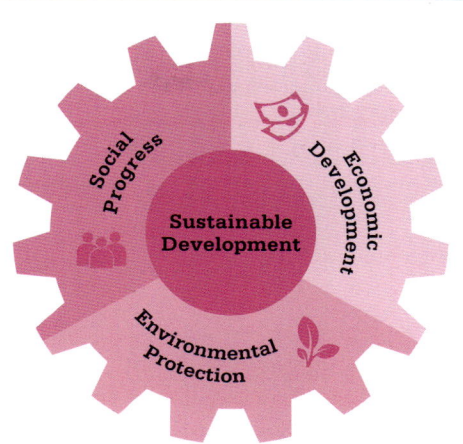

Economic, social, and environmental concerns are often called the three pillars of sustainability (also referred to as the triple bottom line) and provide a framework for evaluating the long-term impacts of decisions. These areas are interconnected and ideally should be kept in balance.

The report covered many topics, such as population growth, food security, energy, and industry. It also detailed how unsustainable consumption in the developed world, and poverty and hunger in the developing world contribute to environmental problems.

In popularizing the concept of sustainable development, *Our Common Future* shifted the world's focus from viewing economic development and the environment as separate issues to recognizing them as intertwined and interdependent at every level.

Many of the development paths of the industrialized nations are clearly unsustainable.
Gro Harlem Brundtland
Our Common Future, 1987

Many of the report's ideas were adopted by governments, academia, and businesses, and began to shape policy in many areas, such as waste-management programmes, renewable energy projects, and urban planning reforms.

"The Brundtland Report" also paved the way for international cooperation. It led to the Rio Earth Summit in 1992, which resulted in the Rio Declaration that outlined 27 principles intended to guide nations toward sustainable development.

Goals for the future

In 2015, all UN member states pledged to work towards 17 sustainable development goals (SDGs) to be achieved by 2030. However, critics argue that SDGs are vague and lack quantifiable targets, making it hard to measure progress and ensure accountability. Progress has been uneven due to a lack of political will, spiralling costs, geopolitical conflicts, and inequalities exacerbated by the COVID-19 pandemic. There have been successes, but the majority of goals are in danger of not being achieved by 2030. ■

Gro Harlem Brundtland

Born in Oslo, Norway, in 1939, Gro Harlem Brundtland trained and worked as a physician in her native city before joining the Labour Party and entering the government in 1974 as Minister of the Environment. In 1981, she became the first woman to serve as prime minister of Norway.

From 1983, Brundtland chaired the UN World Commission on Environment and Development, which published *Our Common Future*. After resigning as prime minister in 1996, she became an international leader in sustainable development and public health. She served as Director-General of the World Health Organization (1998–2003), and became a UN special envoy on Climate Change (2007–10). Brundtland is also a founding member of The Elders, a group of global leaders that was originally convened by Nelson Mandela.

Key work

1987 *Our Common Future*

ALL PEOPLE ARE ENTITLED TO EQUAL ENVIRONMENTAL PROTECTION
THE ENVIRONMENTAL JUSTICE MOVEMENT

IN CONTEXT

KEY FIGURE
Robert D. Bullard (1946–)

BEFORE
19th century In the industrial cities of Europe and the US, poor neighbourhoods are disproportionately affected by unregulated pollution.

Late 19th–20th centuries Colonial powers set up many polluting mining operations in their overseas territories.

1973 In a village in Uttarakhand, India, women lead the Chipko movement of non-violent civil disobedience to halt deforestation.

AFTER
1994 US president Bill Clinton signs the Environmental Justice Executive Order.

2022 The United Nations General Assembly declares that everyone has the human right to a clean, healthy, and sustainable environment.

Since the beginning of the Industrial Revolution, impoverished residents have been forced by circumstance to live close to the polluting effects of industrial sites. In the late 1970s, however, a landmark legal case in the US set in motion a movement to champion environmental justice.

In 1979, Black residents of a suburb of Houston, US, contacted lawyer Linda McKeever Bullard to help them prevent the creation of a solid-waste landfill site in their neighbourhood. Bullard's sociologist husband, Robert, conducted a survey to establish where other landfills and incinerators in the city were located, and discovered that 82 per cent of them were in majority-Black communities, even though Black people made up only 25 per cent of the total population.

The residents' lawsuit (Bean v. Southwestern Waste Management Corporation) was unsuccessful and the landfill went ahead. However, the case was the first to challenge environmental racism using civil rights law. It highlighted how marginalized communities bear the bulk of environmental risk, and how social justice and environmental protection are intertwined.

A growing movement

In 1978, residents of Warren County, a rural, majority-Black community in North Carolina, challenged the right of the state to site a landfill for hazardous polychlorinated biphenyls (PCBs) near their homes. Over seven weeks in 1982, marches and sit-down protests attempted to prevent the dumping of 7,000 truckloads of toxic waste, and gained widespread media attention.

The PCB Landfill protests in Warren County, North Carolina, resulted in the arrest of 500 protesters and national publicity for the growing movement of environmental justice.

See also: Humanistic geography 200–01 ▪ Spatial justice 206–07 ▪ Desertification 228–29 ▪ Sustainable development 244–45 ▪ Climate migration 254–55 ▪ Mapping social injustice 278–79

Robert D. Bullard

The fourth of five children, Robert D. Bullard was born in Elba, Alabama, US, in 1946. He studied business at Alabama Agriculture and Mechanical University, graduating in 1968, before serving for two years in the US Marine Corps. Returning to academia, he gained a doctorate in sociology from Iowa State University in 1976. After his work on the distribution of landfill sites in Houston, he widened his research to study environmental racism across the whole of the US South. In 1994, Bullard became the founding Director of the Environmental Justice Resource Center at Clark Atlanta University, and in 2021, he founded the Bullard Center for Environmental and Climate Justice at Texas Southern University in Houston.

Key works

1990 *Dumping in Dixie*
2005 *The Quest for Environmental Justice*
2009 *Race, Place and Environmental Justice After Hurricane Katrina*

In 1983, Bullard published his analysis of the disproportionate placement of municipal waste facilities in the Houston Black community. In 1987, the United Church of Christ's Commission for Racial Justice, led by Benjamin Chavis, produced a groundbreaking report "Toxic Wastes and Race in the United States", prompting a national debate.

Principles for justice

Bullard pledged to build a movement of "all peoples of color", and in 1991, he was instrumental in organizing The First National People of Color Environmental Leadership Summit in Washington. This brought together over 1,000 delegates and led to the drafting of "17 Principles of Environmental Justice".

These principles focus on four key types of justice: distributive justice focuses on the fair allocation of environmental risks and benefits. Procedural justice emphasizes the right of all to participate in environmental decision-making processes. Corrective justice seeks to rectify past wrongs and ensure accountability for environmental damage; and social justice recognizes the interconnectedness of social and environmental issues.

A worldwide campaign

During the 1990s, the US campaign converged with similar movements in developing countries, particularly Latin America and India, where Anil Agarwal founded the Centre for Science and Environment in 1994. Known as "environmentalism of the poor", these movements also show how environmental inequalities impact marginalized, especially Indigenous, peoples by threatening their livelihood, health, culture, or sovereignty. Activists and scholars from both campaigns began to collaborate on a shared goal: for a clean, healthy, and sustainable environment to be recognized as a universal human right. ■

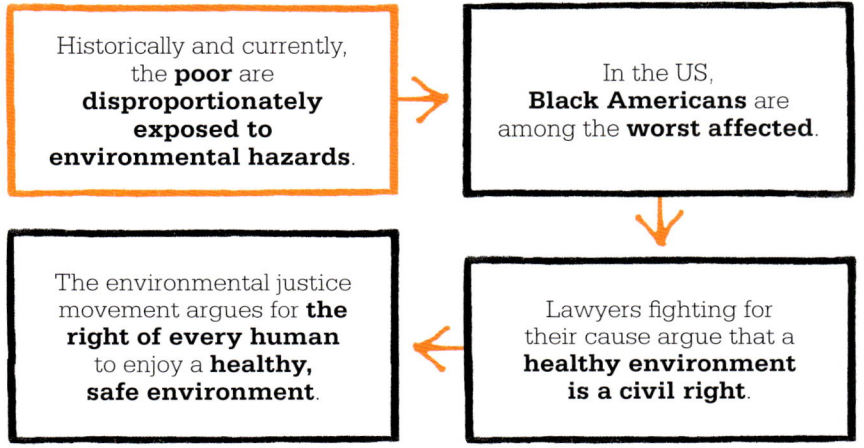

EARTH WAS IN A STATE OF OVERUSE
ECOLOGICAL FOOTPRINT CONCEPT

IN CONTEXT

KEY FIGURES
William Rees (1943–),
Mathis Wackernagel (1962–)

BEFORE
1833 British economist William Forster Lloyd's "Tragedy of the Commons" concept infers that self-interest depletes shared resources.

1940s French botanist André Aubréville highlights human-caused land degradation.

AFTER
1997 Mathis Wackernagel leads the first systematic attempt to calculate nations' ecological footprints.

2006 The GFN (Global Footprint Network) launches Earth Overshoot Day.

2024 Research by the University of Sheffield, UK, forecasts that 16 out of the 20 G20 countries will have a negative ecological footprint by 2050.

Ecological footprint = the area of Earth's resources a nation, enterprise, or individual consumes or requires to assimilate their waste.

Biocapacity = the same area's ability to regenerate its resources and absorb the waste.

Deduct footprint from biocapacity

If the **footprint exceeds** the biocapacity, there is an **environmental deficit**.

If the **biocapacity exceeds** the footprint, there is an **environmental surplus**.

Basic sustainability, or living in a way that doesn't deplete resources for future generations, has been understood for millennia across diverse cultures. However, after two centuries of rapid industrialization in the West, it was evident that natural resources were being consumed far more quickly than they were being replenished.

In 1991, Canadian ecologist and economist William Rees, alongside PhD student Mathis Wackernagel, developed a tool that allowed this imbalance to be quantified.

The 1987 United Nations "Brundtland Report", had prompted UN member states to acknowledge that meeting their country's present needs must not compromise the ability of future generations to meet

ENVIRONMENTAL GEOGRAPHY 249

See also: Geography as human ecology 162–63 ▪ The tragedy of the commons 220–21 ▪ Climate change 232–39 ▪ Globalization 250–53

theirs. However, in order for this commitment to sustainable development to be meaningful, it needed a metric by which it could be calculated. The formula created by Rees and Wackernagel took into account the supply and demand of the planet's ecological resources. Initially, they referred to it as the "appropriated carrying capacity", later adopting the more accessible term "ecological footprint".

A revelatory tool

Rees and Wackernagel started from the premise that an ecological footprint is the area of biologically active land and water, such as grazing pastures and fishing grounds, that an individual, population, or activity requires to support its consumption and its generation of waste. Areas are measured in global hectares (gha), equal to 10,000 sq metres or 2.471 global acres (ga), with 1 gha or 1 ga representing a hectare or acre with world-average biological productivity. An ecological footprint therefore represents the demand that is placed on the natural reserves of Earth. On the supply side of the equation lies the "biocapacity" of the same area of land. This represents an area's capacity to regenerate resources and absorb waste. Quantifying the ecological cost in this way provides a definitive appraisal of the environmental impact on a region. It highlights where populations, regions, or activities are consuming resources at an unsustainable rate and can provide a comparative tool to guide more sustainable decisions.

In recent years, many developed countries have significantly reduced their per capita deficit. This has been achieved through reducing their reliance on fossil fuels, but also through countless sustainable development initiatives. In urban planning, for example, the focus is on integrating parks, gardens, and green roofs into towns and cities, while creating high-density mixed-use buildings prevents the destruction of natural habitats that is the result of urban sprawl.

Earth Overshoot Day

In 2006, Earth Overshoot Day was launched in partnership with the Global Footprint Network. It marks the date each year when the global demand for ecological resources exceeds what the planet is able to regenerate in that year, such as its ability to regenerate fish stocks and forests and absorb carbon dioxide. As such, it marks the date in the calendar when Earth moves into ecological debt and starts to consume resources in a way that will impact their future supply. ■

A once vast biocapacity reserve in Amazonas State, Brazil, was severely depleted in 2024 when illegal fires burned huge swathes of rainforest.

Mathis Wackernagel

Born in Basel, Switzerland, in 1962, Mathis Wackernagel was influenced by the gravity of the 1973 oil crisis to imagine a fossil-free world. Intrigued by solar and wind power technologies and energy-efficient buildings, he studied mechanical engineering at the Swiss Federal Institute of Technology. After graduating, he moved to the University of British Columbia, Canada, where he co-created the footprint concept as the core of his PhD thesis research.

Since establishing the not-for-profit Global Footprint Network in 2003 with American environmentalist Susan Burns, Wackernagel has worked on sustainability projects in Asia, the Americas, Australia, and Europe. In 2011, he received the Zayed International Prize for Environment; in 2008, he was joint recipient of the World Sustainability Award; and in 2024 he was awarded the Nobel Sustainability Award for Leadership in Implementation.

Key work

1996 *Our Ecological Footprint* (with William Rees)

A UNIVERSALIZATION OF HAZARDS
GLOBALIZATION

IN CONTEXT

KEY FIGURES
David Harvey (1935–),
Ulrich Beck (1944–2015)

BEFORE
1944 Hungarian anthropologist Karl Polanyi publishes *The Great Transformation*, showing how capitalist markets commodify natural resources.

1985 Scottish geographer Neil Smith explores the relationship between capitalism and nature in *Uneven Development*.

AFTER
2001 China, then the world's third-largest economy, joins the World Trade Organization, increasing global trade.

2024 The world reaches 1.5°C (2.7°F) of warming above pre-industrial levels, thereby exceeding the threshold set in the 2016 Paris Agreement for limiting climate change.

Globalization is the increased flow of goods, services, capital, people, and ideas across international boundaries. The first wave of globalization began around 1870 but was interrupted in the early 20th century by two world wars. A second wave began in 1945, stimulated by reduced aviation and shipping costs, and a desire to reduce poverty in parts of the world devastated by war. It was given a further boost in the late 1980s when the collapse of the USSR ended the division of the world into separate economic blocs.

The increase in international trade brought many benefits, such as a reduction in tension between

ENVIRONMENTAL GEOGRAPHY 251

See also: Local and global 208–09 ▪ The tragedy of the commons 220–21 ▪ Climate change 232–37 ▪ Ecological footprint concept 248–49

Ulrich Beck

Ulrich Beck was born in 1944 in Slupsk, Pomerania (now part of Poland), and received his PhD in sociology in 1972 from Ludwig-Maximilians University in Munich, where he met his wife Elizabeth Gernsheim, who also became a renowned sociologist. From 1979, he worked as a lecturer at the University of Münster, where he was appointed Professor of Sociology in 1979. A year later, he became co-editor of the German sociological journal *Soziale Welt* and, in 1981, moved to a professorship at Bamberg University. In 1992, Beck returned to Ludwig-Maximilians University. He also served in many temporary international posts, for example at the University of Wales and the London School of Economics in the UK. Beck published 40 books and more than 250 research papers before his death in 2015.

Key works

1986 *Risk Society*
1994 *Reflexive Modernization* (with Anthony Giddens and Scott Lash)

---

nations, and the sharing of ideas and information, which led to innovation. However, it also intensified industrial expansion and resource exploitation on an unprecedented scale.

In 1986, Ulrich Beck, a sociology professor at the University of Munich, published the book *Risikogesellschaft* (*Risk Society*), which argued that environmental risks were not just a manageable side-effect of industrial society, but the predominant product of them. A few years later, in 1989, British-American academic David Harvey published *The Condition of Postmodernity*, stating that globalization exacerbates environmental problems through capitalist expansion and uneven geographical development.

The "Risk Society"

According to Beck's theory, the industrialization of the world could be divided into two phases. In the first, the technological, economic, and social advances comprised a "new modernity"; this aligned itself with the principles of rationality, scientific prediction, and control, and was primarily concerned with the production of wealth. Life chances were determined by one's position within the hierarchy of social classes and according to the logic of wealth distribution.

However, dating from the 1960s, Beck identified a new phase, which he labelled a "second modernity" or "reflexive modernism". By this phase, post-industrial (Western) societies had made use of science and technology in pursuit of progress and to develop solutions to traditional hazards. This had the unintended consequence of creating a world dominated by a range of new risks, often hard to perceive or quantify while far-reaching in impact.

In April 1986, as Beck was preparing to publish his book, a failure at the Chernobyl Nuclear Power Plant in Soviet Ukraine resulted in vast amounts of radioactive material being deposited into the environment, exposing hundreds of thousands of people to the risks of radiation. For Beck, this unexpected disaster provided a dramatic example of the type of incidents that characterized a "risk society".

Nuclear power itself represented human-made technology with immense and unforeseen consequences. When disaster struck, the Soviet government, in spite of their commitment to *glasnost* (a policy of openness), concealed the consequences of »

A US computer simulation showing the significant spread of radioactive particles across the northern hemisphere after the Chernobyl accident.

the fallout, thereby exacerbating the people's exposure to extreme radiation. The spread of radioactive particles in the atmosphere also impacted vast areas, demonstrating how modern risks are not contained by national borders and how social status offers no protection. Furthermore, the invisible nature of radiation made it difficult for individuals to perceive the risks and increased their reliance on experts and individuals to alert them, which they failed to do.

Risk Society was subsequently translated into more than 25 languages, a rare feat for an academic work, and Beck's theory went on to influence many fields of research, such as environmental studies, sociology, and political science. It helped establish "risk" as a central concept in understanding modern society, and especially ecological risks as a direct product of a globalized world.

Multiple crises

In the 1990s, David Harvey's analysis of globalization described how capitalism's drive for endless growth and profit was creating multiple ecological crises – habitat destruction, species extinction, and climate change – and how globalization was accelerating this process. Far from being a force for universal benefit, he argued that globalization was a mechanism through which capital expands at the expense of both human populations and the environment.

The cost of growth

In Harvey's view, globalization evolved in response to capitalism's relentless pursuit of growth and the issue of overaccumulation. If too much capital is produced relative to opportunities to use it profitably, the result is falling profits. Globalization presented a solution to these "crises of overaccumulation" by moving capital into new geographical areas or restructuring existing spaces. Harvey named this concept "spatial fix" and illustrated how it resulted in shifting production and consumption patterns, often at the expense of vulnerable populations and ecosystems.

For example, starting in the 1960s, the Brazilian government launched a campaign to develop the Amazon region for intensive cattle ranching, partly to supply international markets. In just over half a century, the industry increased 17-fold, from 5 million head of cattle to 85 million in 2016, and about 80 per cent of the deforestation in the Amazon basin has been to accommodate pastureland for cattle grazing.

This process has resulted in exports of beef, leather, and dairy products that have generated huge profits for multinational companies. However, the industry has also contributed to the pollution of the air, land, and rivers, and the net impact of deforestation has been the release of 340 million tonnes of carbon annually. Furthermore, the transformation of rainforest to pastureland has negatively impacted biodiversity in the region, not only from tree felling but also by allowing invasive generalist species to spread in from the forest edge.

Relatively small numbers of local people have benefited directly from the Brazilian cattle-ranching economy and Indigenous peoples suffer continued human rights abuses, including illegal displacement. Harvey identified "accumulation by dispossession" as a common feature of a globalized

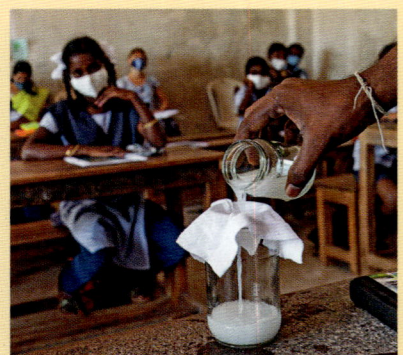

Students detect microplastics in Tamil Nadu, India, by dissolving toothpaste in water and then straining it through filter paper.

Microplastic particles

Since 2004, when British scientist Richard Thompson first detected tiny plastic particles during a beach clean-up, the presence of microplastics in the environment has become a prominent social and political concern. Created through the disposal of everyday household objects, such as cosmetics and clothing fibres, they are often contaminated with poisonous substances and provide a classic example of how human-made risk has become a byproduct of our modern societies.

Microplastics affect everyone, regardless of nationality or social class. Research shows that they pose a potential threat to marine life, soil health, plant growth, and human health through the food chain. They have been found in human tissues and studies suggest links to cardiovascular disease, fertility issues, and other health problems. However, to date, the precise scope of the risk is unknown and many major industries continue to use microplastics despite growing concerns.

ENVIRONMENTAL GEOGRAPHY

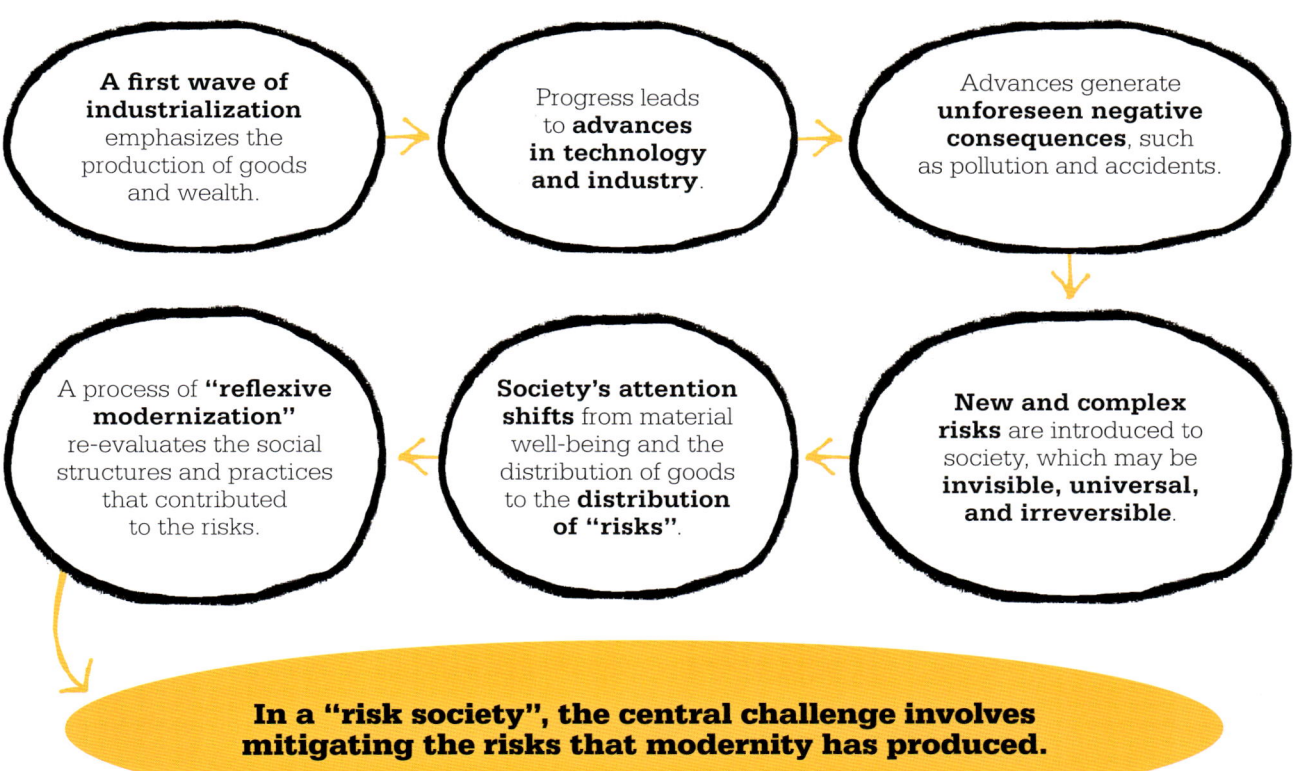

capitalist system that seeks to centralize wealth and power in the hands of a few. In order to facilitate further economic growth, existing assets, often in the form of land and resources, are appropriated from people and public entities through various means, such as land grabbing, privatization, and the displacement of populations.

Mitigation strategies

Throughout the 1980s and '90s, Beck and Harvey contributed to a growing awareness of the impact of globalization on the environment. As Beck predicted, societies in the West became "reflexive" and began to confront the unintended, self-generated consequences and risks of their own modernization processes. In 1992, the UN acknowledged human-caused interference with the climate system with a foundational treaty – The United Nations Framework Convention on Climate Change (UNFCCC) – that established a basis for international cooperation and led to the Kyoto Protocol and the Paris Agreement.

Over the subsequent decades, green economic strategies have increasingly been factored into trade agreements, including the transition to renewable energies and guarantees of environmental protection. The interconnectedness of globalization, which allows for the flow of trade across borders, also enables the rapid spread of energy-efficient and eco-friendly technologies, which are then adapted for local contexts.

However, environmental burden-shifting remains a critical issue. Strict regulations in developed countries have led to companies moving their operations to places with weaker environmental controls. For example, Western countries export significant amounts of plastic and electronic waste to developing countries, along with its pollution and health risks.

In recent years, heightened trade tensions and policy uncertainty has led to a slowdown in the growth of the global economy. The terms of globalization are undergoing a process of reappraisal and its future could potentially be shaped by a broader range of interests. Alongside economic efficiency, social and environmental sustainability have become primary concerns. ∎

THE BEST SINGLE MEASURE OF GLOBAL ENVIRONMENTAL DECLINE
CLIMATE MIGRATION

IN CONTEXT

KEY FIGURE
Norman Myers (1934–2019)

BEFORE
1976 American environmental analyst Lester Brown first proposes the term "environmental refugee".

1985 UN Environment Programme (UNEP) expert Essam El-Hinnawi defines climate refugees as people who have "had to leave their habitat, temporarily or permanently, because of a potential environmental hazard or disruption in their life-supporting ecosystems".

AFTER
2018 French legal academic Benoit Mayer analyses the issue of climate migration in social, political, and environmental contexts.

2021 In a White House report, the US government officially recognizes the link between climate change and migration.

Climate migration is the movement of people from one location to another, either within a country or across international borders, due to the impact of environmental phenomena, such as climate change. These impacts can include sudden onset disasters, which are often the result of extreme weather – hurricanes, floods, or droughts – as well as slow-onset changes, such as desertification and sea-level rise.

In 1993, British environmentalist Norman Myers published an article in *Bioscience* titled "Environmental Exodus", which brought the idea of climate migration to the forefront of international debate. He wrote that "large numbers of environmental refugees could be among the most significant of all upheavals entrained by global warming" and estimated there would be 150 million refugees by 2050, a figure he subsequently updated to 200 million.

The article also identified specific regions as "hotspots", where the impacts of climate change, such

Displaced by devastating floods in Pakistan in 2022, climate refugees were forced to live in temporary shelters or relief sites.

ENVIRONMENTAL GEOGRAPHY

See also: Migration theory 152–53 ▪ Environmental impact assessment 272–73 ▪ Geodemographic analysis 274–77 ▪ Mapping social injustice 278–79 ▪ Monitoring environmental change 284–89

The gravest effects of climate change may be those on human migration as millions are uprooted by shoreline erosion, coastal flooding, and agricultural disruption.
Intergovernmental Panel on Climate Change
First Assessment Report, 1990

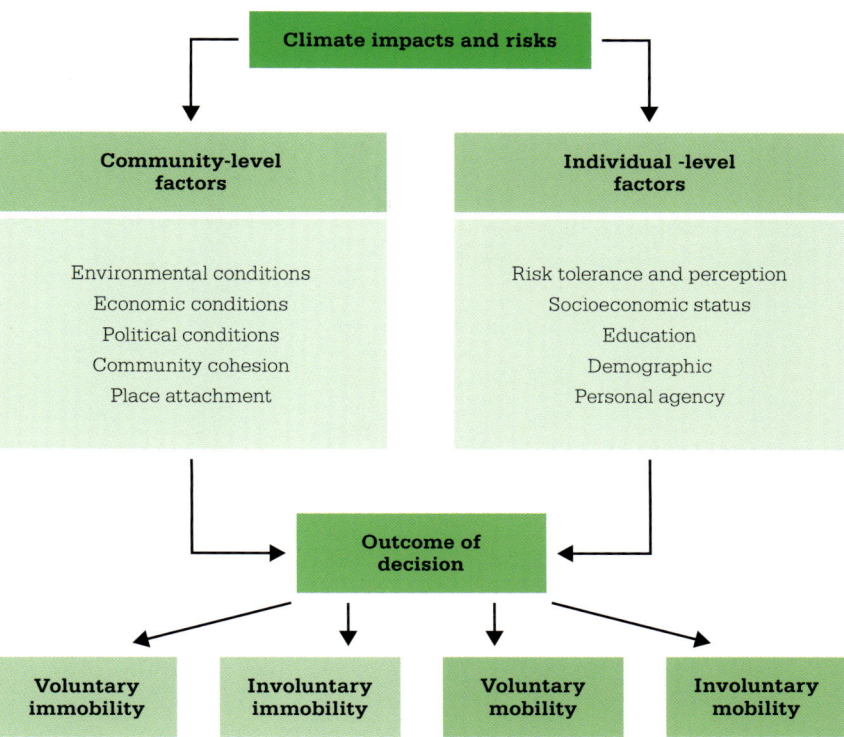

Climate mobility or immobility is influenced by a complex pattern of political, economic, and social ties and all these must be considered when modelling migration patterns.

as water shortages or rising sea levels, were expected to be most severe. These were primarily located in the developing countries of sub-Saharan Africa, South Asia, and Latin America, and Myers theorized that populations in these regions were most vulnerable to large-scale displacement. His migration projections gained instant international media attention and, ever since, have been widely cited in official reports as well as academic literature.

Myers' work significantly raised awareness of the human cost of climate change. However, the methodology he used for modelling migration was soon criticized as simplistic, as it was based on the assumption that populations in "hotspot" regions would inevitably be displaced, and relied on aggregate global forecasts rather than specific case studies. Critics viewed this argument not only as lacking empirical grounding but also as failing to consider the complexities of migration patterns and the potential for communities to adapt to environmental challenges.

Climate mobility

In the years since Myers' report, scientific research has focused on the complex range of factors influencing people's decisions to stay or leave their homes in the face of environmental pressures. A new term – "climate mobility" – represents a more nuanced reality. It encompasses various forms of human movement, including displacement, relocation, and migration, both within and between countries. It also recognizes involuntary immobility, where people may be constrained by poverty or lack of agency, as well as voluntary immobility, where strong ties to culture, identity, place, and community institutions resist the idea of displacement.

A greater understanding of all factors is key to developing policies that can build climate resilience in the most vulnerable places and, where conditions become untenable, to planning relocation with proper support and resources. ■

THE CENTRAL ROLE OF MANKIND IN GEOLOGY AND ECOLOGY
THE ANTHROPOCENE

IN CONTEXT

KEY FIGURES
Paul J. Crutzen (1933–2021),
Eugene F. Stoermer
(1934–2012)

BEFORE
1875 Austrian geologist Eduard Suess describes the parts of Earth where life occurs as the "biosphere".

1920s French Jesuit priest Pierre Teilhard de Chardin uses the term "noosphere" to mean the sphere of human thought on Earth.

1943 Ukrainian geochemist Vladimir Vernadsky writes that in the noosphere, humans are "a large-scale geological force".

AFTER
2025 Disputing a proposed mid-20th century start date, the International Commission on Stratigraphy votes against recognizing the Anthropocene as a new geological epoch.

Earth's current geological epoch – the Holocene, proposed by British geologist Charles Lyell in 1833 – denotes the 11,700-year post-glacial period that encompasses recorded human history. In the 20th century, scientists began to hypothesize about a new geological epoch – the Anthropocene – to reflect the disturbing changes human activity was causing to the planet.

At the University of Michigan, Professor Eugene F. Stoermer used the term Anthropocene informally from around 1980, to describe the adverse effects humans were having on lakes and rivers, based on his studies of microscopic algae.

ENVIRONMENTAL GEOGRAPHY 257

See also: Desertification 228–29 ▪ Climate change 232–39 ▪ Silent Spring 240 ▪ Sustainable development 244–45 ▪ Ecological footprint concept 248–49 ▪ Globalization 250–53

> Human activities have now brought Earth outside of the Holocene's window of environmental variability, giving rise to the proposed Anthropocene epoch.
>
> **Katherine Richardson**
> ***Earth beyond six of nine planetary boundaries,*** 2023

Dutch meteorologist and Nobelist Paul Crutzen introduced the term independently at a meeting of the International Geosphere–Biosphere Programme (IGBP) Scientific Committee in February 2000. In "The Anthropocene", an article published in an IGBP newsletter the same year, he and Stoermer defined the term and proposed it as a new geological epoch.

The human effect

While Earth is some 4.6 billion years old, humans have only been present for the past few hundred thousand years. However, in recent centuries industrialization and an expanding global population have together had a profound effect, as Stoermer and Crutzen noted, pointing to the tenfold increase in population over 300 years, from 600 million to 6 billion humans; the current figure now exceeds 8 billion. The article warned of the consequent drain on natural resources, and the release of harmful greenhouse gases as a result of the burning of fossil fuels. It also explained that up to 50 per cent of Earth's land surface had already been transformed by human activity, increasing the rate of species extinction in tropical rainforests ten thousandfold. The scientists also predicted dramatic climate change as a result of anthropogenic releases of CO_2.

Disappearing species

Other academics worldwide have since expanded on the measurable detrimental effects humans are having on the world's natural systems – physical, chemical, and biological. Their evidence reveals ever-increasing anthropogenic CO_2 emissions, extreme weather events, global warming, deforestation, habitat destruction, pollution, agricultural mismanagement, ocean acidification, reduction in biodiversity, and unprecedented rates of wildlife extinction. American science writer Elizabeth Kolbert's highlighted Earth's accelerating losses of biodiversity in her Pulitzer prize-winning book

Sewage and other human waste produce excess phosphorus, which encourages the growth of cyanobacteria (blue-green algae). This can choke rivers, such as here in Puy-de-Dôme, France.

The Sixth Extinction (2014). She reported that up to half of all plant and animal species could disappear within this century. Earth's earlier five great extinctions were the result of powerful volcanic eruptions or asteroids colliding with our planet; the sixth is due to humans. Amphibians such as »

Eugene F. Stoermer

Born in Iowa, US, in 1934, Eugene F. Stoermer, gained a doctorate in biology at the University of Iowa. Here, he began to research diatoms – microscopic, single-celled algae with cell walls uniquely made up of transparent silica. After further studies in Philadelphia, he returned to Iowa and used newly developed electron microscopy to analyse the algae.

In 1965, Stoermer joined the University of Michigan, where he researched diatom species in the Great Lakes and was appointed professor of biology. His study of aquatic ecosystems contributed greatly to a deeper understanding of anthropogenic eutrophication – how excess phosphorus and nitrogen from sewage, synthetic fertilizers, and industrial wastewater adversely affect lakes and rivers. Stoermer died in 2012.

Key work

1999 *The Diatoms: Applications for the Environmental and Earth Sciences* (with John P. Smol)

Paul Jozef Crutzen

Born in Amsterdam in 1933, Paul Jozef Crutzen gained a love of natural sciences at secondary school but, after failing to qualify for university, initially studied civil engineering instead. In 1958, Crutzen moved to Sweden where he worked as a computer programmer in the Meteorology Institute of Stockholm University.

In 1968, he completed a PhD thesis on ozone in the upper atmosphere. His discovery that chlorofluorocarbon gases deplete Earth's ozone layer earned him and his co-workers the 1995 Nobel Prize for chemistry.

From 1977, Crutzen was director of research at the National Center of Atmospheric Research in Colorado, US, and from 1980 he was director at the Max Planck Institute for Chemistry in Mainz, Germany. He also worked at the Scripps Institution of Oceanography in California and at universities in South Korea, Sweden, and the Netherlands. He died in Mainz in 2021.

Key work

1982 *"The Atmosphere after a Nuclear War"* (with John Birks)

---

frogs, she reported, are the world's most endangered class, their extinction rate up to 45,000 times higher than it was in the past. As a result of anthropogenic global warming, vegetation also dries and dies, animal species starve, and both are consumed by increasing numbers of wildfires.

In a 2015 paper, American and Mexican scientists confirmed that the current vertebrate extinction rate is up to 100 times higher than the historic rate. This, they say, indicates that a sixth mass extinction is already underway. While averting it remains possible, they warned that the window of opportunity "is rapidly closing".

Energy and population

In *The Great Acceleration* (2016), environmental historians John R. McNeill and Peter Engelke argue that the Anthropocene runs from 1945, with two main drivers of its excesses – energy and population.

The radioactivity released into the atmosphere by nuclear bomb tests, such as this at Bikini Atoll in 1946, is one of the human-imposed "novel entities" that have been linked to the Anthropocene.

They reveal how an expanding global population's huge demand for energy – and the mining, drilling, refinement, transportation and burning of polluting fuels that produce it – cause environmental problems at every step.

A related planet-changing human impact has been the creation of petroleum-based plastics, microplastics, and synthetic fibres, and their consequent waste, which litters the land, contaminates waterways, and clogs the marine environment. The Organisation for Economic Cooperation and Development (OECD) has warned that the global build-up of plastic waste will almost triple to 1,014 million tonnes (1,118 million tons) by 2060 unless governments together take preventive steps.

Danger levels

Back in 2008, Johan Rockström, director of the Stockholm Resilience Centre, and a team of international scientists had identified nine processes most heavily influenced by humanity, the stability of which

ENVIRONMENTAL GEOGRAPHY

> Our species has probably used more energy since 1920 than in all of our prior human history.
> **John R. McNeill and Peter Engelke**
> *The Great Acceleration*, 2016

is key to the maintenance of the entire Earth system of atmosphere, land, oceans, and living things. The nine areas of human influence are: climate change, novel entities (synthetic chemicals causing genetic modifications), stratospheric ozone depletion, atmospheric aerosol loading, ocean acidification, biogeochemical flows (extra nitrogen and phosphorus added via industrial and agricultural processes), freshwater change, land-system change (transformation of natural landscapes), and biosphere integrity (the diversity, extent, and health of ecosystems and living organisms). The team aimed to calculate the thresholds or boundaries of human pressure on these areas before the Earth system was threatened.

By 2015, four boundaries had been crossed, and in the 2018 article "Trajectories of the Earth System in the Anthropocene", Rockström and other scientists warned that human actions were setting the planet on a Hothouse Earth pathway leading to possible temperatures 4–5°C (7–9°F) above pre-industrial levels and far higher sea levels than at any previous time in the Holocene. Five years later, 28 scientists noted that six of the boundaries had been exceeded. By this time, trapped radiation and additional CO_2 in the atmosphere were also perceptibly altering climate patterns worldwide.

The team warned that the planet was now "well outside of the safe operating space for humanity". However, anthropogenic changes to the Earth system could still be mitigated, say scientists, if human societies manage their relationship with Earth, accepting their role as an integral part of it and their impact on its stability.

Delayed acceptance

Despite evidence of human-led change to Earth, the Anthropocene has not yet succeeded the Holocene as an official geological epoch. Part of the problem is determining its beginning. Key to the recognition of a new geological epoch is a geochemical marker – a chemical signal preserved in sediment or ice. In 2024, a working group of the International Union of Geological Sciences (IUGS) suggested a start date in the mid-20th century, when atomic weapons tests took place, leaving such a chemical marker in the fossil record. Also required is a stratigraphic marker, a layered deposit in the geological record. Plastics that have fused to rock are one potential example.

Until these criteria are agreed on, the IUGS has proposed that the Anthropocene should be considered a major geophysical event on a par with the Great Oxygenation Event some 2.3 billion years ago or the Cambrian Explosion (the great emergence of organisms) over 500 million years ago. ∎

> The Industrial Revolution significantly increases the **mining and burning of fossil fuels** and its innovations create ever more **destructive weapons**.

> The world's population begins **to rise at an unprecedented rate**, putting pressure on global **food and water resources**.

↓ ↓

> Industrial developments **modify landscapes and waterways**. Atomic and hydrogen bombs **release radiation** and deposit **radioactive fallout**.

> Intensive farming encourages **deforestation**, and causes soil degradation, **water pollution, and biodiversity loss**.

↓ ↓

> **In multiple ways, modern humans are leaving an indelible footprint on the planet – hence the call for the naming of a new geological epoch – the Anthropocene.**

APPLIED GEOGR

D
APHY

INTRODUCTION

Los Angeles, US, introduces the nation's **first zoning restriction**, prohibiting industry in three residential areas.

Syukuro Manabe and Richard Wetherald present a computer-based climate model that **demonstrates the greenhouse effect**.

Richard Webber creates the first version of the **geodemographic segmentation system** Acorn to classify people by their residential neighbourhood.

Luc Anselin develops influential software that expands the **scope of spatial data analysis**.

1904   **1967**   **1970s**   **1980s–1990s**

1960s   **1970**   **1970s**   **1993**

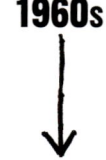 Torsten Hägerstrand develops a method of representing **movement of people** in an environment over time, establishing the field of **time-geography**.

 The US National Environmental Policy Act requires impact assessments to estimate how **developments will affect the natural world**.

 Compton J. Tucker develops the Normalized Difference Vegetation Index to **monitor agriculture globally** using satellite technology.

 Peter Calthorpe popularizes the concept of the urban village, laying the foundations of the **Smart Growth movement**.

In the field of applied geography, the tools and techniques employed by geographers – from surveying methods to data collection using advanced technology – are used to solve social, economic, or environmental problems in the real world.

Urban planning

Some 150 years ago, German engineer Reinhard Baumeister showed how geographic analysis could guide urban development in terms of planning for transport, sanitation, and future growth. His ideas laid the groundwork for building codes, land use, and zoning around the world. Much later, in the 1960s, Swedish geographer Torsten Hägerstrand linked individual behaviour to spatial structures, with the aim of planning amenities to take account of people's daily activities and movements, and make them more time-efficient.

From the 1970s, Belgian geographer Luc Anselim made use of the emerging geographical information systems (GIS) technology to pioneer "spatial data analysis". This uses mathematical and statistical models to analyse information tied to a specific location, which can then be used to optimize public transport routes or map disease spread, for example. The field also advanced the study of geodemographics. British geographer Richard Webber created the first neighbourhood classification system allowing businesses and organizations to create detailed profiles of communities for marketing, urban planning, and service targeting.

Spatial data analysis can also be used to pinpoint social justice inequities. In the 1980s, American geographer Laura Pulido was instrumental in highlighting how environmental burdens such as industrial waste facilities disproportionately affected poorer communities in California, US.

In the 1990s, the Smart Growth movement focused on urban planning. Influenced by American architect Peter Calthorpe, it was a move to combat urban sprawl and create compact neighbourhoods, with essential services accessible to residents either on foot or using sustainable forms of transport.

Monitoring nature

The establishment of the US Environmental Protection Agency in 1970, followed by other national

Global Forest Watch, a pioneering environmental monitoring initiative, is established to **track deforestation**.

Laura Pulido describes how **industrial pollution** disproportionately affects **communities of colour** in Southern California.

Andrew Skidmore pioneers the use of **geographic information system technology** for ecosystem monitoring.

Copernicus, the European Union Space Programme's Earth observation programme, is launched.

1997 — **2000** — **2000s** — **2014**

1999 — **2009** — **2010** — **2020–2021**

The United Nations Office for Disaster Risk Reduction is created to oversee the **prediction and monitoring of natural hazards**.

The International Renewable Energy Agency (IRENA) is formed to promote a transition towards the use of **renewable energy**.

After a large earthquake, volunteers use **OpenStreetMap** to map Haiti and facilitate the **relief operation**.

The Covid-19 Dashboard is developed at Johns Hopkins University to track **cases, deaths, recoveries, and vaccination rates**.

---

agencies, provided a framework for monitoring and regulating pollutants, and setting standards for a cleaner, healthier environment. Geographers and other scientists use remote sensing to monitor ecosystems and the effects of climate change. Global Forest Watch uses satellite data to map the world's forests and track illegal deforestation, supplying its data free to researchers and governments. Similarly, NASA's Earth Observing System (EOS) surveys interactions between land, oceans, atmosphere, and biosphere.

Climate modelling
From the late 1950s, computer power began to transform the science of weather forecasting. American theoretical meteorologist Norman Phillips developed a mathematical model that could track the interactions between thousands of variables in the atmosphere. In 1967, Japanese climatologist Syukuro Manabe produced the first computer-based model of the world's climate system for the US Weather Bureau – the basis for advanced climate-modelling today. In the 21st century, NASA's EOS and the EU's Copernicus Earth observation programme both gather data for use in climate-modelling.

Sustainability
Accurate monitoring and the clear visualization of data can assist with efficiency across numerous areas, ranging from agriculture to resource management. Combined with satellite data, GIS can reveal issues, such as low soil moisture levels or pest attacks, which can then be resolved with targeted precision. The Global Atlas of Renewable Energy is a GIS that uses climate data and remote sensing to pinpoint areas with potential for renewable energy projects and shares its data on a free web platform, open to all.

Disasters and disease
New technologies help combat natural disasters, too, enabling clearer, timelier predictions and alerts. Via open-source software and satellite images, volunteers can help map disasters, aiding responders on the ground. Public health is also more easily tracked; the Covid-19 Dashboard, developed in 2020, was a feat of data-driven disease-mapping. In so many areas, applied geography is proving its worth in the 21st-century world. ∎

ZONING SEEKS TO PROTECT AND STABILIZE WHAT IS GOOD
LAND USE ZONING

IN CONTEXT

KEY FIGURE
Reinhard Baumeister
(1833–1917)

BEFORE
1845 In *The Condition of the Working Class in England*, German socialist philosopher Friedrich Engels highlights the impact of industrialization on urban living and public health.

1853 Emperor Napoleon III commissions French official Georges-Eugène Haussmann to undertake a massive urban renewal project in Paris.

AFTER
1926 The US Supreme Court upholds zoning rules in Euclid, Ohio, against the challenge of Ambler Realty Co., affirming local government power to implement zoning regulations.

2017 Germany introduces a new zoning category, "Urban Areas", which allows a mix of residential and commercial uses within high-density areas.

The concept of regulating how different parts of cities should be set aside for different uses, such as housing or public buildings, dates back at least as far as the Roman Empire of the 1st century BCE. In the 19th century, however, the huge growth of cities during the Industrial Revolution created unplanned and chaotic expansion, with factory chimneys and abattoirs standing in the midst of residential areas. This led to crises in public health and demanded a new approach to urban planning.

In 1876, German engineer Reinhard Baumeister published *Urban Expansion with Respect to Technology, Policing of Building Code and Economy*. Basing his ideas on the Prussian concept of rules, known as "building lines", for street layouts, he observed that industries tended to cluster together in cities. He argued that this should be reinforced through zoning areas into three distinct classes of land use: industry and wholesaling, alongside homes for workers; trades that require direct

---

Unregulated development in the city leads to **poor living conditions** and **public health issues**.

⬇

Dividing the city into **land use zones** enables the **separation of industrial from residential areas**.

⬇

Urban planning regulation results in **healthier, safer, and more efficient urbanization** to the benefit of citizens.

See also: Rank-size rule 160 ▪ Primate city rule 161 ▪ Central place theory 182–83 ▪ The concentric zone model 184–87 ▪ The "smart growth" movement 282–83

APPLIED GEOGRAPHY 265

The green belt

In 1898, in reaction to the squalor produced by 19th-century urbanization, British town planner Ebenezer Howard published the book *To-morrow: A Peaceful Path to Real Reform*. Wanting to combine the positive aspects of town and country life, he proposed the idea of "Garden Cities", where self-contained communities were made up of a balanced mix of residential, industrial, and rural areas, with a surrounding ring of open land – recreational and agricultural. The concept of a "green belt" to prevent urban sprawl – the continued spread of suburbia and industry into the surrounding countryside – soon found support. His ideas were taken up in his home country, where the 1938 Green Belt Act enforced such a zone around London, and the 1947 Town and Country Planning Act empowered local councils to designate their own green belts. The concept of a designated green belt spread globally too, to cities such as Copenhagen, São Paulo, Melbourne, and Tokyo.

A plan of the city of Strasbourg from 1904 reflects the completion of the first major phase of the Neustadt Project, including most of the official buildings.

contact with the public, plus homes for workers; and homes whose owners have no trade and work in a differing range of occupations.

The Neustadt Project

In 1880, Baumeister was made technical advisor on a far-reaching transformation of Strasbourg, a city annexed from France in the Franco-Prussian War (1870–71). Known as the Neustadt Project, the scheme tripled the city's surface area. With tree-lined avenues, green parks, and separated areas for residential, commercial, and industrial use, the undertaking pioneered the development of building codes and zoning practices.

The ambition of the Neustadt Project laid the groundwork for the Frankfurt Planning Acts (1891), a piece of legislation that established modern principles of close control over building types and their physical characteristics, such as height and bulk, all in distinct zones.

Zoning in the US

The Frankfurt Acts became a blueprint for urban development around the world, especially in the US, where rapid industrialization had also led to unregulated construction and public health concerns. In 1904, Los Angeles passed a zoning ordinance prohibiting industry in three residential districts. In New York City, lawyer Edward Bassett – inspired by the German town planning he observed in 1908 – began to advocate urban zoning to control congestion and improve the quality of life. Focusing on business, he argued that zoning reduced disputes between property owners, stabilized property values, and limited land speculation.

In 1916, Bassett wrote the New York City Zoning Resolution, the first all-city zoning code. It regulated the height, size, and use of buildings, ensuring, for example, that skyscrapers did not completely block light and air from the streets below. The code also divided the city into residential, manufacturing, and commercial districts.

The influence of Bassett's code spread as urban planners became more attuned to a formalized approach. Zoning expanded rapidly in the US, and in two years, from 1924 to 1926, the number of zoned cities increased from 62 to 456. By 1929, nearly 800 cities in the US had zoning ordinances. ▪

PEOPLE ARE NOT PATHS, BUT THEY CANNOT AVOID DRAWING THEM IN SPACE-TIME
TIME-GEOGRAPHY

IN CONTEXT

KEY FIGURE
Torsten Hägerstrand
(1916–2004)

BEFORE
1939 "Aereal differentiation", the idea that each space is an interplay of human-made and natural traits, is proposed by American geographer Richard Hartshorne.

1970 American geographer Waldo Tobler publishes his first law of geography: "Everything is related to everything else, but near things are more related than distant things."

AFTER
2016 Whim, the first Mobility as a Service (MaaS) integrated transport app – based on time-geography concepts – is launched in Helsinki, Finland.

2021 During the Covid-19 pandemic, time-geography is used to understand and monitor the spread of the virus over space and time.

Everyone carries out their daily lives within spaces that are defined by factors such as housing, work, travel, education, and recreation. Time-geography is a way of capturing the movements of people through different spaces over time. That data is then analysed and can be presented visually to reveal any links between human movements and their immediate geographic, social, and economic environments.

Humans on the move

Swedish geographer Torsten Hägerstrand developed the space-time concept in the 1960s and '70s. This was in response to what he saw as a misguided approach

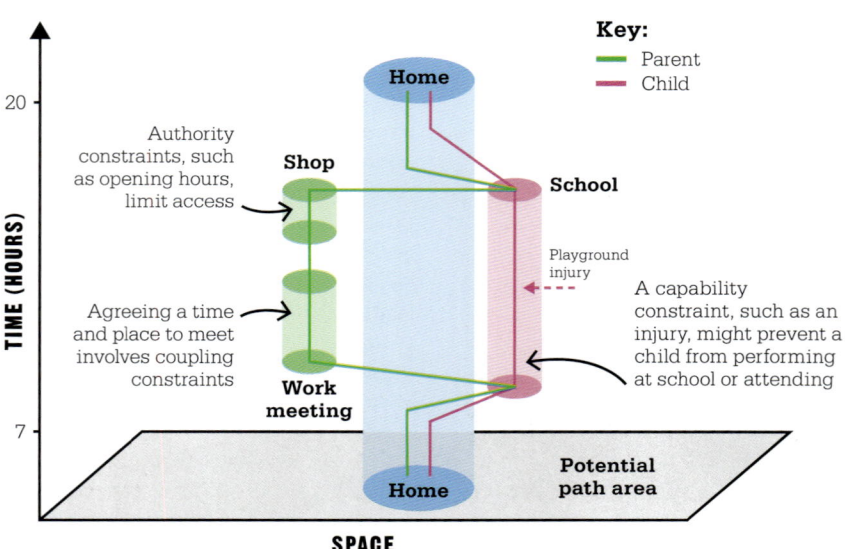

Time-geography data can be presented in three-dimensional graphics, tracking people's life paths through space and time. This graphic represents a family's day, with columns to indicate the space-time constraints that affect their activities.

APPLIED GEOGRAPHY 267

See also: Geographic information systems 52–59 ▪ Spatial interaction theory 192–93 ▪ The "smart growth" movement 282–83 ▪ Early warning systems 296–303 ▪ Tracking disease 310–13

---

from urban planners to developing neighbourhoods and infrastructure, such as transport and health facilities. Hägerstrand agreed with the common principle that the purpose of development is to improve the living standards of residents and boost social and economic activity, making human environments more liveable. However, he realized that little attention was given to why, and when, people moved from one area to another. In Hägerstrand's view, understanding the movements of a population would help to maximize the liveability of places.

Following life paths

Hägerstrand's aim was to create a framework that enabled the reasons for people's many movements to be investigated effectively. To do this, data had to be collected at the individual level, so each person's life was a path plotted through time and space. Setting out his ideas in a 1970 address to the European Congress of the Regional Science Association, he noted that "life paths become captured within a net of constraints,

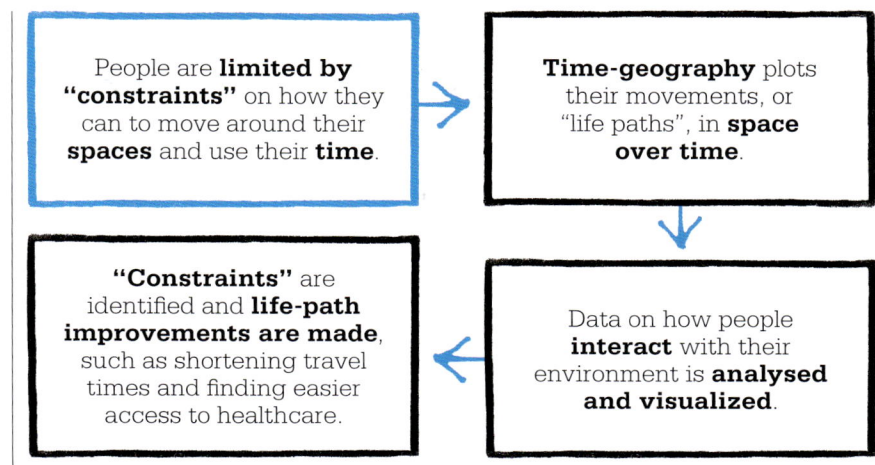

People are **limited by "constraints"** on how they can to move around their **spaces** and use their **time**.

Time-geography plots their movements, or "life paths", in **space over time**.

Data on how people **interact** with their environment is **analysed and visualized**.

"Constraints" are identified and **life-path improvements are made**, such as shortening travel times and finding easier access to healthcare.

… thought does not encounter in its own world the constraints of space and time.
Torsten Hägerstrand
Geography and the Study of Interaction between Nature and Society, 1976

---

some of which are imposed by physiological and physical necessities and some imposed by private and common decisions".

By "constraints", Hägerstrand meant the forces that kept a person in a certain place or on a particular path, and he identified three classes of constraint – capability, authority, and coupling. Capability constraints are limits due to biology, such as health, age, gender, and the need to eat and sleep, plus the distance a person can move on foot in a day. Authority constraints are limits imposed on movement and activity due to decisions made by others, such as traffic laws and transport timetables. Coupling constraints occur when a person is restricted in their movements because they have to be at the same location as another person, object, or event. For example, a sales team might be constrained to attend a daily meeting at 9am in

the same room for 30 minutes, or a parent might have to collect a child from school at 3.15pm.

Practical applications

Today, time-geography principles are studied in tandem with geographic information systems (GIS) to provide insights into patterns of human activity and movement in space and time. This data has been successfully applied to urban planning, especially transport and health provision. ▪

The space-time "constraint" of poor road access in the American city of Portland, between the waterfront and a hilltop university campus, was overcome by an aerial tram linking the two.

THE EXTREME SENSITIVITY OF EARTH'S CLIMATE
CLIMATE MODELLING

IN CONTEXT

KEY FIGURE
Syukuro Manabe (1931–)

BEFORE
1904 Norwegian scientist Vilhelm Bjerknes shows that the way Earth's atmosphere works can be described mathematically, using "primitive equations".

1956 American theoretical meteorologist Norman Philipps creates the first basic computer-driven climate model.

AFTER
2024 Destination Earth (DestinE), a European climate research collaboration, launches digital twins of Earth to help the understanding of future climate changes.

2024 IBM, in collaboration with NASA, introduces Prithi WxC, an open-source AI model that runs large climate data sets on a desktop computer.

Earth's atmosphere, oceans, land, and ice interact with each other and with energy from the Sun to drive changes that create weather systems (short-term atmospheric conditions) and climate (long-term global and regional patterns of weather). A climate model is a computer simulation of these patterns, based on complex mathematical equations that account for all the reactions between matter and energy in the atmosphere. The development of computer-based models was pioneered in the 1960s by Japanese climatologist Syukuro Manabe, who laid the foundations for modern climate science.

See also: The Greenhouse effect 80–83 ■ Climatic zones 96–101 ■ Extreme weather 142–43 ■ Climate change 232–39 ■ Early warning systems 296–303

Climate modelling can speed up the progression of time in simulations, both forwards and backwards, allowing future predictions of climate as well as re-creations of conditions in the past. Climatologists use models over a range of timescales to investigate the impact of specific starting conditions on the global climate system, including the future effects of global warming due to human-made climate change.

Modelling breakthroughs

In 1959, Syukuro Manabe left Japan to join the US Weather Bureau's newly created Geophysical Fluid Dynamics Laboratory (GFDL), where he oversaw work on the computer coding of climate models. Gradually, he added levels of complexity to the models, such as rainfall evaporation and heat exchange across oceans, land, and ice sheets. Then, in 1967, he co-published – with American fellow climatologist Richard Wetherald – an important and influential climate science paper, "Thermal equilibrium of the atmosphere with a given distribution of relative humidity".

In his paper, Manabe described how radiation and clouds interacted to redistribute water vapour and heat through the atmosphere. His model included experiments to see how greenhouse gases – those that trap heat and warm the planet – might change the climate. This led to his estimate that, if carbon dioxide (CO_2) concentrations doubled, Earth would warm by 2.35°C (4.14°F) – a figure still within the range of modern global warming estimates. Manabe also found that, as CO_2 increases, the troposphere (lower atmosphere) warms, but the stratosphere (upper atmosphere) cools, potentially affecting air circulation and other weather patterns. These discoveries created the basis for »

Satellite data can be modelled to predict future climate events at a specific time and date, such as this forecast for the pattern of ocean currents around Antarctica.

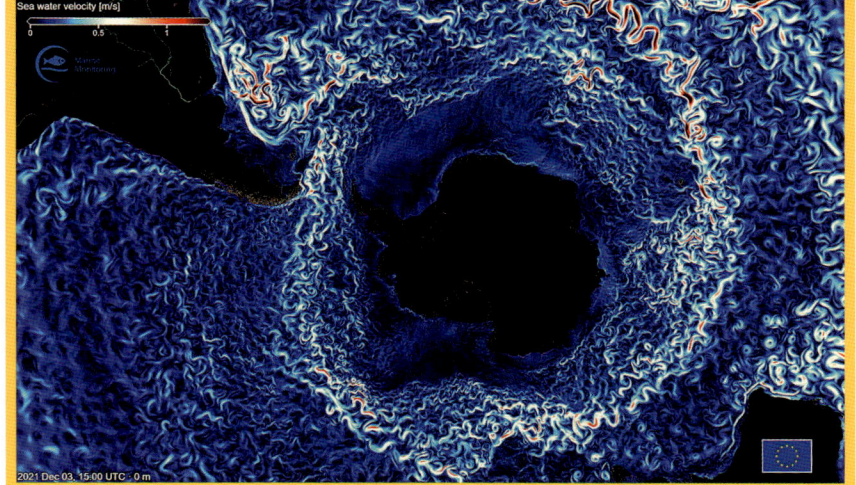

Syukuro Manabe

Born on Shikoku island, Japan, in 1931, Syukuro Manabe rejected his family's wish that he train as a doctor, studying meteorology instead, and gaining his doctorate from the University of Tokyo in 1959. He then moved to the US, where he worked until 1997, and where he produced his first climate model.

From 1997 to 2001, Manabe was Director of the Global Warming Research Division of the Frontier Research System for Global Change in Japan. He returned to the US in 2002 as a research collaborator at the Program in Atmospheric and Oceanic Science at Princeton University, where he also served as senior meteorologist. In 2021, Manabe shared the Nobel Prize in Physics for his work on climate modelling.

Key works

1967 "Thermal equilibrium of the atmosphere with a given distribution of relative humidity" (with Richard Wetherald)
1969 "Climate calculations with a combined ocean-atmosphere model" (with Kirk Bryan)

270 CLIMATE MODELLING

Modelling the planet

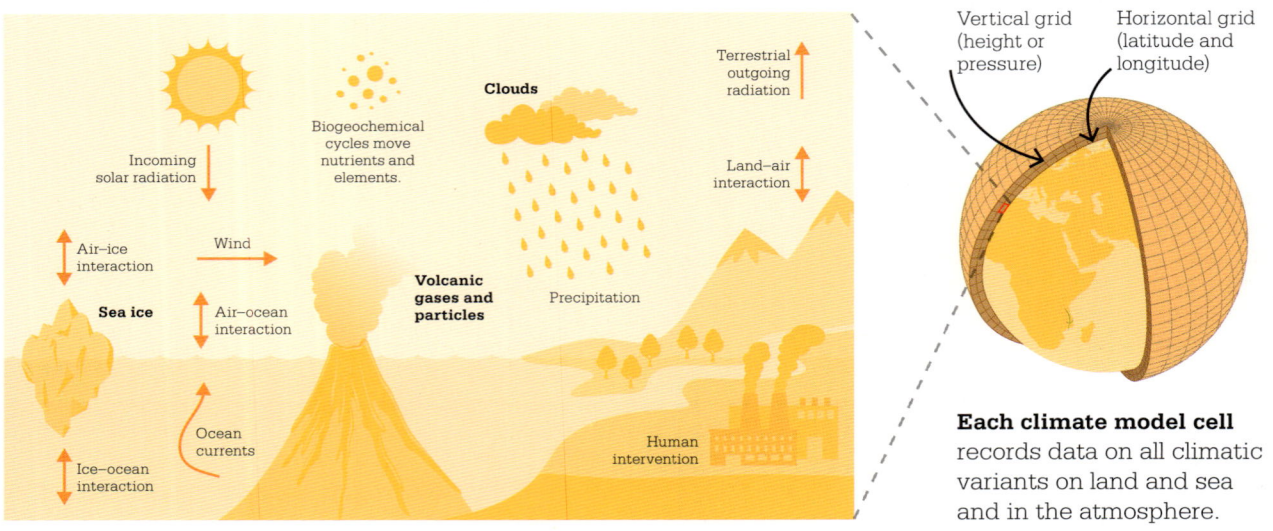

Each climate model cell records data on all climatic variants on land and sea and in the atmosphere.

modelling Earth's climate and predicting the impact that CO_2 might have on it.

In 1969, Manabe made a further advance in climate modelling; with American oceanographer Kirk Bryan, he created the first "coupled" model, combining a model of the ocean (which absorbs most of the solar energy that reaches Earth) with one of the atmosphere. This enabled scientists to model the most complex interactions in the climate system and to make more accurate climate predictions.

How models work
Climate models are based on the circulation of gases – air (mostly oxygen and nitrogen) and water vapour – in Earth's atmosphere. The model recreates the planet's atmosphere, dividing it into a three-dimensional (3-D) grid made up of thousands of cube-shaped cells. Every cell has lateral coordinates – longitude and latitude – as well as a vertical coordinate that gives its altitude.

Each cell is also represented in the model by a series of data points, including values for temperature, pressure, humidity, cloud cover, precipitation (such as rain or snow), solar radiation, and terrestrial radiation – the heat being sent back into the atmosphere from the planet's surface. Based on mathematical equations, these data points create a snapshot of the conditions of the modelled air within the cell.

As the models get ever more complicated… no one person can appreciate what's going on inside them.
Syukuro Manabe
Carbon Brief website, 2015

A cell's coordinates link it to a position above a surface, most commonly that of the ocean but also of land masses. The section of the surface will have its own, distinct set of data. For example, frozen polar seas and dark tropical oceans each have different impacts on the air above them, as do humid forests, arid deserts, and, as is increasingly becoming apparent, urban centres.

Interaction and resolution
Models turn Earth's climate system into millions of data points, simulating the way that energy transfers between air, water, land, and ice. From this starting point, the model moves through time, with the data changing as each new equation is solved. To explore how certain external influences affect the climate system, climatologists introduce a "forcing" into a cell's equation, such as a volcanic eruption or a change in greenhouse gases due to human activity. Results from each cell in the grid are then

APPLIED GEOGRAPHY

passed to neighbouring cells, where new sets of data points will have a further effect on the changes in conditions over time.

The level of accuracy in the predictions of a climate model relies largely on its resolution – defined by the size of the cells. The smaller the area represented by the cell, the greater and more precise the detail. The more grid cells there are, the greater the number of mathematical calculations, which demands huge computer power. As computer technology has developed, so the levels of climate model resolution have increased. At the start of the 1990s, a climate model used square cells with sides of around 500 km (310 miles). By the mid-1990s that measurement had halved to 250 km (155 miles), and by 2007 had been reduced to 110 km (68 miles). During that period, more climatic variations were also added to the calculations, including ocean currents, river flows, and vegetation changes.

Time and testing

Model resolution is also affected by the period of time between each snapshot of evolving data, known as the "time step". This period could be hours, days, or years, depending on the size of the grid cells, with the time step reduced to minutes for the smallest cells. The smaller the steps, the greater the detail and accuracy in the model.

Manabe pioneered a testing process called "hindcasting", a crucial factor in model accuracy. Hindcasting involves running the model backwards from today's starting conditions, with the input of historical data such as known temperatures, greenhouse gas levels, and other atmospheric conditions. A model that accurately simulates the actual observed climate data of the past decades or centuries can, when run forwards, be assumed to generate reliable data about the future of the climate.

Modelling the future

The latest generation of models are called "Earth system models". These embrace the fundamental exchanges of energy between land, air, and ocean, but also factor in the effects of human activity, such as fossil-fuel emissions and agriculture. They also account for biogeochemical cycles, which describe the interactions of chemical elements such as carbon, nitrogen, and phosphorus in Earth's physical system. The complexity of this data demands even greater processing power, such as that provided by the UK Met Office's cloud-based supercomputer, the world's first dedicated to weather and climate science. Launched in 2025, it can perform 14,000 trillion calculations per second and produce model predictions down to 1.5 km (0.9 miles) resolution. Such precision increases climate scientists' ability to predict extreme weather events, such as floods and droughts. ∎

> We need to be adding more processes… modelling finer details so we can better explain the climate.
> **Dr Doug McNeall**
> Climate scientist, UK Met Office, 2018

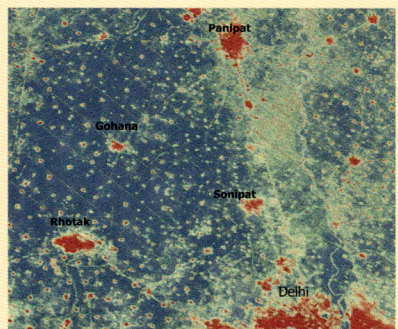

A satellite image taken at night on 5 May 2022 shows heat islands (red) in and around Delhi, India, where the temperature was around 4.4°C (7.9°F) higher than in nearby rural areas.

Heat islands

The impact of climate change on urban centres is especially significant. These are the places where more than half the world's population live, and where the average rise in temperature is higher than in rural or wild places. This creates phenomena called "heat islands", formed when a high density of buildings absorbs and holds heat from the Sun. That is then compounded by the heat generated through consumption of fuels for air-conditioning, transport, and other energy-intensive activities. Since the 1970s, the heat island phenomenon has been the focus of research by Canadian geographer Timothy Oke – an expert on urban microclimates. He has explored the factors that contribute to urban heat islands, such as impervious surfaces, poor building design, and lack of vegetation. In 2012, he helped to introduce the concept of the local climate zone (LCZ). Categorized into 17 classes, these are used to understand and analyse the urban heat island (UHI) effect.

WE MUST DESIGN WITH NATURE
ENVIRONMENTAL IMPACT ASSESSMENT

IN CONTEXT

KEY ORGANIZATON
US Environmental Protection Agency (1970)

BEFORE
1953 American ecologists Eugene P. and Howard T. Odum popularize the idea of the ecosystem and human interconnectedness with the environment.

1962 In the book *Silent Spring*, American conservationist and biologist Rachel Carson raises awareness of environmental dangers posed by pesticides.

AFTER
1972 In Stockholm, the UN adopts 26 principles focusing on sustainable development and the preservation of the human environment.

1989 The World Bank adopts procedures that require EIAs for its funded projects.

1990s Around half of Sub-Saharan African nations develop legislation on EIAs.

In the US, the 1960s saw growing public concern over ecological damage caused by human activities such as farming and industry. Public and political support for environmental legislation was galvanized, and in January 1970 the US Congress passed the National Environmental Policy Act (NEPA). Sponsored by President Richard Nixon, the law promoted protection of the environment against degradation caused by public projects such as roads and power plants. Other nations soon followed suit. Then, in December 1970, Nixon established the US Environmental Protection Agency (EPA), in part to enforce NEPA.

US President Richard Nixon greets William D. Ruckelshaus (right), the first administrator of the EPA, as he is sworn in at a White House ceremony on 4 December 1970.

One of NEPA's stipulations was that an environmental impact assessment (EIA) process should be carried out to estimate the effects any new development might have on the natural world. Based on a review of its findings, a project could be stopped before any environmental damage was done, changed in some way to lessen or negate that impact, or have mitigations added to ensure any losses to the ecological system were restored over time.

Making an assessment

The EIA process is usually carried out over five stages – screening, scoping, reporting, application submission and consultation, and decision-making. Screening happens as a project is designed or planned, with a developer gauging whether the works may have an environmental impact and therefore require approval as part of an EIA process. The developer will then "scope" opinion from relevant authorities and agencies to decide what environmental aspects an assessment needs to cover.

Next, the developer prepares an environmental statement. Its contents may include extensive

APPLIED GEOGRAPHY 273

See also: The tragedy of the commons 220–21 ▪ Silent Spring 240 ▪ Sustainable development 244–45 ▪ The environmental justice movement 246–47 ▪ Globalization 250–53 ▪ Monitoring environmental change 284–89

People protest in Washington, DC, in 2013 against the proposed Keystone XL oil pipeline between the US and Canada. The project was cancelled in 2015 after an EPA environmental review.

primary survey work – carried out by a variety of environmental experts – of landscape, soils, ecology, and traffic. The statement then becomes part of the overall application for project approval.

The statement is submitted to the relevant body, such as a local or government planning authority, to be reviewed and, if necessary, clarified. This stage may also include a public consultation. Based on the statement review and any consultation processes, the deciding body will accept or reject the project, or may request changes, in which case the developer will have to decide whether or not to proceed.

Spreading protection

EIA systems are now present in more than 190 countries. They are, however, mostly limited to the potential impacts of specific projects at specific locations, focusing on single developments, and may not cover the potentially wider environmental impacts of product manufacture, resource extraction, or service provision.

Since the 1980s, interest has grown in life-cycle assessments (LCAs). These are frameworks that allow manufacturers or service providers to assess the environmental impact of a new product or service, tracing it from raw materials, through development, production, and distribution, to recyclable and non-recyclable waste. Although they are mostly voluntary and lack the legal force of an EIA, LCAs are increasingly becoming part of the environment protection methods of governments and organizations. ■

Restoring nature to its natural state is a cause beyond party and beyond factions.
Richard Nixon
State of the Union address, 1970

Human actions **can harm nature** and its ecosystems.

→

To protect the environment, industrial, construction, and engineering **projects must be assessed and monitored**.

↓

EIAs preserve the environment and help to restore natural resources and landscapes.

←

Every new development should be subject to **environmental impact assessment (EIA)**.

THE NEW MAGIC
GEODEMOGRAPHIC ANALYSIS

IN CONTEXT

KEY FIGURES
Jonathan Robbin (1929–2019),
Richard Webber (1947–)

BEFORE
1925 In their book *The City*, American sociologists Robert E. Park and Ernest Burgess emphasize the link between social characteristics and geographical location.

1949 American sociologists Eshref Shevky and Marilyn Williams use three "social axes" to characterize areas and patterns in Los Angeles.

AFTER
1990s The growth of Internet usage allows people's social characteristics and behaviours to be monitored through their Internet Protocol (IP) address.

2017 The launch of Geospatial artificial intelligence (GeoAI) provides new levels of fast, in-depth data analysis.

Geodemographics is a fusion of demography and geography. Demography is the study of human populations through statistics, such as birth rate and ethnicity. It examines how and why populations change and finds meaning in the way they can be divided into subgroups, for example by gender, age, or education. Geodemographics links all this information to where populations live and work, to reach a deeper understanding of people and their neighbourhoods.

Until the mid-20th century, the merging of population detail with place was part of a movement to understand and improve living

APPLIED GEOGRAPHY 275

See also: Thematic mapping 34–37 ▪ Geographic information systems 52–59 ▪ Demographic transition model 176–81 ▪ Censuses and population geography 188–89 ▪ Mapping social injustice 278–79 ▪ Spatial data analysis 280–81

Analysts process data in a US Census Bureau computer room in the 1980s. Computerization of records paved the way for a much deeper understanding of populations.

Many **forms of data** hold information about **households**.

That data links **social, medical, and financial** information to a **specific location**.

Analysis of the data places people in distinct **clusters or segments**.

Understanding clusters and segments allows governments, businesses, and other organizations to **reach people** more effectively.

conditions. In the 1970s, statistical and technological innovations, powered by computer development, took geodemographics in a new direction. This was pioneered especially by American sociologist Jonathan Robbin and British geographer Richard Webber, who created statistical models that enabled populations to be divided into any number of "clusters" and "segments", transforming not only local and national social planning but also how businesses identified customers through marketing.

Social roots and censuses
Geodemographics grew in the 19th century out of the work of social reformers, who wanted to identify and help people living in poverty in the booming industrial cities. In London, for example, British social reformer Charles Booth studied the city street-by-street, recording the social and economic status of the inhabitants on a series of "poverty maps", using colour coding to show the location of all classes of people.

Also in the 19th century, the first national censuses were introduced in countries such as Britain and France to count and locate their surging populations for taxation, food supply, and other government-related purposes. From the early 20th century, census authorities began to expand the data captured in their population surveys, and sociologists used the increasing amount of census data to analyse the problems, such as crime and immigrant poverty, caused by rapid urbanization. From this, it became possible to group, or "segment", people according to place and activity or social feature.

Postal code patterns
From the 1950s, computers began to bring greater speed, efficiency, and analytical power to census data gathering. In the US, the 1970 census was the first to provide easily accessible electronic data files. It also introduced more precise information about the size and location of populations within specified "blocks". Jonathan Robbin, who founded the Claritas market research company in 1971, saw the potential of linking this data to the US Postal Service's Zone Improvement Plan (ZIP), which had introduced a nationwide system of postal (ZIP) codes in 1963. He combined demographic data from more than 200,000 blocks with the 36,000 postal areas. From his analysis of the results, he produced 40 consumer lifestyle categories, or "clusters" – naming his system Potential Rating Index »

276 GEODEMOGRAPHIC ANALYSIS

Richard Webber

Born in England in 1947, Richard Webber gained master's degrees between 1969 and 1973 in economics and transport and town planning. After working as a researcher for the UK-based Centre for Environmental Studies (CES), he set up the micro-marketing division of the CACI company, where, in the late 1970s, he developed the Acorn geodemographics system. He was managing director of Experian from the launch of the Mosaic system in 1985 until 2001.

In 2005, Webber founded OriginsInfo, a geodemographics analysis tool based on people's first or last name, and in 2014 joined the British broadcaster, writer, and former politician Trevor Phillips in founding Webber Phillips, a company that offers analysis of cultural, ethnic, and linguistic characteristics of target groups. Since 2016, he has been visiting professor at the University of Newcastle.

Key work

2018 *The Predictive Postcode: The Geodemographic Classification of British Society* (with Roger Burrows)

for ZIP Markets (PRIZM) in 1974. This allowed businesses, particularly those that sold their products by direct mail, to reach consumers much more effectively.

At the same time, in the UK, Richard Webber was working as a researcher on the Liverpool Social Area Study, which used 1971 census data to analyse how the city could better use government-funded projects to end social deprivation. Examining the various social, economic, and spatial indicators, Webber identified clusters of deprivation previously missed by the local authorities. Webber extended his system of cluster analysis from the local to the national level to produce A Classification of Residential Neighbourhoods (Acorn), a marketing tool, which, from 1975, was further developed by Webber and the American information technology company CACI to segment the UK's neighbourhoods and households into categories, groups, and types. In the 1980s, Webber joined with information services company TRW (later named Experian) to develop the Mosaic geodemographic system, which used a wider range of data in order to identify individual people within households.

Analysing clusters

To identify people who conform to set categories, geodemographic analysis uses an algorithm called "k-means clustering", which gathers sets of data points into meaningful classifications based on how similar or close those points are to each other. The "k" in "k-means" relates to the number of categories, or clusters, that the algorithm is asked to find. For example, an online retailer might want to group its

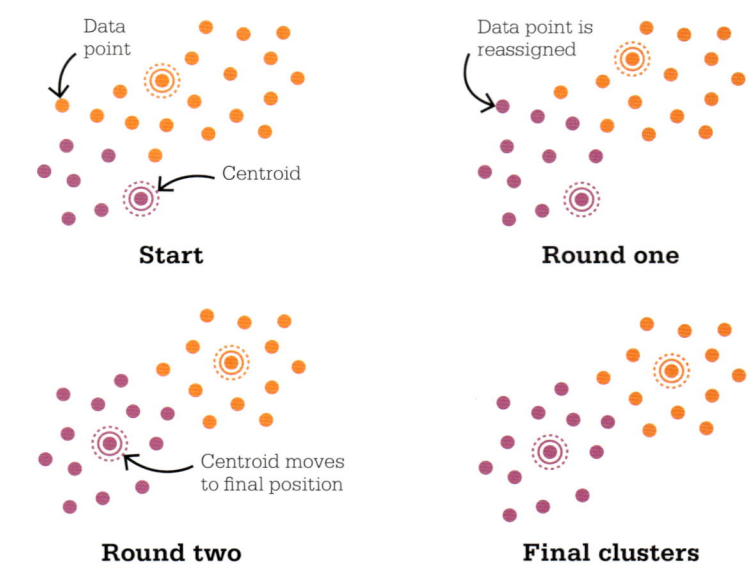

The k-means clustering algorithm starts with sets of data points and random centroids. In round one, the centroids are recalculated (based on the statistical average – the centre – of each set of data points) and the data points assigned to their closest centroid. In round two, the centroids move again to the average position, which in this case becomes the final iteration as any further changes would have no effect.

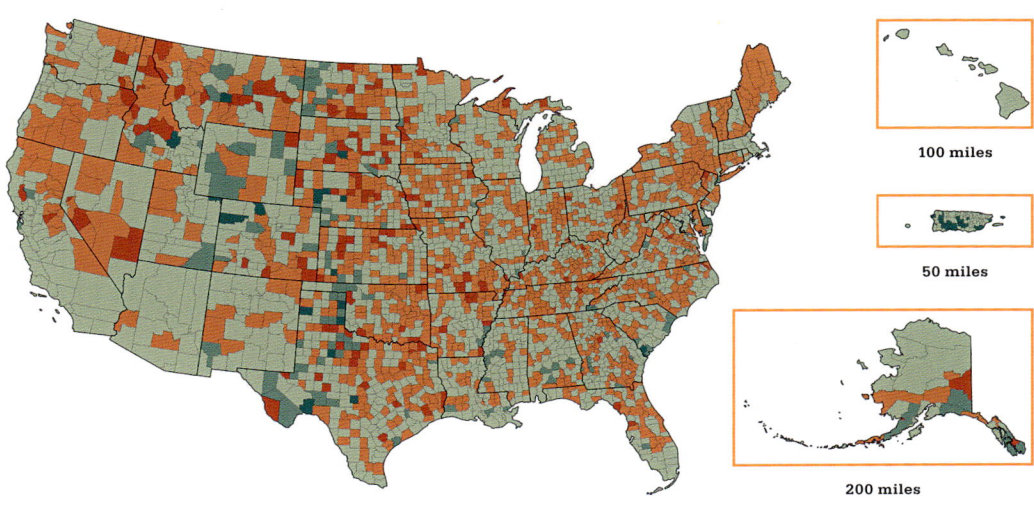

A map produced by the US Census Bureau map shows US counties segmented by changes in median age in 2020–21, providing spatial insights into longevity trends.

**Key:
Change (in years)**
- 1.1 or more
- 0.6 to 1.0
- 0.1 to 0.5
- -0.4 to 0.0
- -0.5 or more

National median age change was 0.3

customers into three clusters – frequent, budget, and big spenders. The algorithm first selects random points, called centroids, and each data point, such as a customer name or location, forms a cluster around the closest centroid. The centroids are then recalculated based on the mean (average) position of the points in each cluster, with the process repeated until the centroids and clusters become fixed. Compact and distinct customer clusters could then allow an online retailer to tailor its marketing strategies to the most relevant customers.

Segmentation

Using geodemographic analysis to classify households is called "segmentation". This has become a powerful marketing, customer service, product development, and strategy tool for businesses and public authorities. The systems created by Jonathan Robbin and Richard Webber – PRIZM, Acorn, and Mosaic – have remained at the forefront of segmentation.

Two PRIZM systems operate in North America. In the US, PRIZM places every adult into one of 68 segments, from "Upper Crust" to "Bedrock America". In Canada, there are 67 categories, from "The A-list" to "Just Getting By". Built from census data, the classifications represent a socioeconomic rank, which indicates both spending power and, to some extent, the interests of the group. The rank is derived from data such as age, income, property, education, number of dependents, and media and technology use.

Acorn, which is specific to the UK, takes a broader approach, adding open-source data to the "official" data from the national census held each decade. It divides households into seven categories, based largely on disposable income, from "Luxury Lifestyles" to "Low Income Living" and "Not Private Households", which covers situations such as prisons, care homes, and student accommodation.

Mosaic, like Acorn, uses census and open-source data, but covers a wider range of consumer, financial, and lifestyle information to produce more detailed household and individual segmentation. Although developed for the UK, the system has been adapted by Experian for many other countries, including Australia, Denmark, France, Germany, and Spain. With access to geodemographic data from multiple countries, Experian can then segment populations across the world, allowing even further marketing reach.

Data developments

Since the start of the 21st century, many new data sources have been integrated into geodemographic analysis, generated particularly by the World Wide Web. Vast amounts of complex digital information – "big data" – can be collected from social media, mobile-phone records, credit card transactions, and other real-time data, and tied to specific locations. Much of the segmentation and micro-marketing pioneered by PRIZM and Acorn is now used by personalization algorithms to tailor online experiences to every computer and smartphone user.

The development of artificial intelligence (AI) has also added machine-learning techniques to geodemographics. This allows systems to analyse data sets "unsupervised" – without being explicitly programmed – to find clusters and spatial relationships. ∎

JUSTICE HAS A GEOGRAPHY
MAPPING SOCIAL INJUSTICE

IN CONTEXT

KEY FIGURE
Laura Pulido (1962–)

BEFORE
1880s British social reformer Charles Booth uses census data to create colour-coded maps that highlight wealthy and poorer areas of London.

1900 In charts, graphs, and maps, Black American intellectual W.E.B. Du Bois and his team illustrate the reality of Black lives in America for the Paris Exposition in France.

AFTER
2022 In *Abolition Geography*, American scholar Ruth Wilson Gilmore studies links between where people live and their likelihood of being imprisoned.

2024 Urban sociologist Eric Klinenberg publishes *2020: One City, Seven People, and the Year Everything Changed*, examining social disruption experienced by New Yorkers during the Covid-19 outbreak.

In the early 1970s, British academic Richard Webber was commissioned to use census data in Liverpool, UK, to help those in most need of state aid. His pioneering work marked the start of geodemography and led to the development of tools in the UK and US that analyse the demographics of residents based on their postcodes.

In the late 1980s and 1990s, a new wave of critical geographers began to delve beyond home addresses to uncover cultural and ideological power structures, racism, and other forms of prejudice that determine the characteristics of underprivileged neighbourhoods. Since the 1990s, American geographer Laura Pulido has been a prominent figure in this approach.

Prejudice and power
Pulido focused initially on environmental racism in Los Angeles. Her paper "Rethinking Environmental Racism: White Privilege and Urban Development in Southern California" (2000) noted how industrial pollution and waste facilities have been historically sited close to poor Latino and Black communities. Pulido showed that this was not accidental but due to a long history of discriminatory planning, zoning, and economic policies, proving that inequality is not a natural phenomenon but created by territoriality and the exercise of power unjustly – often along racial and cultural lines.

Analytical tools
Many organizations and some governments have developed tools that can map complex inequalities using geographic information systems (GIS). In the US, for example, the Spatial Equity Data Tool, developed by the Urban Institute in Washington, DC, shows

Studying environmental racism is important for an additional reason: it helps us understand racism.
Laura Pulido
"Rethinking Environmental Racism", 2000

See also: Possibilism 157 ▪ Territoriality theory 202–03 ▪ The environmental justice movement 246–47 ▪ Geodemographic analysis 274–77

Demographic data can be used to **map poverty and deprivation**.

⬇

Such **maps do not explain why** areas have **different levels** of services and support.

⬇

Critical geography analyses factors such as racism, class, and power **to draw conclusions**.

⬇

Deprivation and inequality are due to social injustice.

---

how resources are distributed across different areas. It permits anyone to upload data, which can then be analysed and presented in map form or via charts. The US Environmental Systems Research Institute (Esri) has a variety of GIS software that can map topics such as health needs and voter access, while the Voting Rights Lab monitors polling access, new voter ID laws, and instances of gerrymandering (the manipulation of electoral boundaries).

Beneficial geodemography that can identify environmental social injustice is now widespread. For example, Canada maps the health facility access of its Indigenous First Nations peoples, while bodies such as the Observatório das Metrópolis in Rio de Janeiro and the Brazilian NGO Instituto Pólis in São Paulo produce GIS-driven social justice research and focus on metropolitan challenges. GIS maps on NITI Aayog, India's national portal, are designed to improve the allocation of resources to areas in need. In Africa, Nigeria's GRID3 project uses high-resolution satellite imagery and local data to identify vulnerable communities and pinpoint their education, nutrition, and infrastructure requirements.

Accessible visual data

Multiple classification systems and geodemographic tools can now collect and visualize population data. However, while potential social injustices may be more easily identified, as Pulido points out, it takes political will to resolve them. ■

Laura Pulido

Laura Pulido was born into a working-class family in Los Angeles, US in 1962. She has talked of her fascination with maps and landscapes from an early age and her increasing awareness of the impact of race, class, and gender politics. After her 1991 PhD in Urban Planning, she taught for many years at the University of Southern California, where she pursued her research into environmental justice and racism. In 2020, Pulido moved to the University of Oregon to become the Collins Chair and Professor of Indigenous, Race, and Ethnic Studies and Geography at the University of Oregon. There she has continued to study the relationship between race, place, and environmental processes, recently examining how white supremacy and white nationalism impact climate change.

Key works

1996 *Environmentalism and Economic Justice: Two Chicano Struggles in the Southwest*
2006 *Black, Brown, Yellow, and Left: Radical Activism in Los Angeles*

THE IMPORTANCE OF "WHERE"
SPATIAL DATA ANALYSIS

IN CONTEXT

KEY FIGURE
Luc Anselin (1953–)

BEFORE
1950 Australian statistician Patrick Moran introduces Moran's I, a way of measuring the clustering or dispersal of data variables across locations.

1969 British geographers Andrew Cliff and Keith Ord set out a framework for calculating "spatial autocorrelation".

AFTER
2006 Amazon Web Services (AWS) cloud platform vastly expands access to the storage and processing of spatial data.

2010s Geospatial artificial intelligence (GeoAI) accelerates the speed and precision of spatial data analysis.

2021 British geographer James Cheshire and American designer Oliver Uberti publish *Atlas of the Invisible*, an infographic world atlas based on spatial data analysis.

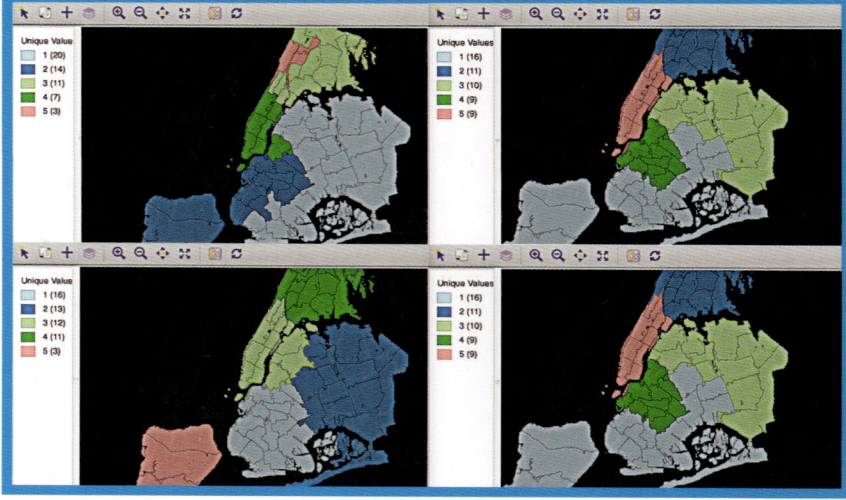

Spatial data is any kind of data that also has a location component, measured by geographic coordinates. Spatial data analysis evolved in the 1960s with the development of the data-mapping geographical information system (GIS) and advances in the understanding and measurement of spatial autocorrelation – the concept that values for data variables near to each other tend to be more similar than those further apart.

In the 1980s, Belgian geographer Luc Anselin saw the potential of bringing statistical data and

GeoDa analysis software can impose different spatial constraints on data sets to produce data visualizations, such as different clusters of property renters in New York City, above.

geographical information systems (GIS) data together to examine the influences of location and spatial interrelation in economics and beyond.

A new way to view data

Spatial data analysis (SDA), or geospatial analytics, takes data – the "what" and the "where" – and,

APPLIED GEOGRAPHY 281

See also: Thematic maps 34–37 ▪ The global positioning system 48–51 ▪ Geographic information systems 52–59 ▪ Time-geography 266–67

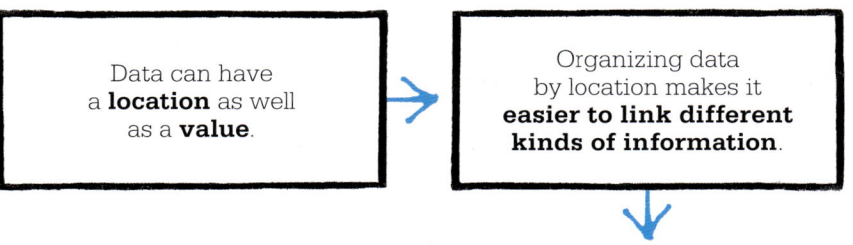

Location-linked data can be analysed to reveal insights and help decision-making.

using software and other analytical tools, interprets it to extract actionable information. Belgian economist Jean Paelinck was the first to see the value in close study of such data. In 1974, he introduced the term "spatial econometrics" to describe how economic factors, such as real estate values, transportation accessibility, and agricultural practices, are affected by spatial relationships and geographical location. This became known as neighbourhood effects.

By the 1980s, GIS capabilities had expanded. Luc Anselin identified a lack of specialized software able to take advantage of complex datasets, so developed his own. Released in 1991, SpaceStat allowed users to visualize data, display it on maps and graphs, and perform statistical analyses to identify spatial patterns. This evolved in 2003 into the free, open-source GeoDa software, which by 2023 had more than 630,000 users.

The power of place

With GeoDa, data is collected from historical and real-time sources, including maps and surveys, remote-sensing imagery, and the global positioning system (GPS). The data is then analysed with a variety of computer-based tools, such as GIS software, statistical software, and visualization programmes.

Analysts then employ a range of techniques to find patterns and relationships in the data. For example, sets of data can be overlaid to find interactions, such as comparing flood-prone areas to population densities to identify communities most at risk. Zones can be created around specific features – known as buffer analysis – to see how they relate to neighbouring areas, such as the distances between a new housing estate and vital medical amenities. Areas with particularly high data values, known as hotspots, can also be studied to gain relevant insights, such as high footfalls that could influence the siting of shops.

In recent years, spatial data analysis has become an increasingly important source of information for policymakers and those involved in fields such as urban planning, health provision, and agriculture. As artificial intelligence (AI) aids the rapid development of data collection, storage, and visualization technology, spatial data analysis will become even more influential. ▪

Infographics

Spatial data analysis has reached a wide public audience through computer-generated thematic maps known as infographics. Marrying complex spatial data with inventive graphic design, they convey information on socially relevant subjects such as climate change, population, agriculture, tourism, and consumer behaviour in an accessible and visually striking way.

The concept of visualizing data dates back to the 19th century, but advances in graphics software in the 1980s heralded a rapid popularization of infographics, informed by design principles set down by American statistician Edward Tufte in his 1983 book *The Visual Display of Quantitative Information*. Following Tufte's emphasis on clarity, precision, and efficiency in data visualization – and the avoidance of what he called "chartjunk" – infographics have become an integral part of data communication.

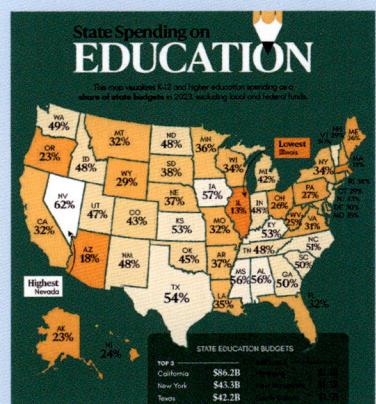

An infographic map of the US shows the budget percentage that each state spends on education, making obvious the spatial differences among states.

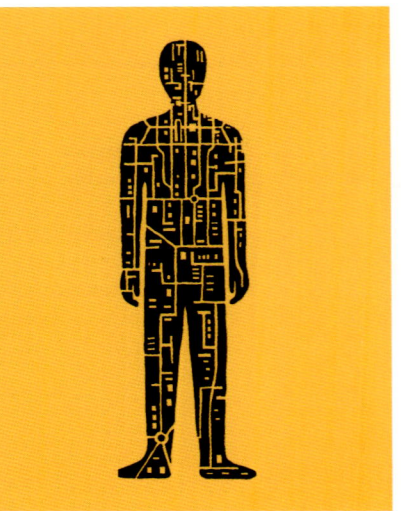

FIRST WE SHAPE THE CITIES – THEN THEY SHAPE US
THE "SMART GROWTH" MOVEMENT

IN CONTEXT

KEY FIGURE
Peter Calthorpe (1949–)

BEFORE
1958 American engineer John E. Arnold launches Stanford University's design programme and advocates for "creative thinking" in design that prioritizes human needs.

1975 The Oregon Experiment proves the success of an organic, user-driven approach to architectural planning.

AFTER
2002 Smart Growth America is formed – a national organization that promotes Smart Growth initiatives across the US.

2007 Smart Growth UK, an informal coalition, is formed by a group of national NGOs and individuals interested in promoting Smart Growth principles for planning, transportation, and community development in the UK.

New Urbanism is a theory of urban planning that seeks to create compact neighbourhoods, where people live and work, and where amenities, such as shops, schools, and leisure areas, are all within easy reach on foot, by bike, or using public transport. The idea began to emerge in the US from the late-1970s and became known as "Smart Growth" in the 1990s. It was conceived as a response to the issue of post-war suburban sprawl, which had spread over vast areas of land and necessitated the use of private cars for almost every journey, even local ones. Urban planners and architects began to see the 19th-century model applied to European cities as providing a solution.

In 1991, the Local Government Commission in Sacramento, California, invited a group of architects, urban planners, and elected officials to a conference at Yosemite's Ahwahnee Hotel to discuss ideas for making cities as clean, walkable, and livable as

Seaside, Florida, one of the first New Urbanist towns, prioritized public spaces, such as a town centre and pedestrian-oriented streets.

APPLIED GEOGRAPHY 283

See also: Sustainable development 244–45 ▪ Environmental impact assessment 272–73 ▪ Land use zoning 264–65

Smart Growth City: Copenhagen

Known as one of the most bicycle-friendly metropolises in the world, Copenhagen is often cited as an exemplary model for Smart Growth. While the city missed its ambitious target to become carbon-neutral by 2025, it still aims to reach this goal in the near future.

Copenhagen's infrastructure utilizes technology to optimize efficiency and sustainability. The city has implemented initiatives not only with citizen engagement but also in collaboration with universities, businesses, and government. Investment in public transport, alongside prioritizing cycling as a mode of transport and preserving green spaces, has reduced air pollution and enhanced the quality of life for residents. In the modern district of Ørestad, the UN17 Village, comprising 400 homes and shared co-living areas, incorporates all 17 of the UN's Sustainable Development Goals into one housing project, the first of its kind.

Copenhagen's cycle lane network contributes to cleaner air by reducing reliance on cars and their associated emissions.

---

possible. This resulted in the Ahwahnee Principles for Resource-Efficient Communities.

Many professionals who had attended the conference, such as Peter Calthorpe, Andres Duany, and Elizabeth Plater-Zyberk, set up the Congress for the New Urbanism only two years later, in 1993. The same year, Peter Calthorpe set out his vision for transit-oriented development (TOD) in his book *The Next American Metropolis*, defining it as "a mixed-use community that encourages people to live near transit services and to decrease their dependence on driving".

Plans for "Smart Growth"

Throughout the 1990s, a consensus formed in the US regarding the need for sustainable and efficient urban development. As such, it absorbed many of the core tenets of New Urbanism. The US Environmental Protection Agency partnered with several non-profit and government organizations to create the Smart Growth Network, which published its "Smart Growth Principles" in 1996 and distributed the "Growing Smart Legislative Guidebook" on behalf of the American Planning Association.

In 1997, the Governor of Maryland, Parris N. Glendening, initiated "Smart Growth and Neighborhood Conservation", which involved a series of laws and budgetary measures aimed at directing state resources towards revitalizing older neighbourhoods, reducing sprawl, protecting green spaces, and enhancing communities. This legislation established a framework that was soon taken up nationwide.

The concept of Smart Growth soon spread through the influence of international conferences and publications, and as a result of the global focus on sustainability. In 2016, the Paris Agreement was ratified by 196 parties at the UN Climate Change Conference (COP21). In its commitment to achieve emissions reduction and climate resilience it promoted many Smart Growth principles. ∎

IT'S HARD TO MANAGE WHAT YOU CAN'T MEASURE

MONITORING ENVIRONMENTAL CHANGE

MONITORING ENVIRONMENTAL CHANGE

IN CONTEXT

KEY ORGANIZATION
Global Forest Watch (1997)

BEFORE
1948 The UN Food and Agriculture Organization (FAO) publishes the first survey of the world's forests.

1972 NASA launches the Earth-observing Landsat 1 satellite, allowing the close monitoring of deforestation.

1972 The UN Environment Programme (UNEP) is founded to monitor deforestation and other environmental concerns.

AFTER
2008 The UN Programme on Reducing Emissions from Deforestation and Forest Degradation (UN-REDD) is launched to help tackle climate change due to forest activities.

2021 Leaders of more than 140 countries sign the Glasgow Leaders' Declaration on Forests and Land Use to halt and reverse forest loss by 2030.

As well as retaining massive quantities of clean, fresh water and having the richest biodiversity on Earth, the world's rainforests, particularly those in the Amazon and Congo basins, both absorb and store vast amounts of carbon dioxide (CO_2). Where large areas of forest are lost, known as deforestation, the land's capacity for absorbing CO_2 is reduced, and CO_2 stored in dead trees is released. As a result, deforestation has become one of the world's main drivers of increased atmospheric greenhouse gases and climate change.

To monitor the scale, location, and causes of deforestation, scientists need to gather accurate, independent, and verifiable data. Some of the most crucial data is provided by the Global Forest Watch (GFW), established in 1997 by the World Resources Institute. A non-governmental organization, it surveys deforestation, especially in the tropics – the region of Earth surrounding the equator – where 94 per cent of unsustainable forest clearance occurs.

Satellite monitoring

Earth's most important tracts of tropical forest are often in remote places, so obtaining accurate data

> From now on, the bad guys cannot hide and the good guys will be recognized for their stewardship.
> **Dr Andrew Steer**
> Former President and CEO, World Resources Institute, 2014

about deforestation was, for a long time, almost impossible. This problem was mitigated in the 1970s by the launch of the NASA Landsat satellite programme, which began to capture detailed images of Earth's surface from space, a process known as remote sensing. Further satellite data resources were added with the introduction in 2014 of the Sentinel programme by the European Space Agency (ESA).

GFW harnessed satellite data to produce global and regional forestry maps, including, in 2006, the first global map of intact forest areas, produced in conjunction with the environmental campaign group Greenpeace. Since 2014, GFW has used both Landsat and Sentinel satellite data to monitor forests and track changes, analysing the information through the cloud-based Google Earth mapping platform and making it available in near real time on an interactive online platform. Two of GFW's data

Forest clearance can be easily detected by satellite. This image from Sentinel-2 shows large-scale agricultural deforestation in 2020, in the Amazon region of Pará, Brazil.

See also: Aerial photography and remote sensing 40–45 ▪ Geographic information systems 52–59 ▪ Greenhouse effect 80–83 ▪ Glaciation and ice ages 86–89 ▪ Desertification 228–29 ▪ Climate change 232–39 ▪ Climate modelling 268–71

Tropical primary forest loss

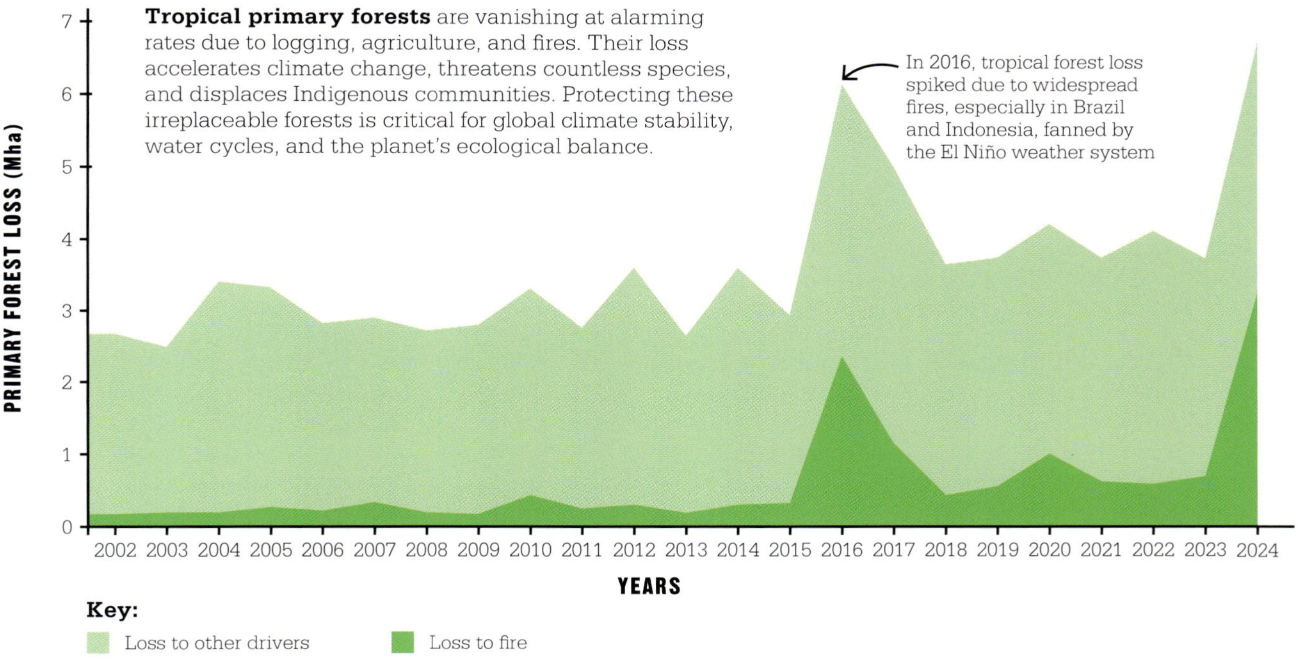
Tropical primary forests are vanishing at alarming rates due to logging, agriculture, and fires. Their loss accelerates climate change, threatens countless species, and displaces Indigenous communities. Protecting these irreplaceable forests is critical for global climate stability, water cycles, and the planet's ecological balance.
In 2016, tropical forest loss spiked due to widespread fires, especially in Brazil and Indonesia, fanned by the El Niño weather system
PRIMARY FOREST LOSS (Mha)
0
1
2
3
4
5
6
7
2002 2003 2004 2005 2006 2007 2008 2009 2010 2011 2012 2013 2014 2015 2016 2017 2018 2019 2020 2021 2022 2023 2024
YEARS
Key:
Loss to other drivers
Loss to fire

sets are maintained in the US by the University of Maryland's Global Land Analysis and Discovery (GLAD) laboratory. One data set monitors tree canopy cover, the other analyses tree cover height, both at an image resolution of 30 m (98 ft), meaning each pixel making up the satellite image represents a 30 m x 30 m (98 ft x 98 ft) area.

GFW has also worked with the GLAD laboratory to develop a system of deforestation alerts, based on Landsat data. This has allowed GFW to produce weekly or monthly warnings of disturbances, natural or human-made, in forest cover, so that local communities and environmental organizations can take action. In 2017, GFW launched Forest Watcher, a mobile software application that allows offline access to forest-monitoring data and tools.

In 2021, GFW launched a third dataset, Tropical Tree Cover (TTC), based on data from the Landsat 8 and Sentinel-2 satellites. It has a ground resolution of 10 m (33 ft), which means it can detect small patches of tree cover, even in urban areas. As is the case with all GFW platforms, the data is provided free for governments, researchers, and the public, enabling rapid analysis of deforestation and wildfires, and land use more generally.

Mapping with lasers

As well as utilizing satellite images, the GFW datasets also incorporate information provided by the NASA Global Ecosystem Dynamics Investigation (GEDI) programme, launched from the International Space Station in 2018. GEDI produces high-resolution observations and measurements of forest structure using LiDAR (light detection and ranging) technology, which allows analysts to "see" what lies beneath the forest canopy.

The GEDI LiDAR system emits laser pulses and notes the time it takes for the pulse to be reflected back from a forest area, providing an extremely accurate measurement of distance. Thousands of pulses per second are emitted, creating a three-dimensional (3-D) image of the forest shape. The faster the pulse rate, the more detailed the image created. LiDAR can provide information on » tree canopy height and density, as well as terrain features important to reforestation projects. It can also estimate the CO_2 or potential of trees for storage of biomass (energy-producing organic matter), monitor logging activity, and assess wildfire risks and post-fire recovery.

LiDAR can also be operated from aircraft, such as planes and drones. In 2024, for example, environmentalists ran a project to appraise the state of rare dry forest in Ecuador's Anna Lotta Biosphere Reserve, which has lost 70 per cent of its original extent, due largely to climate change and human activity. Researchers employed high-resolution LiDAR on board a light aircraft to map the forest and create a vertical profile of its structure. The survey recorded around 1.62 billion laser points, detailed enough to clearly show the canopy, sub-canopy, understorey, and ground surface, and to calculate the above-ground biomass and the CO_2 storage of its trees.

Seeing through cloud

Rainforests are often covered in thick cloud cover, which presents monitoring challenges. While clouds do not affect LiDAR technology on low-flying aircraft, it interferes with the gathering of information by optical sensors on Landsat and Sentinel satellites, which must wait for breaks in the cloud cover. Sporadic monitoring of persistently cloud-covered tracts of forest has meant some alerts were not raised or arrived too late to stop illegal forest clearance or fire breakouts.

Since 2021, GFW has overcome cloud-cover visibility problems by utilizing the Radar for Detecting Deforestation (RADD) system, which is carried by the ESA's Sentinel-1 satellite. RADD sends out radio waves with a wavelength long enough to penetrate even thick cloud cover, so its data can be used to issue alerts to forest monitors in close to real time. Also, RADD's high 10 m (33 ft) resolution means it can detect the harvesting of trees of high value, often felled while the surrounding forest is left alone.

Losses and gains

Since the start of this century, the destruction of the world's rainforests has been largely rising. Data collected by GFW shows that 6.7 million hectares (16.6 million acres)

One-third of forest lost this century is likely gone for good.
World Resources Institute
Press release, 2025

of tropical primary rainforest were destroyed in 2024, an area roughly the size of Panama. This was the biggest annual total for more than two decades, with the largest areas of forest loss in Brazil, Bolivia, and Democratic Republic of the Congo. Massive fires were responsible for much of the destruction, often the result of a changing, hotter climate, although many were started deliberately to clear forest. The global loss of tropical primary forest in 2024 equated to 3.1 gigatonnes (3.4 billion tons) of greenhouse gas emissions, about the same as India's annual fossil-fuel CO_2 emissions.

Skyscrapers in Kuala Lumpur, the capital of Malaysia, are shrouded by a thick cloud of pollution in September 2015 during the "heat haze" crisis.

Environmental detective work

Heat hazes, caused by forest and peat fires, are a regular occurrence in Southeast Asia, usually peaking in the dry season, between July and October. In 2015, a particularly intense "haze crisis" hit the region, when extreme air pollution impacted Indonesia, Malaysia, and Singapore. More than 500,000 cases of acute respiratory tract infection were reported, and as many as 100,000 people died – the majority Indonesians – as a result of the pollution. Many of the fires were started deliberately to clear the land for palm oil and other plantations. Detailed maps produced by GFW revealed that the fires were often linked to specific businesses, especially in the palm oil and pulp and paper sectors. Such real-time data made it possible for legal action and boycotts to be directed at more than 50 plantation companies. This would not have been possible without the real-time data produced by GFW.

Retreating bodies of ice can be followed using satellite images. Here, the Upsala Glacier in Argentina's Los Glaciares National Park has shrunk between 2001 (left) and 2021 (right) due to climate change.

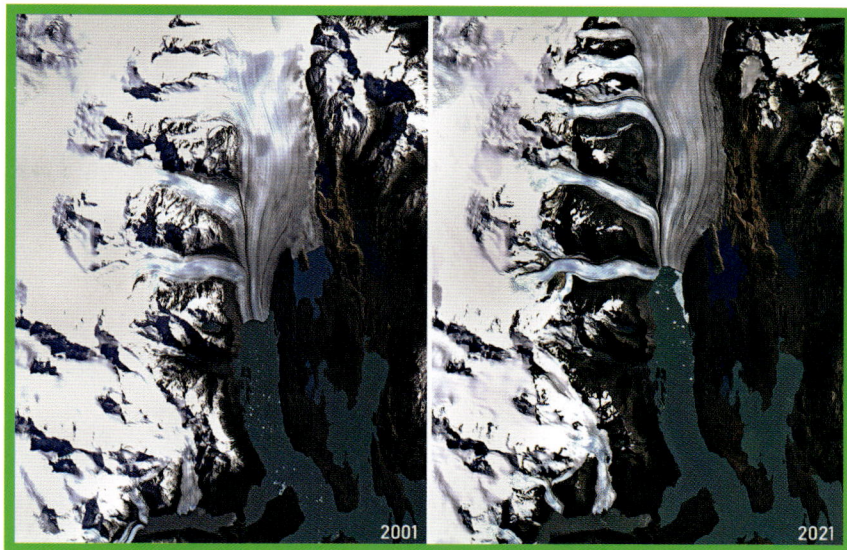

While the pace of deforestation has continued to increase, so has the emergence of forest restoration projects in countries such as Brazil, Bolivia, Mexico, Kenya, Tanzania, Laos, Indonesia, and Papua New Guinea. Since 2011, more than 60 nations have joined the Bonn Challenge – set up by the German government and the International Union for Conservation of Nature (IUCN) – to restore 350 million hectares (865 million acres) of degraded and deforested landscape by 2030. The initiative reached a milestone of 150 million hectares (371 million acres) of pledged restoration in 2017.

Glaciers and ice sheets

The gradual warming of Earth's atmosphere, including the major contribution to greenhouse gases made by deforestation, has dramatically reduced the planet's large bodies of ice since the late 20th century. Arctic sea ice covers a smaller area each year, and the mass of the huge terrestrial ice sheets of Antarctica and Greenland steadily diminishes. Mountain glaciers have retreated over the same period, losing a mass of ice equivalent to a depth of around 27 m (89 ft) of water since 1970 – the same as taking a slice about 30 m (98 ft) deep from each glacier. This in turn influences climate change, since the albedo effect, whereby ice and snow reflect a large proportion of solar radiation back into space, is weakened, adding to atmospheric warming. Melting ice also causes sea levels to rise, threatening to flood low-lying regions. To mitigate flooding and other dangers, scientists monitor changes to Earth's ice sheets and glaciers through a combination of satellite-based remote sensing, aircraft surveys, and surface-based sampling, such as the extraction of ice cores, providing information about past climate conditions.

Some of the most sophisticated ice- and glacier-tracking is done by NASA's Advanced Spaceborne Thermal Emission and Reflection Radiometer (ASTER), launched in 1999 in conjunction with the Japanese government and carried on the Terra satellite – part of NASA's Earth Observing System for exploring the complex climate connections between the planet's atmosphere, land, snow, ice, and oceans. ASTER captures high-resolution images of Earth across a range of wavelengths to produce detailed maps of glaciers and ice sheets. Images and maps are then analysed by NASA's Global Land Ice Measurements from Space (GLIMS) project to create a constantly updated inventory of the world's glaciers and other bodies of ice.

NASA's Ice, Cloud, and Land Elevation Satellite 2, or ICESat-2, launched in 2018, provides further space data. Through its Advanced Topographic Laser Altimeter System (ATLAS), ICESat-2 uses laser light fired at 10,000 pulses per second to produce high-resolution 3-D pictures that allow scientists to measure the elevation of ice sheets, glaciers, sea ice, and oceans. ∎

If all glaciers and ice sheets melted, global sea level would rise by more than 60 m (195 ft).
NASA
NASA Earth Data, 2025

THE SUN AND WIND WON'T BE SENDING YOU A BILL
RENEWABLE ENERGY GEOGRAPHY

IN CONTEXT

KEY FIGURE AND ORGANIZATION
Hermann Scheer (1944–2010), **International Renewable Energy Agency** (2009)

BEFORE
1887–88 American engineer Charles F. Brush builds an early automated wind turbine. The electricity it produces powers 350 lamps in his home.

1903 Wind-power pioneer Poul la Cour establishes the Danish Wind Electricity Society.

AFTER
2029 The year in which, according to the International Energy Agency (IEA), global solar photo-voltaic electricity will overtake hydropower to become the largest global renewable power source.

2030 The year when renewable energy sources are, according to the IEA, expected to account for 46 per cent of global electricity generation.

In the 21st century, many nations have recognized that renewable energy, such as wind- and solar-generated power, is an essential alternative to fossil fuels. Proponents of the technology know, however, that achieving its full potential requires worldwide participation. German politician Herman Scheer, an early key advocate, helped to found the International Renewable Energy Agency (IRENA) in 2009. IRENA's members soon envisioned a global platform for renewable energy that would identify resources by natural geographic and climatic conditions at locations worldwide, with shared data accessible to all. The Clean Energy Ministerial (CEM), founded in 2010 by the US Department of Energy, partnered with IRENA and helped to fund the project. Together they launched a conceptual platform for solar and wind energy in 2012 and the fuller Global Atlas for Renewable Energy a year later.

Worldwide reach
The platform seeks to bridge the gap between wealthier countries that have the means to assess their

Renewable energy sources, such as wind and solar power, are essential alternatives to **harmful fossil fuels**.

→ To promote awareness and encourage investment, **geographic locations** for different energy types **must be identified** worldwide.

↓

A geographic information system (GIS), fed with data, can indicate the best location for **renewable energy resources**.

← **This is renewable energy geography.**

See also: Geographic information systems 52–59 ▪ Biomes and ecological zones 136–39 ▪ Sustainable development 244–45 ▪ Environmental impact assessment 272–73 ▪ Monitoring environmental change 284–89

Herman Scheer

Though born in Wehrheim, Germany, in 1944, Herman Scheer grew up in Berlin. After studying at Heidelberg University, he graduated as a doctor of political science in Berlin in 1979. A long-time Social Democrat, he joined the German Bundestag (parliament) in 1980, representing Baden-Württemberg.

Eight years later, Scheer founded Eurosolar, the European Association for Renewable Energy, which lobbies for worldwide renewable energy sources to replace nuclear and fossil fuels. A driving force in the founding of IRENA, he also continued to campaign in the Bundestag for clean energy. In 2000, he spearheaded the German Renewable Energy Act (EEG), which includes the Feed-in Tariff (FIT) law, enabling those who generate renewable energy to sell it back to the grid at favourable prices. In 2010, Scheer died suddenly in Berlin.

Key works

2005 *A Solar Manifesto*
2011 *The Energy Imperative: 100 Percent Renewable Now*

---

renewable energy potential and those that lack it. As importantly, it aims to promote greater awareness of the world's renewable energy potential, and guide policymakers and investors in these new markets. At its launch in January 2013 at the IRENA Assembly in Abu Dhabi, UAE, the countries showing interest quickly expanded from 22 to 37.

The platform is essentially a geographic information system (GIS) that combines climate data with remote-sensing information about land type, and overlays it all on a world map of current renewable energy infrastructure. It indicates, for example, where winds are constant and have a consistently high speed, such as regions with long coasts facing the prevailing ocean winds, and pinpoints hot subtropical deserts with a high solar power potential due to their many hours of sunshine and aridity.

In 2014, the platform added its first geothermal layer, and from 2018 to 2021, it expanded into hydropower, bioenergy, and marine (tidal) energy. It uses more than 1,000 data sets, upgraded regularly from a range of global sources, including NASA, European meteorological agencies, academic models, and specialized studies. Potential future projects are also identified and planned, such as the African Clean Energy Corridor, with networks of wind, hydro, and geothermal power stretching from Ethiopia to South Africa.

Other atlases

Similar GIS resources now also focus on single forms of energy. The Global Wind Atlas, developed by Denmark's Technical University and funded initially by the CEM and later the World Bank, was launched in 2015. Bratislava-based Solargis developed the Global Solar Atlas, funded by the World Bank Group and launched in 2017. Both are supported by the World Bank's Energy Sector Management

Workers started to build South Africa's first municipally owned solar plant in Atlantis in 2025. It has more than 12,500 solar photo-voltaic panels.

Assistance Program (ESMAP), which helps lower-income countries to develop their energy sectors.

In 2024, renewable energy sources, such as solar and wind power, supplied almost a third of global electricity. A 2022 study by Oxford University in the UK estimated that moving from fossil fuels to solar, wind, and other renewable sources could save the world £10.2 trillion by 2050. ∎

YOU CANNOT LIVE WITHOUT SPACE

REMOTE-SENSING AGRICULTURE

IN CONTEXT

KEY FIGURE
Compton J. Tucker

BEFORE
1930s The US Soil Conservation Service uses aerial photography to monitor soil erosion and land use.

1960s American scientist Robert Colwell helps to develop Landsat, the first Earth remote-sensing satellite.

1983 NASA develops the Airborne Imaging Spectrometer, a breakthrough in hyperspectral imaging.

AFTER
2011 G20 governments launch GEOGLAM, or Group on Earth Observations Global Agricultural Monitoring.

2025 After 26 years in orbit collecting more than 3 million images of Earth, Landsat-7 is decommissioned.

Remote sensing is the process of gathering information about the surface of Earth without being in direct contact with it. The data, which is collected by sensors on Earth-observation satellites, can help to improve agricultural productivity, especially in light of challenges such as limited resources or extreme-weather events driven by climate change.

A major development in this field was the creation in the 1970s of the normalized difference vegetation index (NDVI) by Earth scientist Compton J. Tucker. This metric enabled the global monitoring of plant health and

APPLIED GEOGRAPHY 293

See also: Aerial photography and remote sensing 40–45 ▪ Geographic information systems 52–59 ▪ Sustainable development 244–45 ▪ Monitoring environmental change 284–89 ▪ Global Earth observation programmes 308–09

changes in land cover (as a result of deforestation or expanding urban centres, for example) using satellite data. Tracking variations in the NDVI over time allows farmers to gauge crop development, detect any early signs of water stress, disease outbreaks, pest damage, or nutrient deficiencies – and take action before serious damage is done.

Eyes in the sky

The first Earth-observation satellite for agricultural applications – Landsat 1 – was launched in 1972 by NASA in collaboration with the

Improved crop yields could help to cut the number of people experiencing hunger, which in 2024 was 8.2 per cent of the global population, according to the World Health Organization.

US Geological Service. It was equipped with sensors that monitored the visible and near-infrared (thermal) bands of the electromagnetic spectrum. Since then, eight more Landsats have been launched, with the next mission planned for 2030. NASA's Earth Observing System (EOS) also includes the Terra and Aqua satellites. They carry sophisticated instruments such as MODIS (moderate resolution imaging spectroradiometer), which is able to capture data in 36 spectral bands.

The European Space Agency's satellite Sentinel 2, launched in 2015, has been providing multi-spectral imagery with a resolution of up to 10 m (33 ft) – ideal for collecting field-level data.

Drones, or unmanned aerial vehicles, offer high-resolution images as well as great flexibility.

[Earth science is about] understanding how this planet works and helping people make… informed decisions.
Karen St Germain
Director of NASA's Earth Science Division, 2022

Farmers can use drones themselves to scan their fields in real time, identifying specific areas that require attention, allowing for much more targeted agricultural management practices.

Gathering remote data

Remote sensing is achieved by measuring energy – such as the amount of light – that is reflected and emitted from Earth's surface. »

Compton James Tucker

A leading scientist in Earth-observation research, Compton J. Tucker was born in Carlsbad, New Mexico, US. After earning a degree in biology in 1969, he studied for a master of science and a PhD in systems ecology – all at Colorado State University.

In the mid-1970s, Tucker joined NASA's Goddard Space Flight Center. His pioneering work on the normalized difference vegetation index (NDVI) is often credited with establishing a worldwide standard for measuring not only the amount of vegetation on Earth, but also how well it is performing.

Tucker has received many awards, including the NASA Exceptional Scientific Achievement Medal two years running (1986 and 1987). In 2025, he was elected a member of the National Academy of Sciences in the US.

Key work

1986 "Satellite Remote Sensing of Primary Production"

Some satellite sensors measure long-wave infrared radiation, which is particularly useful for estimating surface temperatures. Other sensor types include radar (radio detection and ranging) and LiDAR (light detection and ranging), both of which send out electromagnetic energy and measure the amount that is reflected. LiDAR is generally more precise than radar; however, radar's cloud-penetrating capability is useful in regions with persistent cloud cover – for instance, in tropical rainforests.

After a thorough check, any inaccuracies caused by clouds or Earth's atmosphere are adjusted. The resulting spatial and geographic information is fed into a geographic information system (GIS) database for storage, management, and analysis. GIS creates a detailed picture of the land, providing accurate data of soil conditions, crop growth, and weather patterns, as well as calculating indices like NDVI. This information is then processed and translated into graphs to assist in farm management.

Precision agriculture

The application of remote-sensing technology is particularly useful in giving early warning about potential problems. Where water is a scarce resource, GIS combined with other satellite data can visualize soil moisture levels, which helps to make irrigation systems more efficient by ensuring water is applied only where it is needed. This reduces water consumption and improves productivity.

Soil mapping also facilitates the collection of detailed information about the variability of nutrients in different areas. This process helps to identify the location of soils with different fertility levels and is not only a guide for farmers in making decisions about the types of crops that can be planted, but also where fertilizer should be applied.

Satellite data can also offer accurate predictions of yields and optimal harvesting times, allowing farmers to plan when to gather their crop. Monitoring local weather conditions such as temperature, humidity, and windspeed provides more accurate predictions than relying on broader regional weather reports. This allows farmers to, for example, schedule pesticide or fungicide applications when they will be most effective and will not be washed away by rain.

Precision agriculture makes use of data and technology to optimize field-level management of crop farming. It allows farmers to observe, measure, and respond to variability within and between

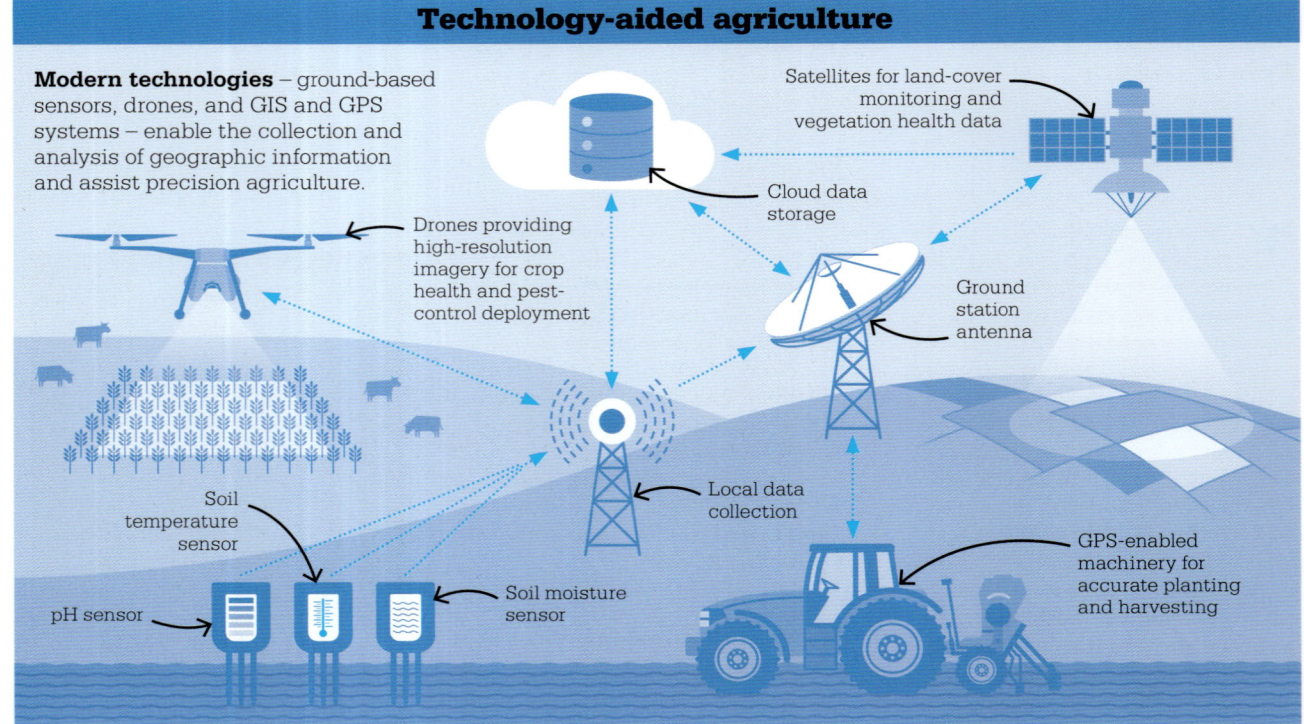

Technology-aided agriculture

Modern technologies – ground-based sensors, drones, and GIS and GPS systems – enable the collection and analysis of geographic information and assist precision agriculture.

- Satellites for land-cover monitoring and vegetation health data
- Cloud data storage
- Drones providing high-resolution imagery for crop health and pest-control deployment
- Ground station antenna
- Local data collection
- Soil temperature sensor
- pH sensor
- Soil moisture sensor
- GPS-enabled machinery for accurate planting and harvesting

APPLIED GEOGRAPHY 295

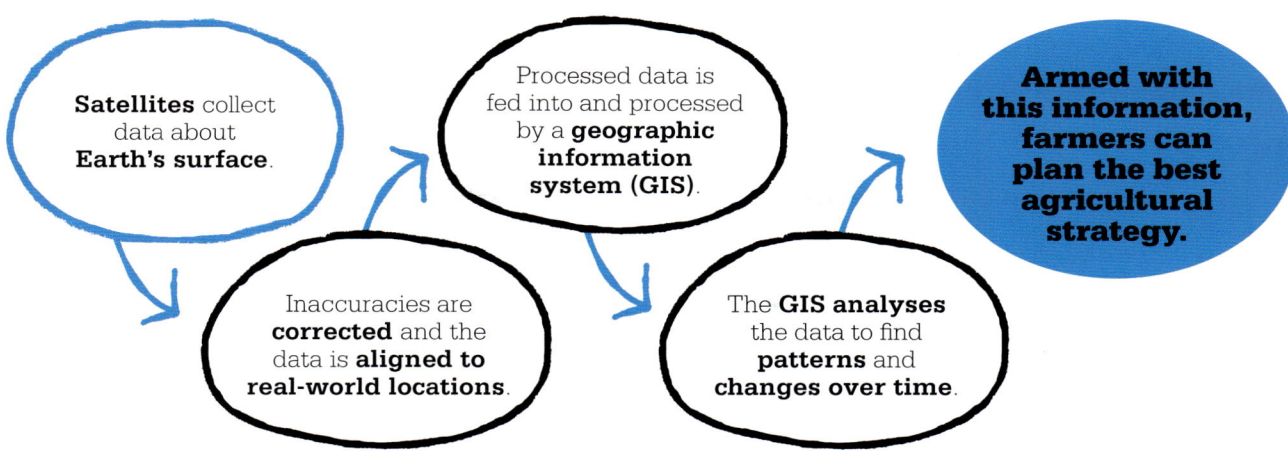

fields. For example, rather than applying the same amount of fertilizer across an entire field, precision agriculture enables the use of variable rate technologies. This means farmers can adjust the amount of fertilizer they apply in different areas based on need, reducing waste and increasing yields at the same time.

Combine harvesters and tractors guided by a global positioning system (GPS) ensure that operations such as planting, fertilizing, and harvesting are carried out with pinpoint accuracy. Sensors mounted on equipment monitor yield, moisture levels, and soil conditions, feeding data back into management systems.

High-tech science

Technology has transformed agriculture into a high-tech science, although this brings its own problems. Data accessibility and affordability prevents many farmers from making use of these new tools. Not all have the training or resources to use them effectively. There are also questions about how the data gathered is controlled and by whom. Initiatives like NASA's Harvest programme aim to make technology such as remote-sensing data more accessible to end users.

Remote-sensing and precision agriculture are about producing more food, but they are also about producing food more sustainably and efficiently. As climate change alters weather patterns and puts unprecedented stresses on natural resources, these tools will become even more important in providing farmers with the resilience they need to meet these challenges. ∎

Remote-sensing studies of vegetation normally use specific wavelengths.
Compton J. Tucker & Piers J. Sellers
"Satellite Remote Sensing of Primary Production", 1986

Reading the NDVI

The normalized difference vegetation index (NDVI) is based on observations of reflected radiation in the red and near-infrared regions of the spectrum. A healthy plant reflects a lot of radiation in the near infrared and much less in the red region, in which instance the NDVI is high. In the case of an unhealthy plant, the opposite is true.

NDVI values vary between -1 and 1. The closer to 1 the NDVI value is, the greater the amount of healthy green vegetation present. Negative values of -1 to -0.1 usually indicate an absence of vegetation, such as on open water, snow, or ice. Values of 0 to 0.1 are typical of bare soil, rock areas, or deserts, while values of 0.2 to 0.5 suggest sparse vegetation, such as grasslands, shrub, or drought-stressed fields. High values of 0.6 to 1 generally point to lush, healthy vegetation, such as crop fields or mature forests.

Drought, disease outbreaks, pest damage, and nutrient deficiencies can all result in declining NDVI values.

THERE'S NO SUCH THING AS A NATURAL DISASTER

EARLY WARNING SYSTEMS

EARLY WARNING SYSTEMS

IN CONTEXT

KEY FIGURE AND ORGANIZATIONS
Susan Cutter (1950–)
World Meteorological Organization (1950), **United Nations Office for Disaster Risk Reduction** (1999)

BEFORE
1935 British meteorologist Robert Watson-Watt develops the earliest radar alert system.

1941 Japan creates an early earthquake-centric tsunami warning system at its Sendai Meteorological Observatory.

1955 The National Hurricane Center is founded in Miami, Florida, US, to track storms and issue public warnings.

AFTER
2027 The UNDRR and WMO seek to implement the Early Warnings for All initiative by the year's end, scaling up multi-hazard early warning systems (MHEWS) in future years.

Technological advances between World War I and World War II prompted the development of the first widespread early warning alerts for extreme weather. While the military use of radar was to detect enemy aircraft, scientists soon discovered that it could also identify precipitation such as rain, snow, or hail. More accurate weather forecasts made it possible to detect extreme weather in advance, and wealthier nations began to develop effective early warning systems. From the 1990s, geographers such as American academic Susan Cutter also began to produce detailed studies of what makes certain areas vulnerable to disasters and how relief resources are best targeted.

In the 21st century, as climate change increased the frequency of extreme weather events, the World Meteorological Organization (WMO), United Nations Office for Disaster Risk Reduction (UNDRR), and other UN bodies began to plan a global early-warning system.

A system of stages

Multi-hazard warning systems are part of a nation's wider climate information service that produces timely forecasts in more general terms over longer timescales. Disaster planners assess what hazards are likely in their area, how frequently the region experiences these events, and what specific vulnerabilities exist. Since 2009, the UNDRR has produced a Global Assessment Report (GAR), published every two years. Its information includes case studies of past disasters, maps of global risks, and the latest scientific research into risk reduction.

After assessing which extreme weather events are most likely to occur in their area and require warnings, planners can set up monitoring and forecasting systems to help detect them. Once an event is detected that is extreme enough to require a warning, an alert is issued via various media, such as TV broadcasts, news notifications, and weather apps. Warnings, which must be kept consistent across all platforms, will generally be graded by threat level, such as yellow, orange, or red. In advanced early warning systems, the highest level of threat triggers an audible push notification delivered through the cellular network to every connected device in an affected region.

Extreme weather and other natural hazards **can be forecast** by satellite and other advanced technology.

→ New technology also facilitates **early warning systems**, which **save lives** and reduce damage.

→ **Most people have access** to effective early alerts, but at least **30 per cent of the world's population still do not**.

↓

The people at greatest risk must receive early warnings as climate change increases the frequency and intensity of extreme events.

APPLIED GEOGRAPHY 299

See also: Aerial photography and remote sensing 40–41 ▪ Seismology and earthquake magnitude 132–35 ▪ Extreme weather 142–43 ▪ Floodplain management 230–32 ▪ Climate modelling 268–71 ▪ Crisis mapping with crowdsourcing 304–05

Key stages that early warning systems should include, as shown here, are based on advice from the United Nations Office for Disaster Risk Reduction (UNDRR) as its Early Warnings for All (EW4ALL) initiative is rolled out.

Risk awareness	Monitoring and alerts	Response capacity	Communication
Hazard	Observation	Resources	Access
Exposure	Analysis	Plans	Understanding
Vulnerability	Trigger	Rehearsal	Action

The final stage in developing an early warning system is to put in place the means to respond to a threat. This means ensuring that emergency services have effective training and resources, preparing action plans for the evacuation of residents, or offering guidance on how people can shelter in place with adequate provisions.

Seismic activity

While it is currently impossible to predict the location and magnitude of earthquakes before they occur, worldwide networks of seismographs can register their earliest tremors. Operations such as the US-operated Global Seismographic Network, a digital seismic network with 152 stations, are now part of a global cooperative system that constantly monitors seismic activity.

An earthquake's primary seismic waves (P waves) have a top speed of about 8 km (5 miles) per second, and the more destructive secondary waves (S-waves) travel at half this speed. As that is slower than the signal speed from seismic detectors, there is a small but significant window for alerting people further from the epicentre to prepare for the imminent arrival of an earthquake. The ShakeAlert system operating on the West Coast of the US uses a network of sensors to detect the magnitude and direction of an earthquake's primary waves. If this wave data indicates that a nearby population centre is threatened, the system orders a warning to go out to subscribers. For example, the train network in San Francisco brakes all trains when an alert is received. ShakeAlert has been functioning in its present form since 2018. Typically, the system provides messages 45 seconds ahead of an earthquake's damaging secondary waves. Earlier versions often sent warnings during or after.

Earthquake alerts are broadcast to the general public in Japan, South Korea, Taiwan, and Mexico. Google has also developed a global system in which motion sensors »

When a magnitude 7.8 earthquake hit Türkiye in 2023, Google's warning system failed to give a highest-level alert. In 2025, Google admitted the failure.

EARLY WARNING SYSTEMS

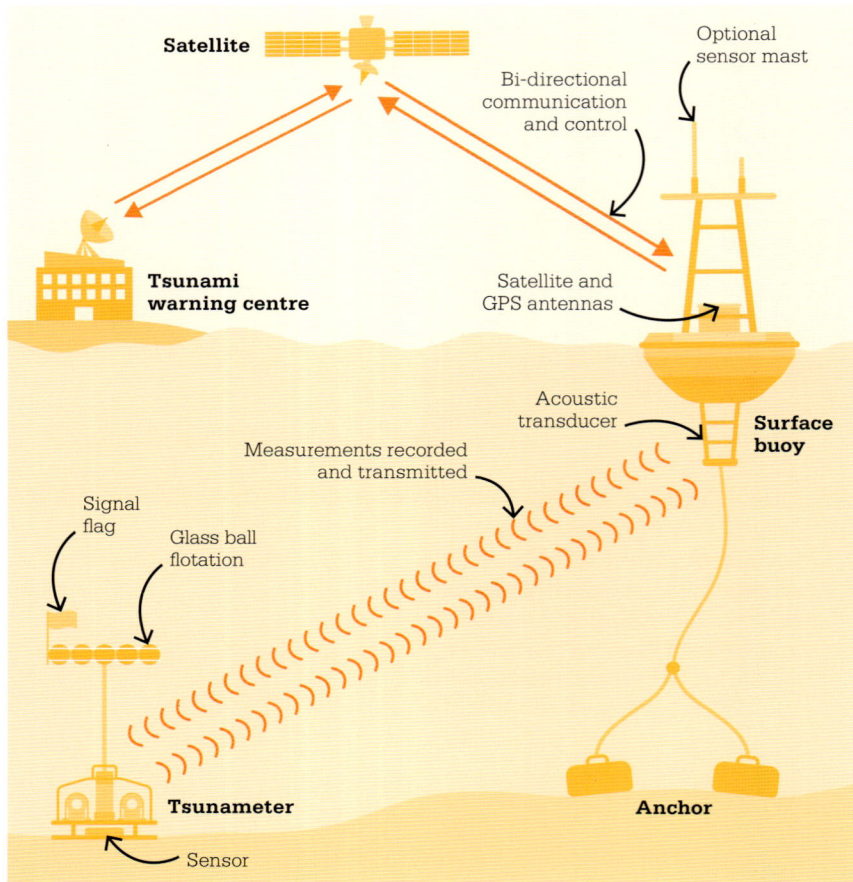

The DART II system consists of a tsunameter on the sea floor, which sends signals to a buoy at the surface. These are then transmitted via satellite to a tsunami warning centre.

in users' Android devices detect seismic action and early warnings are delivered via the network.

Tsunami alerts

Earthquakes and tsunamis are among the most destructive natural events. Tsunamis, the great waves that subocean earthquakes can create, are rare. However, as just over one in seven people – 15 per cent of the world's population – lives within 10 km (6.2 miles) of a coast, giant waves can potentially endanger millions of people in a matter of hours.

Tsunamis are created by a section of the sea floor shifting up or down. The displaced water ripples out in all directions, and when the wave meets the coast, land can be inundated under many metres of water. In 2004, the Indian Ocean tsunami and earthquake killed nearly 230,000 people. At that time, only one tsunami early warning system (TEWS) existed in the Pacific Ocean. Subsequently, TEWS were established in the Indian and Atlantic oceans, and Caribbean and Mediterranean seas. TEWS are designed to detect earthquakes that could generate a tsunami and provide timely warnings.

Following a devastating tsunami in 2011, Japan's early-warning system was upgraded. An extensive earthquake-detection network was built on the seabed, extending east from its main island Honshu to the Japan Trench. The network includes some 5,700 km (3,540 miles) of cable that connects 150 seafloor observatories, each equipped with sensitive instruments such as accelerometers and seismometers. The network is spread across a deep, seismically active region of the seabed a few hundred kilometres off the eastern coast of Japan – the primary source of any tsunami that threatens the country. A tsunami can travel from the trench to the mainland in less than an hour, and much more quickly if the epicentre is closer.

The epicentre of the powerful 2011 earthquake was only 70 km (43 miles) offshore. The new seismic detection system, the first part of a larger network completed in 2025, offers a better chance of quickly detecting and analysing the threat of a similar event and adds crucial time for emergency responses. A magnitude 6.0 earthquake in 2018 proved the efficacy of the system. Alerts reached cities before the first jolt was felt and seconds before an alert from the closest land-based seismometer. With the completion in 2025 of a smaller network to its southwest, beneath the Philippine Sea along the geologically active Nankai Trough, Japan now has a comprehensive earthquake and tsunami detection system.

DART and other systems

Building networks like Japan's across every seismically active ocean region is not feasible.

Sea-floor earthquakes can be detected by seismographs, but as they are far out to sea and deep below the surface, it is difficult to estimate how the water above has been displaced. Most tsunami warning systems now use the Deep-ocean Assessment and Reporting of Tsunamis (DART) system, operated by the US National Oceanic and Atmospheric Administration (NOAA). Completed in 2008, the network includes 39 systems located in geologically active areas of the Pacific, Atlantic, and Caribbean basins, and similar compatible systems exist in the Indian Ocean.

The DART systems consist of tsunameters – essentially pressure detectors – anchored to the seabed, which sense when a tsunami wave passes, then gather details of its speed and direction, and estimate its size. The tsunameters relay data via acoustic signals up to buoys floating on the surface. These send on the information via satellite links to a warning centre, which can then issue alerts to regions in the path of the wave, and warn coastal populations to move to high ground.

Volcanic eruptions

Compared to earthquakes, volcanic eruptions are easier to predict, so there are now few surprise eruptions near to where people live. Of some 1,300 active volcanoes, mostly in clusters around the planet, 500 are highly active and frequently erupt. Vulcanologists monitoring the seismic activity of mountains study changes in the gases emitted from vents and fumeroles for signs of eruption. When magma inside a volcano pushes upwards, it raises the temperature of a mountain, which swells due to its internal pressure. Remote-sensing satellites regularly capture the temperature and shape of target volcanoes to detect potential eruptions.

Next, the threat posed by an imminent eruption is evaluated. Most volcanoes are in remote areas far from cities. Early warnings trigger the evacuation of anyone living nearby. Ash and dust from eruptions can spread over a wide area, so warnings to people further afield and downwind of any ash clouds may allow them to insulate their buildings and shelter in place until the skies have cleared.

Storm warnings

Tropical storms that form over the ocean not only pose a hazard to shipping but also cause death and destruction when they hit land. The storms grow in strength as warm ocean waters feed them with rising columns of water vapour, ultimately creating a vast spiral of cloud banks and strong winds. Once the storm hits land, its energy disappates, but high winds damage houses, and heavy rain causes flooding. Most deadly of all, the storm's low pressure »

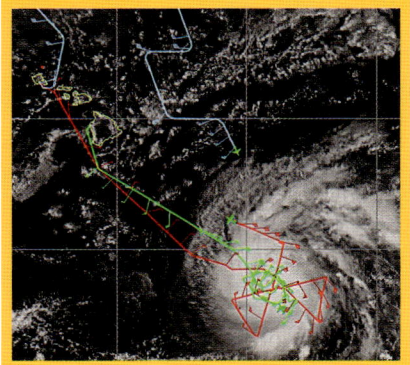

A satellite image of a hurricane over Hawaii in 2018 shows superimposed routes of NOAA hurricane-hunter planes that gathered data in an operation with the Joint Typhoon Warning Center.

Wildfires

As summers become hotter and vegetation dries, the intensity and frequency of wildfires increases. Forest fires emitted a record 8.6 billion tonnes (9.5 billion tons) of CO_2 in 2023, and destroyed 6.7 million hectares (16.5 million acres) of primary tropical forest, mainly in South America, in 2024.

Since 2010, the Global Early Warning System for Wildland Fires (Global EWS-Fire) has linked national and local systems. From satellite-detected hotspots and weather forecasts, it can provide alerts up to 14 days in advance. Countries and regions, such as Canada and Europe, also have their own fire-warning systems that include a fire weather index to assess forest conditions, and fire behaviour maps. A new machine-learning tool at the European Centre for Medium-range Weather Forecasts uses multiple data sources and remote-sensing to monitor the state of vegetation in different areas of Earth. It aims to predict the likelihood of fire in square kilometre units worldwide up to 10 days in advance.

A new network of fire alarm sirens was established in Paradise, California in 2013, following the failure of electronic warning systems during the 2018 wildfire.

EARLY WARNING SYSTEMS

A house is submerged as heavy monsoon rains cause major floods in 2021 on Mindoro Island, Philippines, a vulnerable nation that is strengthening its early warning systems.

forms a bulge in the ocean surface, creating a tall, deadly surge that washes over the coast.

Hurricanes, typhoons, or tropical cyclones, the largest tropical storms, take several days to develop, and are tracked by storm warning centres using weather satellites. They are most frequent in summer and autumn months. Hurricane hunters – specialized aircraft linked to weather forecasting centres – fly into problematic storms, drop wind sensors, and gather key data. Using this data, hurricane centres can predict where a storm will make landfall, and estimate how powerful and destructive it will be. Warnings are issued to affected regions for all large storms that are likely to progress over land. Although island communities may only receive a few hours' warning before a major storm strikes, timely alerts generally give communities a few days to prepare.

Similar weather warning systems operate at middle and higher latitudes, where low-pressure weather systems can trigger hail, thunderstorms, blizzards, and tornadoes. Forecasters can usually predict more extreme events days in advance, refining their advice as a weather system approaches. During storms and high winds, the principal dangers are from falling trees and flying debris, so warnings generally attempt to dissuade people from travelling until the threats have passed.

Spirals of wind

The world's fastest wind systems – tornadoes – strike only small areas over short periods but can deliver winds that are more destructive and faster than any hurricane. They form when large thunderclouds that enclose tight vertical spirals of wind, called a mesocyclone, make contact with the ground. The resulting whirlwind is powerful enough to suck material up into the sky.

Since the 1970s, doppler radar has been used to search for mesocyclones in thunderstorms developing in tornado-prone regions, such as the southern and eastern states of the US, the world's biggest hotspot. If they are found, "watch signals" are issued, asking local residents to be prepared for an alert; a tornado warning, as opposed to a tornado watch, means a tornado has been spotted and is imminent. Horizontal radar pulses, which use a system called dual-polarization, can also detect solid debris picked up by a thunderstorm, a further clear indication that a tornado has formed. Warnings are issued to communities in the path of the storm – often only a small area as tornadoes seldom travel far. The general advice is to take shelter in the centre of an internal room on the lowest floor of a resilient building.

Flood warnings

Sources of floodwater include extreme rainfall that can create flash floods, where huge torrents form suddenly in valleys that are normally more arid. Large quantities of rain can also cause permanent water courses to swell and break their banks. Predicting floods is complex, however, as the quantity of rain falling is not directly linked to flood risk. Rain can percolate into soil and add to groundwater, and only when ground is waterlogged will the rain flow along the surface and

All people on Earth must be protected by early warning systems within five years.
António Guterres,
UN Secretary-General,
2022

feed directly into river systems. To gain a better idea of potentially imminent floods, hydrologists often use a series of flow gauges, set up in a watershed, to measure how much water is entering the system. This allows them to calculate when and where the water levels will pose a threat. Where these measures are in place, flood warnings can be given days or hours in advance so that people can prepare, and authorities can enact plans to erect temporary barriers to divert floodwater away from residential areas. Yet only a small proportion of the world's watersheds are gauged and, when devastating floods strike in some parts of the world, those living in poorer communities struggle to save their lives. Thousands of people die in floods every year.

Helping the vulnerable

From the 1970s, interdisciplinary disaster research led by British academics Phil O'Keefe, Ken Westgate, and Ben Wisner challenged the idea that disasters are purely natural events and highlighted the role of socio-economic factors. The concept of social vulnerability became integral to the study of natural disasters and the question of how best to measure, monitor, and assess vulnerability and resilience has been the life work of American geographer Susan Cutter, who founded the Hazards Vulnerability and Resilience Institute at the University of South Carolina.

Using censuses and other socioeconomic and demographic data, Cutter developed two important indexes in 2003 and 2010 to measure and evaluate vulnerability and resilience and to guide authorities and planners seeking to protect against disasters or manage their aftermath. Her Social Vulnerability Index (SoVI®) sets criteria for assessing which areas are most at risk, together with social factors, such as poverty or age, which affect how people prepare for, tolerate, and recover from hazards.

Cutter's Baseline Resilience Indicators for Communities (BRIC) index helps planners to evaluate the capacity – such as early warning systems and defensive infrastructure – that communities have for withstanding hazards. Both indexes have been widely used in the US, as well as Portugal, Brazil, China, and elsewhere.

Protecting the most vulnerable is also the goal of the UNDRR and WMO, who have developed an Early Warnings for All (EW4ALL) framework to alert all nations to extreme events. According to WMO, an alert given 24 hours in advance could reduce the impact of a disaster by a third.

Unequal provision

According to the UN, by 2024 more than half of all countries had multi-hazard alert systems, an increase of over 50 per cent in less than ten years. Poorer nations, some of which are highly vulnerable to extreme events, remain the least equipped to confront them. If communities are unprepared, the devastation and suffering caused is inevitably greater.

WMO estimates that EW4ALL could save up to £12 billion ($16 billion) a year, as well as thousands of lives. In 2022, UN Secretary-General António Guterres set a five-year target – global access to early warnings systems by 2027. ∎

> Vulnerability… the potential for loss, is an essential concept in hazards research and is central to the development of hazard mitigation strategies.
> **Susan Cutter**
> *Hazards, Vulnerability, and Environmental Justice*, 2006

Susan Cutter

A native of Ohio, US, Susan Cutter was born in Cincinnati in 1950. She studied geography at California State University, then completed postgraduate studies at the University of Chicago in the 1970s. After holding several academic positions, she joined the University of South Carolina's geography faculty in 1993, where she founded the Hazards Vulnerability and Resilience Institute. Cutter has pioneered research in disaster vulnerability/resilience science throughout her career and served on many national advisory boards and committees. She received the American Association of Geographers' Lifetime Achievement Award in 2010, and in 2024 she was elected to the National Academy of Science.

Key work

2014 *Hurricane Katrina and the Forgotten Coast of Mississippi* (co-author)

EVERYONE, EVERYWHERE
CRISIS MAPPING WITH CROWDSOURCING

IN CONTEXT

KEY FIGURE
Steve Coast (1980–)

BEFORE
Early 1960s British-Canadian geographer Roger Tomlinson creates the Canada Geographic Information System (CGIS), the first functional digital mapping system.

1963 American architect Howard Fisher releases his Synagraphic Mapping Program (SYMAP), which enables users to print thematic maps generated from digital data.

AFTER
2012 The launch of the Digital Humanitarian Network links relief agencies with volunteer digital providers so they can work together to supply aid.

2014 The Euro-Mediterranean Seismological Centre releases its LastQuake app, designed to spread earthquake alerts and crowdsource information about injuries and damage.

The initiative of one person, plus the wide availability of digital technology, has democratized cartography in the 21st century, with huge benefits for crisis mapping in disaster areas.

OpenStreetMap (OSM) was a pioneering project launched in 2004 by British computer scientist Steve Coast while he was still a student. Open-source mapping software did already exist, but Coast's idea was to crowdsource and create a global map database built entirely by volunteers, who need only an internet connection and user account to contribute their work.

Saving lives in Haiti
In 2010, when a powerful magnitude 7.1 earthquake struck Haiti and destroyed much of the Caribbean island's infrastructure, OSM proved

The Ushahidi platform, used to tackle social and public health challenges, is maintained by the Ushahidi team, pictured in their Nairobi office in 2013.

APPLIED GEOGRAPHY

See also: Aerial photography and remote sensing 40–45 ▪ Seismology and earthquake magnitude 132–35 ▪ Extreme weather 142–43 ▪ Early warning systems 296–303

> You can't protect what you can't map.
> **Patrick Meier**
> American academic, 2016

a vital crisis-mapping tool. It was aided by the open-source software platform Ushahidi, launched in 2008 by technicians and bloggers in Kenya to map violence triggered by controversial elections.

During Haiti's earthquake relief operation, responders had to find and rescue residents, battle a cholera epidemic, and restore utilities. OSM volunteers, working at the scene or remotely, used current satellite images to update maps for rescue workers, initially highlighting damaged buildings and blocked roads. Volunteers also used Ushahidi software to supply first-hand reports, crowdsourced from text messages and open-source channels, such as news stations and social media. Other data could then be added to maps, such as the location of trapped survivors, casualty counts, the prevalence of disease, and where pharmacies could supply medicines.

Effective collaboration

The humanitarian response to the Haiti earthquake was a notably collaborative initiative. American academic and entrepreneur Patrick Meier, an early pioneer of crisis mapping, coordinated volunteers working with Ushahidi, which he had helped adapt for humanitarian use. Meier co-founded the International Network of Crisis Mappers in 2009. In 2015, he published *Digital Humanitarians*, which discusses how wider access to digital data has influenced the response to extreme events.

Crisis mappers worldwide

During the Haiti earthquake relief operations, the Humanitarian OpenStreetMap Team (HOT) also took shape. HOT became the first community-driven, disaster-mapping organization to formally liaise with aid agencies, providing crisis map services to governments and NGOs, and making large-scale printable maps of a disaster zone freely available.

Within months of the Haiti disaster, HOT again invited the help of OSM mappers as massive floods affected Pakistan. Here, satellite imagery of the vast flooded area was still limited and users achieved less mapping than hoped for. However, the efficacy of crisis mapping had been shown in Haiti, and it soon became an integral part of humanitarian responses. Aided by ever-clearer and more accessible satellite images, such maps have proved their worth in epidemics, wars, and extreme weather events. By 2025, HOT had mapped an area of more than 900 million people. ▪

Open-source digital mapping enables ordinary people to **work together and add detail** to maps.

The potential for **crisis mapping** is soon identified.

In disasters such as the 2010 Haiti earthquake, crisis mapping **proves its worth and saves lives**.

Crowdsourced maps are now used in **disasters worldwide**.

Organizations such as HOT **link digital volunteers with humanitarian organizations**, improving data-use efficiency.

BIODIVERSITY CAN BE MONITORED ON A GLOBAL SCALE
MONITORING BIODIVERSITY

IN CONTEXT

KEY FIGURE
Andrew K. Skidmore
(1959–)

BEFORE
1985 American biologist Michael E. Soulé founds the Society for Conservation Biology, dedicated to preserving global biodiversity.

2000 British environmentalist Norman Myers identifies 10 deeply threatened tropical areas rich in plant and animal species, introducing the idea of biodiversity hotspots.

2001 The Global Biodiversity Information Facility (GBIF) is launched, with the aim of providing open access to data about all types of life on Earth.

AFTER
2023 The Alan Turing Institute and the UK Centre for Ecology & Hydrology launch the AMBER project, which uses AI tools and citizen science to monitor wildlife diversity.

For decades, scientists have noted alarming biodiversity losses, but a key challenge has been measuring the scale of the problem and pinpointing specific areas of concern. The emergence of satellite technology, geographic information systems (GIS), and satellite-based sensors has revolutionized such research, supplying an ever-clearer view of changes on Earth's surface.

Australian geographer and ecologist Andrew Skidmore was among the first to press for and promote the use of high-resolution, satellite-based hyperspectral sensors for monitoring biodiversity. By capturing hundreds of narrow bands of light, these instruments allow researchers to assess vegetation types based on their unique spectral signatures (how they absorb, reflect, and transmit light), and thus detect individual plant species and monitor ecosystem traits.

Effective new tools
Complementary remote-sensing technology includes multispectral imaging, LiDAR (Light Detection and Ranging), and radar. These tools perform many functions, such as assessing vegetation and measuring soil moisture and sea surface temperatures, all of which are indicators of biodiversity and ecosystem health. Essential Biodiversity Variables (EBVs), such as vegetation height, leaf area, and

Emperor penguins stand on sea ice in Halley Bay, Antarctica. Satellite images of guano stains revealed a previously unknown emperor penguin colony here.

See also: Biomes and ecological zones 136–39 ▪ Natural resource management and conservation 222–27 ▪ Environmental impact assessment 272–73

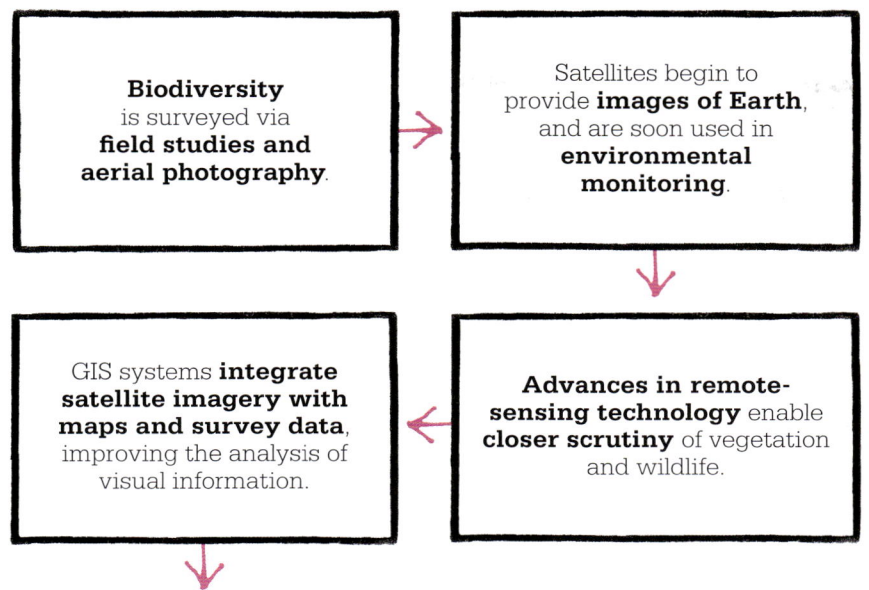

Combined ground and satellite data enables scientists to monitor ecosystem health and implement conservation strategies.

Andrew K. Skidmore

Born in Australia in 1959, geographer Andrew Skidmore graduated from the Australian National University and, after working in forestry, undertook a PhD in remote sensing and GIS. In the 1990s, he moved to the University of Twente in the Netherlands, and from 1997 to 2016, he headed its Department of Natural Resources. There he became a professor in its Faculty of Geo-Information Science and Earth Observation, and also Professor of Spatial Ecology at Macquarie University in Sydney, Australia.

Skidmore has developed an international reputation for applying remote sensing and geospatial technologies to ecological research. He has published widely on species mapping and loss, invasive species, migration, and how climate and land-use changes affect biodiversity.

Key works

2015 "Agree on biodiversity metrics to track from space" (et al.)
2002 "Taxonomy of environmental models in the spatial sciences"

even primary productivity (the rate at which organisms harness and convert energy) can also now be reliably monitored from space.

Remote sensing is also used to assess fire damage, the impact of agriculture, mining, and deforestation, and to detect unusual vehicle movements or fires that might indicate illegal hunting. In marine environments, satellites track ocean temperature, algal blooms, and coral bleaching, which helps in managing fisheries and the safeguarding of reef biodiversity.

Integrating data

To monitor biodiversity at a local and regional level, researchers may use aircraft and drones, equipped with LiDAR and hyperspectral, thermal, and fluorescence sensors. On the ground, tools include motion-triggered camera traps and environmental DNA (eDNA) – samples taken from soil, water, and air. The equipping of wildlife with GPS trackers, which transmit location data via satellite or cellular networks, enables scientists to follow large wildlife migrations, whether of birds, land mammals, or marine life.

Organizations such as the Group on Earth Observations Biodiversity Observation Network (GEO BON) collect data across countries and integrate it into a unified framework. It then becomes possible to gauge biodiversity levels over vast, remote regions, such as the Siberian tundra or the Congo Basin, which are too large and inaccessible to be directly observed in any other way.

With technological advances, the goal is to integrate satellite, aerial, and ground-based remote sensing with AI-powered data analysis to provide granular (detailed) data that can inform targeted conservation efforts. ∎

TO TAKE THE PULSE OF OUR PLANET
GLOBAL EARTH OBSERVATION PROGRAMMES

IN CONTEXT

KEY PROGRAMME
Copernicus (2014–)

BEFORE
1960 NASA launches TIROS 1, the first operational weather satellite; during the course of 75 days, it sends back 23,000 photos of Earth.

1991 The European Space Agency launches its ERS-1 (European Remote Sensing) satellite, thus inaugurating its Earth observation programme.

AFTER
2017 The National Oceanic and Atmospheric Administration in the US launches the NOAA-20 satellite to observe Earth's oceans and polar regions.

2025 In April, the European Space Agency launches its Biomass satellite with a five-year mission to map Earth's forests and investigate their role in the carbon cycle.

The backbone of modern environmental monitoring, Earth observation programmes rely on orbiting satellites to collect data about Earth's surface, atmosphere, and oceans. This can take the form of optical imagery, radar signals, thermal data, and more. The goal is to observe environmental changes, monitor natural resources, support agriculture, and develop disaster responses. Information is processed and made available to governments, scientists, businesses, and the public.

Images of the Netherlands were among the first sent by the Sentinel-1 mission. Collected data helps monitor water levels around farming land that has been reclaimed from the sea.

The Copernicus programme, launched by the European Union in cooperation with the European Space Agency (ESA) in 2014, is the world's largest and most advanced Earth observation initiative. At the heart of Copernicus is the Sentinel series of satellites.

A range of Sentinels

The Sentinel observation missions comprise constellations of satellites equipped with a range of sensors. Sentinel-1 uses radar to monitor ice movements, land deformation, and natural disasters – even through cloud cover or at night.

The satellites of the Sentinel-2 constellation provide high-resolution optical imagery for vegetation monitoring, land use, and water quality, while the Sentinel-3 satellites focus on sea surface temperatures, ocean colour, and land altimetry (the height of Earth's surface), essential for climate and marine studies.

Sentinel-4 is an ultraviolet, visible, and near-infrared spectrometer that provides data on a wide range of atmospheric pollutants in order to forecast and monitor air quality over Europe. Sentinel-5P monitors air pollution

See also: Aerial photography and remote sensing 40–45 ▪ Geographic information systems 52–59 ▪ Environmental impact assessment 272–73 ▪ Monitoring environmental change 284–89 ▪ Remote-sensing agriculture 292–95

The Sentinel-6 mission measures ocean currents and sea-surface heights. Orbiting at about 1,335 km (830 miles), it covers 95 per cent of Earth's oceans every 10 days.

and greenhouse gas concentrations in the atmosphere, while Sentinel-6 measures sea level rises.

EOS and GEO

NASA's Earth Observing System (EOS), initiated in the early 1990s, is one of the longest-running Earth observation programmes. It aims to provide a long-term view of the planet using a coordinated group of satellites (collectively known as the A-train) to gather data. Among them is Aqua (launched in 2002), which focuses on Earth's water cycle – from clouds and precipitation to soil moisture and ocean salinity. Aura (2004) is dedicated to atmospheric chemistry and air quality, while OCO 2 (2014) measures CO_2.

A partnership of more than 100 countries and participating organizations, the Group on Earth Observations (GEO) was formed in 2005 to bring together data from satellites, ground-based sensors, ocean buoys, and citizen science projects into a unified framework. GEO is dedicated to creating a Global Earth Observation System of Systems (GEOSS) that links data across various disciplines, including climate, agriculture, health, and ecosystems.

A valuable tool

Earth observation programmes enable the tracking of sea levels, glacier melt, and greenhouse gas concentrations, which is essential for understanding the impact of climate change. Other benefits include the rapid assessment of floods, earthquakes, wildfires, and hurricanes, saving lives and guiding recovery efforts. Monitoring soil moisture, crop health, and rainfall helps to improve food security and manage water resources, while habitat mapping, deforestation tracking, and marine ecosystem monitoring can all help to preserve biodiversity. Cities use this data to map growth, monitor air pollution, and design greener infrastructure. ■

Earth observation has changed the way we comprehend our profound impact on the environment.
Josef Aschbacher
ESA's Director of Earth Observation Programmes, 2020

A brief history of global monitoring

Environmental change used to be something observed locally, by people on the ground. That began to change in the 1960s as weather satellites sent back the first images of Earth from space. As the technology improved, those first grainy pictures became powerful data streams. By the 1970s, NASA's Landsat programme was monitoring forests, cities, and coastlines in what would become the longest continuous record of Earth's changing surface. With the rise of environmental awareness, satellites also began to track deforestation, desertification, and biodiversity loss.

Thanks to the ESA's Copernicus programme, with its free, high-resolution data, and organizations such as GEO BON (Biodiversity Observation Network), Earth observations are now more accessible. With the help of artificial intelligence and real-time satellite feeds, there is a clearer picture of life on Earth than ever before.

INFORMATION HAS TO MOVE EVEN FASTER
TRACKING DISEASE

IN CONTEXT

KEY FIGURES
Peter Haggett (1933–2025),
Lauren Gardner (1984–)

BEFORE
1854 English physician John Snow creates a meticulous map to record deaths in a London cholera outbreak.

1947 The World Health Organization establishes the first global disease-tracking service, with information transmitted via telex.

AFTER
2023 British epidemiologist Verity Hill tracks the genomic evolution of variants of SARS-CoV-2 virus (cause of Covid-19 disease) to assess the potential public-health impact.

2024 The European Centre for Disease Prevention and Control implements a "One Health" strategy for dealing with communicable diseases.

The idea that geography plays a role in disease patterns dates back to the mid-19th century. However, a new branch of "spatial epidemiology" emerged in the late 20th century in line with digital technological advances. By using maps and geographic information systems (GIS) to incorporate and assess a wide range of data, this field examined how factors such as environment, demography, socioeconomic status, and behaviours can influence where and how diseases occur, and also how they develop.

From creating malaria risk maps in the tropics to managing the global coronavirus pandemic via the

See also: Thematic mapping 34–37 ▪ Geographic information systems 52–59 ▪ Early warning systems 296–303 ▪ Crisis mapping with crowdsourcing 304–05

Covid-19 dashboard (developed by Lauren Gardner at Johns Hopkins University), data-driven disease tracking has become one of the most important pillars of global public health.

Spatial epidemiology

In the mid-20th century, the "quantitative revolution" in the study of geography emphasized the use of scientific methods, statistical techniques, and mathematical models to analyse spatial patterns and processes. One of the most influential figures in this field was British geographer Peter Haggett, who published *Locational Analysis in Human Geography* in 1965.

A few years later, the chief statistician at the World Health Organization (WHO) invited Haggett to advise on incorporating spatial elements into their disease-diffusion models. This sparked Haggett's interest in medical geography and he began to investigate how geographical factors influence disease spread, including diffusion through contact between adjacent areas (contiguity) and spread through social networks (hierarchical diffusion). In his book *The Geographer's Art* (1982), Haggett describes geography as "the art of the mappable". His work explored how disease-diffusion maps reveal patterns of geographic clustering and variation that can be investigated using spatial statistics, and it paved the way for today's advanced disease-surveillance systems.

Data-driven public health

Modern disease-tracking collects spatial data from multiple sources. Satellites equipped with various sensors are a key component as they monitor environmental factors, such as temperature, humidity, vegetation patterns, and water quality, which are often linked to disease outbreaks. For example, warmer temperatures and increased humidity often extend seasons and geographical ranges for disease vectors, such as mosquitoes, while also providing optimal conditions for bacterial and parasitic growth. Rainfall and vegetation data can indicate where mosquito populations are likely to flourish

Personal health can also be monitored using biosensors, such as smartwatches and fitness trackers. These are able to track continuously individual health metrics, such as temperature and blood oxygen levels, and detect potential issues. Some wearable devices are also designed for contact tracing, helping to identify individuals who may have been exposed to a contagious disease.

In recent years, public engagement has become an increasingly important part of disease tracking. During flu seasons, for example, mobile apps enable users to report symptoms anonymously, helping researchers detect outbreaks faster than using traditional methods. In the same way, citizen-science platforms engage the public in research with data-collecting projects, such as monitoring air and water quality. »

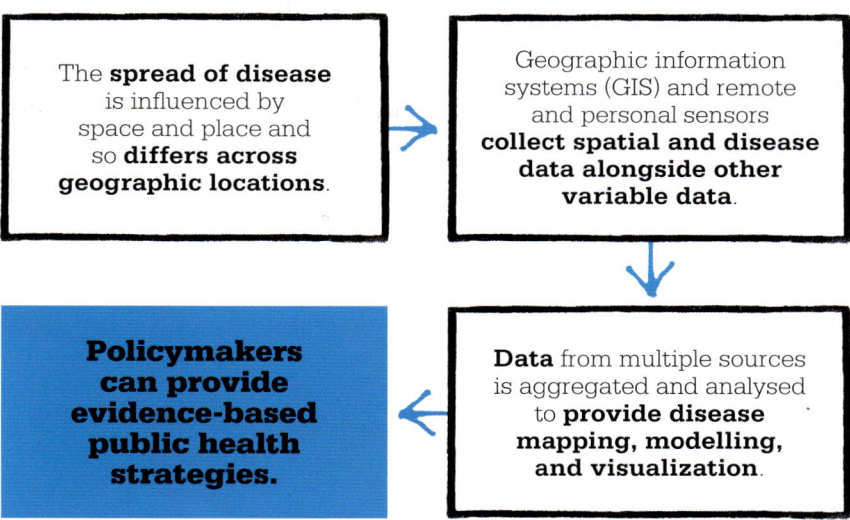

… the essential question is pragmatic: How do we delay or stop epidemic waves spreading in geographical space?
Peter Haggett
***The Geographical Structure of Epidemics*, 2000**

TRACKING DISEASE

Demographic, environmental, and behavioural data is collated using technologies such as geographic information systems (GIS) to create maps that represent population health in a specific region. One of the leading providers of GIS software is the American multinational company Environmental Systems Research Institute (Esri). Founded in 1969 by Jack and Laura Dangermond, its technology has become a vital tool in understanding and managing spatial information. GIS maps can reveal patterns, clusters, and trends that might not be apparent from raw data alone and can help to create risk maps that identify areas with a high probability of disease transmission.

GIS technology is used to map everything from vector-borne diseases, such as malaria, to chronic illnesses, such as diabetes. In sub-Saharan Africa, for example, malaria risk maps have been instrumental in guiding mosquito control programmes, predicting seasonal outbreaks, and directing distribution of protective bed nets to the most vulnerable areas. By combining climate data (such as rainfall and temperature) with population mobility and density, and health records, these maps help make targeted interventions where most effective.

> There will be another pandemic, and these types of frameworks will be crucial for supporting public health response.
> **Lauren Gardner**
> *Nature Computational Science*, 2025

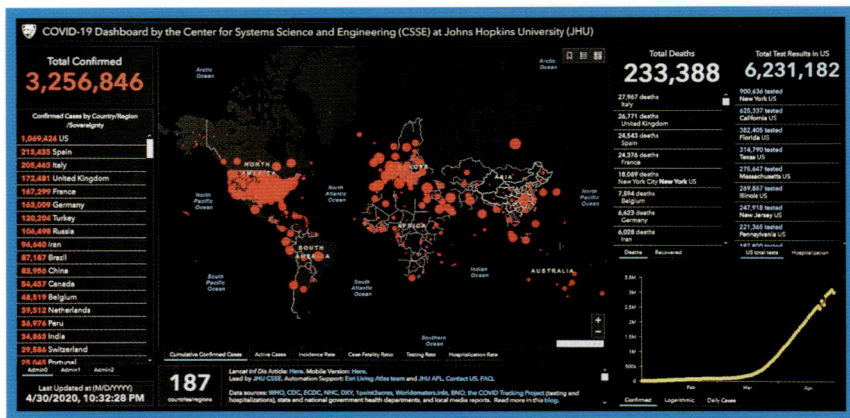

Data visualization

In the mid-1990s, the first digital dashboards for public health were designed to aggregate real-time data from multiple sources and present it in intuitive graphics, such as charts and graphs. By providing a clear overview of health data, public health officials, healthcare providers, and policymakers could identify trends and risks, and make timely and evidence-based decisions.

Public-facing dashboards were also created to provide information about local health risks, promote healthy behaviours, and encourage participation in preventative measures. During significant public health events, such as epidemics and pandemics, dashboards have been crucial for increasing public awareness, sharing vital information, and empowering people to make informed decisions.

The Covid-19 pandemic

When a previously unseen form of coronavirus, later named SARS-CoV-2, was identified in December 2019, the world was unprepared for the speed and scale of its spread.

The Covid-19 Dashboard was used by news organizations and government agencies around the world, who relied on its authoritative data.

Within weeks, cases of the disease, which became known as Covid-19, were being reported worldwide, sparking a global health emergency.

The race to understand this new disease became a test of the world's ability to track, contain, and respond to rapidly changing public health data. Scientists, data experts, public health officials, and technology companies worked together to build tools that could track infections, inform the public, and guide governments in their responses. As a result, Covid-19 became one of the most closely monitored diseases in history.

During the pandemic, Esri provided governments, hospitals, and non-governmental organizations (NGOs) with dashboards and spatial analysis tools that helped monitor outbreaks, manage testing sites, and plan vaccine distribution. Esri's Covid-19 Response Hub became a central resource for GIS professionals and public health officials around the world.

American engineer Lauren Gardner led the development of the now-famous Covid-19 Dashboard at

Johns Hopkins University. Created by the university's Center for Systems Science and Engineering, it launched in January 2020, the day after the US reported its first coronavirus case, and it tracked global cases, deaths, recoveries, and vaccination rates in near real time.

Gardner's team worked in collaboration with the Johns Hopkins Coronavirus Resource Center, which integrated the dashboard with epidemiological data, policy data, and public health guidance. At its peak, the site received over a billion requests a day, highlighting the enormous public appetite for accurate health data during a crisis. By making data transparent and widely accessible, the dashboard empowered health officials to make evidence-based decisions and gave individuals the tools to understand their personal risk.

Disease modelling

While data dashboards helped to track the pandemic in real time, epidemiologists and mathematical modellers worked to predict how the virus would spread, and how public health measures could slow it down. Researchers such as Dr Neil Ferguson of Imperial College, London, produced computer models that warned of the potential for hundreds of thousands of deaths without strict public health measures. These played a major role in shaping lockdown strategies in many countries.

In the US, computational biologist Dr Trevor Bedford, an expert in genomic epidemiology at the Fred Hutchinson Cancer Center in Seattle, analysed the genetic sequencing of viruses to track mutations and understand how the virus was evolving and spreading. His team helped identify early transmission patterns in the US and later contributed to tracking emerging variants of the virus, such as Alpha, Delta, and Omicron.

The data debate

During the Covid-19 pandemic, scientists around the world relied on public dashboards, open-access datasets, and real-time application programming interfaces (APIs) to track variants, assess policy impacts, and guide treatment strategies. Institutions such as the Johns Hopkins Coronavirus Resource Center and the European Centre for Disease Prevention and Control played key roles in curating and sharing this information. This level of transparency not only improved scientific research but also built public trust.

However, collecting personal data raises concerns about privacy and surveillance. Public dashboards can be misinterpreted or misused if not accompanied by clear explanations and context, while false information can spread rapidly on social media with detrimental results. During the Covid-19 pandemic, many people reportedly experienced "data fatigue" as they struggled to make sense of constantly changing numbers.

Implementing disease measures requires a high level of trust in public health. Climate change impacts, antimicrobial resistance, and zoonotic diseases – those that are naturally transmissible from animals to humans and vice-versa – all represent significant threats to the global population. Therefore, addressing these issues requires careful design, clear communication, and investment in both technology and education. ∎

Lauren Gardner

Texas-born, Lauren Gardner earned a BSc in architectural engineering in 2006, followed by an MSc in civil engineering in 2008, and a PhD in transportation engineering in 2011, all from the University of Texas. Her doctoral research focused on the interaction between transportation networks and infectious-disease spread.

In 2011, Gardner became a lecturer at the University of New South Wales in Sydney and a researcher with Australia's National Health and Medical Research Council. She returned to the US in 2019, where she joined Johns Hopkins University. In 2020, together with graduate student Ensheng Dong, Gardener launched the Johns Hopkins Covid-19 Dashboard and, in 2022, Gardner was recognized with a Lasker-Bloomberg Public Service Award. In 2025, with a team of researchers from John Hopkins University, Gardner launched an interactive dashboard that tracks publicly reported measles cases across the US.

DIRECTO

RY

DIRECTORY

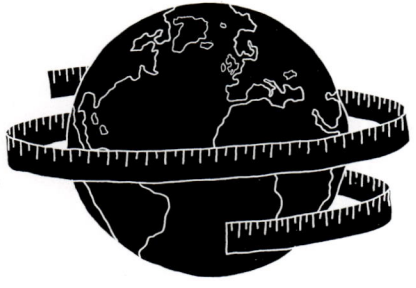

The most important ideas in geography and some of the most prominent people in the field have been presented in this book, but inevitably there has not been space to include all who have shaped the way we think about Earth and its features. This directory, although by no means exhaustive, gives some information on a selection of those figures who have not been introduced elsewhere, including their achievements and the ideas for which they are best known. It also cross-references the pages in the book that discuss the ideas, theories, and geographers they have been associated with or who have influenced their thinking, as well as those they have inspired.

ARISTOTLE
384–322 BCE

Born in Greece, Aristotle was a prominent philosopher and scientist who also tutored Alexander the Great. He made major contributions to biology, ethics, politics, and geography. Through observations of lunar eclipses and the changing position of stars with latitude, he was among the first to posit that Earth is spherical. Aristotle also explored the concept of climate zones, dividing Earth into torrid (hot), temperate, and frigid regions. His systematic approach had a lasting impact on Western science.
See also: The development of cartography 20–21

PTOLEMY
c. 100–170 CE

Claudius Ptolemy, a Greco-Roman scholar who lived in Alexandria, Egypt, was the leading proponent of a model of the Universe that had Earth at its centre. This geocentric model dominated Western science for 1,300 years. Ptolemy also had a profound influence in the field of geography. In his eight-volume treatise *Geography*, he catalogued more than 8,000 places, giving each a precise location using coordinates on a grid of latitude and longitude. These maps contained significant errors, such as underestimating Earth's circumference and distorting the size of Asia; however, Ptolemy's methods offered the first step in transforming map-making from an artistic description of land to an empirical representation of it.
See also: A geographical coordinates system 22–23

PEI XIU
224–271 CE

A founding figure in Chinese cartography, Pei Xiu (also known as Jiyan) was the first geographer in the region to develop some key principles of cartography, such as a graduated scale, a geometric grid reference, and consistent directional measurements. He showed the errors of the maps produced during the preceding Han dynasty – for example, exaggerated distances (though modern analysis has since proved that some Han-period maps were quite accurate). During his tenure as a government minister, Pei Xiu also oversaw the creation of improved geographical records.
See also: The development of cartography 20–21 ▪ A geographical coordinates system 22–23 ▪ Map projections 24–31

ABU ZAYD AL-BALKHI
850–934 CE

Persian scholar Abu Zayd al-Balkhi founded the Balkhi School of Mapping, which focused on conceptual map-making with a regional rather than a global scope, accompanied by explanatory texts and a focus on elements of human geography such as religion and language. His *Suwar al-Aqalim* (*Figures of the Regions*) was one of the earliest attempts to organize geographic knowledge. Al-Balkhi's integration of cartography with cultural and environmental details influenced later Islamic geographers and map-makers.
See also: The development of cartography 20–21

MUHAMMAD AL-IDRISI
1100–65

Muhammad al-Idrisi is remembered for his magnum opus, *The Book of Pleasant Journeys into Faraway Lands*, also known as the *Tabula Rogeriana* because it was commissioned by King Roger II of Sicily. Completed in 1154 after 18 years of work, it was the most comprehensive record of world geography at the time, including political, socioeconomic, and cultural information about different regions. The book's world map was drawn as a circle, or planisphere. In keeping with Islamic traditions, the map had south at the top and featured stylized mountains and rivers; it also included the Indian Ocean as an open body of water with no known far shore.
See also: The development of cartography 20–21 ▪ Map projections 24–31

IBN BATTUTA
1304–69

Moroccan explorer Ibn Battuta is said to have spent nearly 30 years travelling across vast regions of Asia, Europe, and Africa. He chronicled his adventures in a travelogue, *The Rihla* (*rihla* meaning "journey" in Arabic). Despite some questions over its accuracy, *The Rihla* provides a detailed account of the political, social, and cultural landscapes of the 14th-century Islamic world and beyond, with descriptions of the natural environment, trade routes, and cities and their differing customs.
See also: The development of cartography 20–21 ▪ Thematic mapping 34–37

ZHENG HE
1371–1433

A towering figure in Chinese naval history, Zheng He led seven large maritime expeditions for the Ming dynasty between 1405 and 1433. His fleets included vast "treasure ships" that were five times larger than contemporary European ocean-going vessels. The fleets travelled through the South China Sea, the Indian Ocean, the Arabian Sea, and along the east coast of Africa, reaching as far as modern-day Kenya. The voyages expanded China's knowledge of the world beyond its borders, collecting information on foreign cultures, trade routes, and geographical features. The expeditions resulted in detailed maps of land and accurate navigational charts based on star positions and compass bearings.
See also: A geographical coordinates system 22–23

PIRI REIS
1465–1554

An Ottoman admiral and cartographer, Piri Reis was born in Gallipoli (modern-day Türkiye). He is known for the map that he compiled in 1513 by studying older charts, copying coastal outlines, and consulting the reports of experienced sailors he met during military campaigns. One of his sources was a map made by Italian navigator Christopher Columbus. By combining observation with inherited knowledge, Reis created a surprisingly accurate representation of parts of the Americas long before Europeans fully explored them.
See also: A geographical coordinates system 22–23

PEDRO NUNES
1502–78

Portuguese mathematician Pedro Nunes, who was appointed royal cosmographer in 1529, was the first to apply numerical principles to navigation and cartography. His work allowed Portuguese navigators to sail across vast oceans and helped to increase his country's maritime power. Nunes's most significant contribution to navigation was the loxodrome, also known as the rhumb line. This is the path taken by a ship sailing on a constant compass bearing. The path does not follow the shortest route between two points, nor is it straight; instead, it follows a spiral course. Nunes envisaged a nautical chart where the loxodromes would be shown as straight lines. This was achieved by Gerardus Mercator in 1569.
See also: The development of cartography 20–21 ▪ A geographical coordinates system 22–23 ▪ Map projections 24–31

JEAN-ÉTIENNE GUETTARD
1715–86

French naturalist Jean-Étienne Guettard created the first mineralogical map of his country in 1780. As part of his detailed field observations, he documented the distribution of different rock types. He also surveyed the geographical features that linked England and France. By ignoring political boundaries and focusing solely on surface and subsurface features, Guettard laid the groundwork for scientific geography. He was also among the first to recognize the relationship between a region's

geology and its botanical life, noting how plant communities correlated with specific soil types.
See also: Thematic mapping 34–37
- Steno's law of superposition 70–71
- Bioregionalism 242–43

JOSEPH PRIESTLEY
1733–1804

An English chemist and theologian, Joseph Priestley is best known for his discovery of oxygen. His experiments with gases also contributed to early ideas about the carbon cycle. Priestley observed that plants could "restore" air made impure by candles burning or animals breathing, further noting that plants and animals had complementary roles in maintaining air quality. His discoveries laid the groundwork for the modern understanding of the carbon cycle – the continuous movement of carbon among the atmosphere, living organisms, oceans, and Earth through processes such as photosynthesis, respiration, decomposition, and combustion.
See also: The greenhouse effect 80–83

TADATAKA INŌ
1745–1818

Tadataka Inō spent most of his life working at the family sake-brewing business, retiring at the age of 50 to pursue his passion for astronomy and mathematics. In 1800, he was commissioned to map the entire Japanese coastline. Inō embarked on a series of walking expeditions, making meticulous measurements, using astronomical observations to determine latitude (an advanced technique at the time), and developing a precise step-counting method for gauging distance. His *Maps of Japan's Coastal Area*, published posthumously in 1821, became the standard maps of Japan for the next 100 years.
See also: A geographical coordinates system 22–23
- Triangulation 32–33

ABRAHAM WERNER
1749–1817

A prominent figure in geology and mineralogy, and an influential voice at the Freiberg Mining Academy in his native Germany, Abraham Werner is best remembered for developing the theory of neptunism. This posited that all rocks – including those of volcanic origin, such as basalt – were formed as sediments from an ocean that once covered the entire Earth. Although this theory was later disproved, Werner's teaching of the systematic observation and classification of rocks was important groundwork for the scientific study of geology.
See also: Isostasy 94–95 ▪ The cycle of erosion 102–07 ▪ The structure of Earth 112–13

ALEXANDER VON HUMBOLDT
1769–1859

Dozens of geographical features and plant and animal species have been named after Alexander von Humboldt, which is testament to this Prussian explorer's influence on modern science. Humboldt travelled extensively, especially around South America, and he pioneered the field of biogeography, the study of how the distribution of wildlife is linked to geography. His depiction of the Chimborazo volcano, illustrating the distribution of plant life at different elevations, is a classic example of the field.
See also: Thematic mapping 34–37 ▪ Biomes and ecological zones 136–39 ▪ Global ocean circulation 140–41

CARL RITTER
1779–1859

Born in Prussia, Carl Ritter was a professor of geography at the University of Berlin. A contemporary of geographer Alexander von Humboldt, he emphasized the interconnection between the natural environment and human activity. His multi-volume work *Die Erdkunde* (*Geography*) detailed the physical and cultural characteristics of different regions, highlighting the role of geography in shaping history and society. He promoted a regional approach – stressing systematic study – and his teachings helped establish geography as an academic discipline.
See also: Environmental determinism 156 ▪ Geography as human ecology 162–63

MARY SOMERVILLE
1780–1872

Scottish science writer Mary Somerville's main contribution to the field is her 1848 textbook *Physical Geography*. It was one of the first English-language books to gather knowledge from a range of related disciplines – including geology, climatology, and oceanography – for the purpose of describing Earth's surface. Somerville's book helped to establish physical geography as a discipline and remained a standard

textbook for several decades. Somerville was also an advocate for women's rights.
See also: The hydrological cycle 66–69 ▪ Steno's law of superposition 70–71 ▪ The cycle of erosion 102–07

ARNOLD HENRY GUYOT
1807–84

Early in his career, Swiss-American geologist Arnold Guyot studied glaciers in the Alps as a supporter of Louis Agassiz's theory that ice ages had shaped those rugged landscapes. Guyot suggested that glaciers flowed like rivers, only much more slowly. This theory was a big step in understanding how glaciers erode rocks to create distinctive valleys. Later, Guyot became a professor at Princeton University and wrote numerous atlases and textbooks that were widely used in North American schools.
See also: The hydrological cycle 66–69 ▪ Glaciation and ice ages 86–89 ▪ Palaeoclimatology 126–27

EDUARD SUESS
1831–1914

In 1875, London-born Austrian geologist Eduard Suess suggested in his book *The Origin of the Alps* that mountain ranges formed as a result of Earth contracting as it cooled – a view known as contractionism. In the same book, he also coined the term "biosphere" to describe a "zone at the surface of the lithosphere" where life exists. Suess also suggested that South America, Africa, the Indian subcontinent, and Australia were once connected, forming a supercontinent that he named Gondwanaland. As support for his theory, he pointed to the distribution of glossopteris fern fossils across the oceans.
See also: Uniformitarianism 75 ▪ Orogeny 76–79 ▪ Biomes and ecological zones 136–39

FERDINAND VON RICHTHOFEN
1833–1905

In 1860, German geographer Ferdinand von Richthofen joined a two-year expedition to South and East Asia, but all the records he collected on that mission were lost. On a subsequent trip, he travelled widely around China and the surrounding countries. This time he successfully compiled his findings into a five-volume work. Von Richthofen was a pioneer of chorology, or the study of place. He also coined the term "Silk Road", which is now synonymous with the ancient overland trade routes connecting East Asia and the Mediterranean.
See also: Cultural diffusion 164–71

CLARENCE DUTTON
1841–1912

American geologist Clarence Dutton helped to set up the US Geological Survey with John Wesley Powell in 1879. Dutton's area of expertise was documenting the geology and landforms of the Colorado Plateau and the Grand Canyon. In 1889, he coined the term "isostasy" for his theory – now proved – about the nature of Earth's crust. Dutton proposed that the crust maintains a gravitational equilibrium, with lighter, low-density continental crust effectively "floating" higher on the denser mantle. This proved a fundamental principle for understanding mountain building and other vertical movements of Earth's crust, and it was a precursor to the theory of plate tectonics.
See also: Orogeny 76–79 ▪ Isostasy 94–95 ▪ Continental drift 110–11 ▪ The structure of Earth 112–13 ▪ Plate tectonics 114–21

GROVE KARL GILBERT
1843–1918

Known by his initials GK, Gilbert was a leading figure in the early years of the US Geological Survey and is considered one of the founders of geomorphology, the scientific study of landforms and the processes that shape them. He is remembered for his definition of a graded river – that is, a river that over time has achieved an equilibrium between the forces of erosion and the water's sediment load. In other words, the current delivers fresh sediment at the same rate that it washes it away.
See also: Orogeny 76–79 ▪ The cycle of erosion 102–07

VASILY DOKUCHAEV
1846–1903

A Russian geologist, Vasily Dokuchaev is widely regarded as the founder of pedology, or the scientific study of soils. His research transformed the understanding of soil from a simple geological deposit into a complex, dynamic structure. Dokuchaev was able to map the distribution of distinct soil types correlated to the interaction of five key factors: climate, the parent material, topography, the organisms present, and time.
See also: Pedology 128–31

ÉDOUARD-ALFRED MARTEL
1859–1938

French lawyer Édouard-Alfred Martel conducted more than 1,500 cave expeditions, mostly in the karst landscapes of southern France. His adventures included a pioneering journey by canoe along the underground river in the Gouffre de Padirac chasm in 1889. As well as making accurate maps of caves and theorizing about the formation of subterranean features, Martel helped to transform caving into a scientific pursuit. He is regarded as the founder of speleology, the scientific study of caves.
See also: Karst landscapes 108–09

VILHELM BJERKNES
1862–1951

Norwegian physicist Vilhelm Bjerknes proposed that weather could be predicted by using a set of hydrodynamic and thermodynamic equations. In 1917, he established the Bergen School of Meteorology, which was actually a collective of like-minded researchers. The work of the school revolutionized the understanding of atmospheric circulation, leading to high-speed computer-based forecasting techniques, as well as longer-term climate modelling.
See also: Extreme weather 142–43 ▪ Climate change 232–39 ▪ Climate modelling 268–71

MARION I. NEWBIGIN
1869–1934

Combining her interest in geography and biology, Scottish scientist Marion Newbigin became a leading figure in animal geography, a sub-branch of biogeography. Her book *Animal Geography* (1913) applied scientific principles to the study of animal distribution. In *Geographical Aspects of Balkan Problems* (1915) and *The Mediterranean Lands* (1924), Newbigin explored the links between regional geography and politics, a practice that has since fallen out of favour. In 1923, Newbigin became the first female recipient of the Livingstone Gold Medal, awarded by the Royal Scottish Geographical Society.
See also: Biomes and ecological zones 136–39

ALEXANDER DU TOIT
1878–1948

South African geologist Alexander du Toit was key in advancing the theory of continental drift. An early supporter of Alfred Wegener's ideas, du Toit dedicated much of his career to proving the southern continents once formed a supercontinent called Gondwanaland. He took part in several expeditions to South America with the aim of comparing the rock formations, fossils, and glacial activity there with those of his native continent. Du Toit's results formed the core of his work *Our Wandering Continents* (1937).
See also: Continental drift 110–11 ▪ Plate tectonics 114–21

EVA TAYLOR
1879–1966

In 1930, Britain's Eva Taylor became the first woman to hold an academic chair in geography in the UK, at Birkbeck College, University of London. Before this appointment, she had made a name for herself by writing textbooks that influenced a generation of geographers. However, Taylor's main expertise lay in the history of cartography, exploration, and navigation. Her work *Tudor Geography* (1930) described how the English state leveraged the new geographical knowledge of the time to gain power on the world stage.
See also: The development of cartography 20–21 ▪ A geographical coordinates system 22–23

HENRI LEFEBVRE
1901–91

A French Marxist philosopher and sociologist, Henri Lefebvre urged geographers to do more than just describe locations and, instead, to examine the power dynamics that produce them. He is best known for his concept of the "social space", which is actively created and shaped by social, political, and economic forces. Lefebvre argued that capitalism transforms space into a commodity, which then alienates people from their own surroundings. His essay "The Right to the City", which advocated a more democratic and participatory approach to urban life, was published in March 1968 and is said to have influenced the organizers of the student revolts that shook France two months later.
See also: The urbanization of capital 204–05 ▪ Spatial justice 206–07 ▪ The "Smart Growth" movement 282–83

LESLIE HOLDRIDGE
1907–99

During Word War II, US botanist and climatologist Leslie Holdridge spent time in South America,

researching how to produce the antimalarial drug quinine for the US government. Then, in 1947, he published his life zones system as a way to classify land areas based on the relationship between climate and plant growth. Holdridge's system became a cornerstone of biogeography, and it has been particularly influential in tropical forestry and conservation, because it helps to predict how climate change might impact biodiversity and plant communities. In 1954, Holdridge set up La Selva Biological Station in Costa Rica, which remains a world-leading conservation research centre.
See also: Biomes and ecological zones 136–39 ▪ Sustainable development 244–45 ▪ Globalization 250–53

ESTHER BOSERUP
1910–99

Danish economist Esther Boserup is best known for a groundbreaking theory she presented in *The Conditions of Agricultural Growth: The Economics of Agrarian Change under Population Pressure* (1965). In this book, Boserup challenged the Malthusian view that population growth would inevitably lead to a global famine, arguing instead that population pressure is a powerful stimulus for agricultural innovation. Boserup's optimistic perspective emphasized the role of human agency in the face of resource constraints; however, this idea is proving more controversial as the climate crisis continues.
See also: Malthusian theory 148–49 ▪ The tragedy of the commons 220–21 ▪ Sustainable development 244–45

JANE JACOBS
1916–2006

Jane Jacobs's influential book *The Death and Life of Great American Cities* (1961) challenged the post-war trend of large-scale urban renewal projects, arguing that they destroyed vibrant existing communities. An American journalist with no training in city planning, Jacobs based her ideas on careful observation of city life, particularly in New York City's Greenwich Village, where she lived. She reasoned that safe and economically healthy communities formed in places where socially mixed populations lived close together. Jacobs believed that neighbourhoods should maintain a balance of older buildings and new developments. In many ways, these ideas remain a cornerstone of modern city planning.
See also: Gentrification 196–97 ▪ Land use zoning 264–65 ▪ The "Smart Growth" movement 282–83

RAYMOND DASMANN
1919–2002

The American author of the influential textbook *Environmental Conservation* (1959), Raymond Dasmann served in the US Army during World War II. In 1954, he earned his PhD in zoology from the University of California, Berkeley, demonstrating a particular interest in conservation. He became well known as an early advocate for sustainable development. Dasmann was also a champion for the rights of Indigenous peoples, arguing that their cultures are crucial to preserving natural landscapes. In 1971, he helped develop UNESCO's Man and the Biosphere Programme, which seeks to encourage global communities to value their environment and work to protect it.
See also: Sustainable development 244–45 ▪ The environmental justice movement 246–47

TETSUYA "TED" FUJITA
1920–98

Born in Japan, Ted Fujita was involved in studying the aftermath of the atomic bombs at Nagasaki and Hiroshima. He later moved to the US and joined the University of Chicago. A meteorologist, he created the Fujita scale (F-scale), a system for classifying tornado intensity and damage. His analysis of nuclear explosions and a plane crash site played a role in his discovery of downbursts and microbursts, powerful localized downward-moving winds from severe thunderstorms that pose major threats to aviation.
See also: Extreme weather 142–43

MILTON SANTOS
1926–2001

One of the most important thinkers on urbanization in developing countries, Brazilian geographer Milton Santos was a qualified lawyer. However, after being exiled by Brazil's dictatorship, he chose to pursue an academic career in Europe and North America. He wrote more than 40 books, the most influential of which was *The Shared Space* (1979), which contains his theory of urban development. Santos was a vocal critic of globalization, which he said exacerbated social inequalities and marginalized the poor,

particularly in the Global South. In 1994, he was awarded the Vautrin Lud Prize, the most prestigious award in the field of geography.
See also: The urbanization of capital 204–05 ▪ The environmental justice movement 246–47 ▪ Globalization 250–53

ELINOR OSTROM
1933–2012

In 2009, US political scientist Elinor Ostrom became the first woman to win the Nobel Prize in Economics. Her research focused on economic governance, particularly regarding the management of shared resources, known as "the commons". Ostrom challenged the prevailing theory that common resources inevitably face overuse without government regulation or private ownership. Her studies showed that local communities often develop perfectly effective self-governing mechanisms to manage these resources sustainably. Her work emphasized "bottom-up" solutions created by the users themselves.
See also: The tragedy of the commons 220–21 ▪ Natural resource management and conservation 222–27

SUSAN HANSON
1943–

Susan Hanson is an American human geographer known for her pioneering work on gender, transportation, and everyday mobility. A professor at Clark University, Massachusetts, Hanson integrates time-geography and feminist perspectives to explore how spatial and social factors shape people's opportunities in the urban built environment. She shows that men and women experience cities and travel differently due to unequal access to resources, such as transport, work, and time, and their different social responsibilities. Her research continues to influence thinking about sustainable transport in urban planning and social equity in geography.
See also: Gender and space theory 210–13 ▪ Time-geography 266–67

DENIS WOOD
1945–

A former professor of design with a PhD in geography, American cartographer Denis Wood is known for challenging traditional ideas about maps. In 1975, his *Narrative Atlas of Boylan Heights* project started mapping features such as street lights and Halloween pumpkins in a historic neighbourhood of Raleigh, North Carolina, presenting a different view of the area than just its roads and buildings. Wood's *The Power of Maps* (1992) suggests that, far from being objective representations, maps present an argument about the world through the careful selection and arrangement of content.
See also: Thematic mapping 34–37

MAUDE BARLOW
1947–

Known for her activism, Maude Barlow, the Canadian author of *Blue Gold: The Battle Against Corporate Theft of the World's Water* (2002), helped to found the Blue Planet Project, which advocates for the human right to water. Her activism led her to serve as a senior advisor on water to the president of the United Nations General Assembly in 2008–09, and she helped in the campaign to have water recognized as a human right by the UN. Barlow is a vocal critic of free trade agreements, arguing that they undermine local democratic control.
See also: The environmental justice movement 246–47 ▪ Globalization 250–53

NIGEL THRIFT
1949–

British academic Nigel Thrift has had a varied career in the social sciences. His stand-out contribution to geography is non-representational theory, which focuses on practices and interactions that form through social relationships – from children's play, to activities such as dancing. A co-founder of the *Environment and Planning D: Society and Space* journal, Thrift has received many awards, including the Royal Geographical Society's Victoria Medal. In 2015, he was knighted for services to higher education.
See also: Time-geography 266–67

LINDA TUHIWAI SMITH
1950–

A Maori scholar and distinguished professor, Linda Tuhiwai Smith is known for her groundbreaking book *Decolonizing Methodologies* (1999), which critiques how traditional Western research has served colonial interests. Indigenous peoples and lands have historically been represented through Western perspectives, while the Indigenous worldview has been transformed and Westernized. Smith's book has led to the reframing of scholarly

approaches in environmental and political human geography by centring Indigenous knowledge and promoting more collaborative, community-led research.
See also: Spatial justice 206–07

CINDI KATZ
1954–

A professor of Earth and environmental sciences at the CUNY Graduate Center of New York, Cindi Katz is known for her work on social reproduction, which explores how social structures are maintained through the generations by different means. She has also researched the relationship between children and the environment. A founding editor of the *Social and Cultural Geography* journal, Katz was the recipient of the American Association of Geographers' 2024 Lifetime Achievement Honor.
See also: Local and global 208–09
- Gender and space theory 210–13
- Globalization 250–53

JONI SEAGER
1954–

Known for her activism on gender and environment, Joni Seager is a professor of global studies at Bentley University, Massachusetts. Her work challenges traditional approaches to environmental policy and data by highlighting the role of gender. Seager is the author of several influential books, including *The Women's Atlas* (2018), which maps the status of women around the world. Her research also focuses on climate change, militarism and the environment, and illegal wildlife trafficking. Seager has served as a consultant for the United Nations and other international organizations, playing a key role in developing gender-sensitive statistical protocols.
See also: Gender and space theory 210–13

ASH AMIN
1955–

Born in Uganda, Ash Amin lived in Kenya before his family emigrated to Britain when he was 16. He gained a PhD in geography from the University of Reading and has become known for his work on urban and regional development. Amin is a prominent figure in human geography, researching topics such as cities as social and technical assemblages, the relationship between everyday life and urban infrastructure, and the intersections of urban encounters and national belonging. Along with Nigel Thrift, he is a co-founder of the journal *Environment and Planning D: Society and Space*.
See also: Cultural landscape theory 172–75 ▪ Humanistic geography 200–01

SARAH WHATMORE
1959–

British geographer Sarah Whatmore is a pioneer in "more-than-human" geography, a field that does not separate human culture from nature and looks at both the social and ecological dimensions of life. After university, Whatmore became a town planner, but she returned to academia to research the division of labour on farms by gender. A professor at the University of Oxford since 2004, she was appointed Dame Commander of the Order of the British Empire in 2020 for her services to the study of environmental policy.
See also: Gender and space theory 210–13

TIM CRESSWELL
1965–

British human geographer Tim Cresswell studied geography at University College London and earned his PhD from the University of Wisconsin–Madison, where Yi-Fu Tuan was his supervisor. Today, Cresswell is a leading figure in the "mobilities paradigm", which explores how movement through a place shapes social and cultural life. Cresswell has held professorships at several universities, and he currently holds the Ogilvie Chair of Human Geography at the University of Edinburgh.
See also: Humanistic geography 200–01

KAREN BAKKER
1971–2023

A Canadian geographer and author, Karen Bakker was a leading thinker on water governance and the role of digital tools in environmental sustainability. She explored how politics and technology shape access to natural resources, particularly water, and she also suggested that digital innovation in ecology could help combat climate change. Bakker's book *Gaia's Web* helped popularize the concept of digital environmentalism and promote new methods of environmental governance.
See also: Sustainable development 244–45 ▪ The environmental justice movement 246–47

GLOSSARY

Aggregate To form or group into a class or cluster.

Albedo effect The ability of a surface to reflect sunlight, with lighter surfaces having a higher albedo and darker surfaces having a lower albedo.

Anthropocene A proposed geological age, defined as the period during which human activity has been the dominant influence on climate and the environment.

Asthenosphere The semi-solid layer of Earth's **mantle**, below the **lithosphere**.

Biodiversity The variety of plants, animals, fungi, and microorganisms that make up the natural world.

Bioenergy Renewable energy derived from **biomass**, through processes such as direct combustion, thermochemical conversion, or biological conversion.

Biogeography The scientific study of the distribution of living organisms across space and time, and the factors that influence these patterns.

Biome A geographical area classified according to the species of plant and animal life within it.

Biomass The total quantity of a given organism within a habitat, generally expressed as weight or volume. Also a type of fuel made from organic matter, usually burned to generate electricity.

Biosensor An analytical device that measures biological or chemical reactions. Used in applications such as health and disease monitoring, drug discovery, and detection of pollutants.

Biosphere The layer of Earth in which life can exist, situated between the atmosphere and **lithosphere**; the sum of all **ecosystems** on the planet.

Boreal A cold temperate climate zone located in the northern hemisphere, between roughly 50° N and 65° N **latitude**, and dominated by extensive conifer forests (taiga).

Cartogram A thematic map (also called a value-area map or an anamorphic map) on which the size of geographic entities is altered to be proportional to a statistical value, such as population.

Census An official survey of the population of a country, conducted to obtain data that can help a government track societal change, and inform their policy decisions.

Chorochromatic map A type of thematic map (also known as an area-class, qualitative area, or mosaic map) that uses distinct colours and patterns to represent data and data-defined boundaries.

Choropleth map A type of thematic map that uses varying colours to indicate the average values of a particular quantity in predefined geographic areas, such as countries, states, or districts.

Computer simulation A computer-run digital model of a real-world system or phenomenon, designed to study its behaviour under variable conditions.

Conservation The protection and preservation of animal life, plant life, and natural resources.

Conurbation An extended urban area, where the expansion of towns and cities has caused them to form one merged, continuous urban or industrially developed region.

Convection In plate tectonics, the circular motion of Earth's mantle, due to heat from the core, which moves tectonic plates.

Cosmography The scientific study of the origin and structure of the universe as a whole and of its related parts.

Data set A structured collection of related data points, organized for analysis, processing, or training machine-learning models.

Deltaic Relating to a river delta – a landform created at the mouth of a river, where accumulated sediment has resulted in fertile, low-lying plains.

Demographic A particular sector of a population, defined in terms of distinct factors, such as age, income, or background.

Deposition The geologic process by which sediments, carried by agents like wind, water, ice,

GLOSSARY

or gravity, settle and accumulate in a new location, forming layers of material that build up over time.

Desertification The process by which fertile land becomes desert, typically as a result of drought, deforestation, or unsustainable agricultural practices.

Diffusion In anthropology, the dissemination of elements of culture to another region or people.

Ecosystem A community of living organisms in a given environment that interact with and affect one another.

Electromagnetic radiation The transfer of energy through oscillating electric and magnetic fields, travelling at the speed of light, which interacts with Earth's surface and atmosphere.

Emigration The act of leaving one's own country to settle permanently in another.

Epidemiology The study of how diseases spread through populations, and the impact this has on the wider **ecosystem**.

Epoch A division of geologic time, less than a period and greater than an age.

Equatorial Pertaining to the region near Earth's equator – the circle of **latitude** that divides Earth into the northern and southern hemispheres.

Erosion The geological process where Earth materials, such as soil and rock, are gradually worn away and moved by natural forces, such as wind, water, or ice.

Forestry The science or practice of planting, managing, and caring for forests.

Fossil fuels Non-renewable fuels formed over millions of years from plant and animal remains.

Geocoding The computational process of converting a location into a set of coordinates for mapping and spatial analysis.

Geodemography The practice of grouping neighbourhoods or localities into segments based on shared **demographic** and geographic characteristics.

Geodesy The scientific study of measuring and representing the Earth's geometric shape, gravity field, and orientation in space, and how these change over time.

Geomagnetism The branch of geology concerned with Earth's magnetic properties.

Geomorphology The scientific study of Earth's landforms and the processes that create, shape, and change them over time.

Geophysics The study of Earth's physical properties and processes.

Geopolitics The study of the effects of Earth's geography on politics and international relations.

Geosphere Any part of the physical, solid surface and interior of Earth – including all rocks, minerals, and sediments.

Geostatistics A branch of statistics focusing on spatially related data, such as soil moisture or air pollution levels.

Geothermal Relating to or utilizing heat energy from Earth's interior.

Gerrymandering The political manipulation of electoral district boundaries to advantage a particular party, group, or socioeconomic class within the constituency.

Greenhouse effect The way in which gases in Earth's atmosphere trap heat. The buildup of these gases leads to global warming.

Hemisphere A half of a sphere, such as Earth, which is divided into northern and southern halves by the equator, or into western and eastern halves by an imaginary line passing through the poles.

Holocene The name given to the current period of Earth's history, dated from the end of the last major glacial epoch, or "ice age".

Homogenization The process where diverse local characteristics become increasingly uniform or similar across different places.

Hyperbolic line A curve in hyperbolic geometry that represents the shortest distance between two points.

Igneous rock A type of rock formed from the cooling and solidification of molten rock (**magma** or lava).

Immigration The act of entering a foreign country with the intention to live there permanently.

Industrialization The process of transforming an agrarian-based economy into one focused on mass manufacturing and industry.

Infrastructure The essential organizational systems and services that enable an economy to function effectively.

Interpolation The process of using a set of known data points to estimate or predict unknown data points that fall between those known points.

Isostasy The principle of gravitational equilibrium where Earth's crust "floats" on the underlying **mantle**.

Lines of latitude Imaginary lines, also called **parallels**, that measure distance in degrees north or south of the equator.

Lines of longitude Imaginary lines, also called **meridians**, that run from pole to pole and measure distance east or west of the **prime meridian**.

Lithosphere The rigid, rocky outer layer of Earth, consisting of the crust and the solid outermost layer of the upper **mantle**.

Loxodrome An imaginary line (also called a **rhumb line**) on the surface of a sphere that crosses all **meridians** at the same angle.

Magma Hot fluid or semi-fluid material below or within Earth's crust from which lava and other **igneous** rock is formed on cooling.

Magnetometer An instrument for measuring magnetic forces, especially Earth's magnetism.

Mantle The region of Earth's interior between the crust and the core, believed to consist of hot, dense silicate rocks.

Map projection A mathematical method of transforming the 3-D, curved surface of Earth onto a 2-D plane.

Meridians Imaginary lines (also known as **lines of longitude**) between the North and South Pole, to measure distance, in degrees, east or west from a reference point.

Metamorphic Denoting or relating to rock that has undergone transformation by heat, pressure, or other natural agencies.

Meteorology The scientific study of Earth's atmosphere and short-term atmospheric phenomena (weather), with a focus on weather forecasting.

Methodology A system of methods used in a particular area of research.

Microclimate A local set of atmospheric conditions that differ from those in the surrounding areas.

Moraine A mass of rocks and sediment carried and deposited by a glacier, typically as ridges at its edges or extremity.

Multispectral Operating in or involving several regions of the electromagnetic spectrum.

Open-source Denoting software for which the original source code is made freely available via the internet and may be redistributed and modified.

Orogeny A geological process caused by the collision of two tectonic plates, in which a section of Earth's crust is folded and deformed by lateral compression to form a mountain range.

Parallels Imaginary concentric circles (also known as **lines of latitude**) that run around Earth parallel to the equator, to measure distance, in degrees, north or south from a reference point.

Pedology The scientific study of the formation, nature, **ecology**, and classification of soil.

Photogrammetry The process of obtaining accurate, real-world measurements, maps, and 3-D models by taking a series of overlapping photographs.

Polygon A 2-D, closed shape with straight sides that form a complete boundary.

Precipitation Any form of water – liquid or solid – that falls from clouds in the atmosphere and reaches Earth's surface.

Prime meridian The line of 0 degrees longitude that runs from the north pole to the south pole and divides the Earth into the eastern and western hemispheres.

Quantitative geography The use of scientific, statistical, mathematical, and computational methods to analyse, model, and explain geographical phenomena.

Radar An electronic system that detects and determines the location, speed, and other characteristics of objects by emitting radio pulses and analysing their reflected echoes.

Raster data A grid of pixels (or cells) representing **spatial** data, where each pixel holds a specific value representing a geographic feature or phenomenon, such as elevation, temperature, or land cover.

GLOSSARY 327

Real-time computing A system in which input data is processed within microseconds.

Remote sensing The process of using sensors, often mounted on aircraft or satellites, to detect and record energy that is reflected or emitted by a target to thereby gather information about an object or phenomenon from afar.

Rhumb line A navigational path on Earth's surface that crosses all meridians of longitude at a constant angle, meaning a vessel can maintain a fixed compass bearing relative to true north for the entire journey.

Salinity The saltiness or amount of salt dissolved in a body of water.

Satellite Any object that orbits another, larger object, such as a planet or a star, or a machine launched into space for data-collecting purposes.

Scatterometer A **radar** instrument that measures the amount of radar energy scattered back from a surface, particularly in order to determine ocean surface wind speeds and directions.

Seismic waves Mechanical waves of acoustic energy that travel through Earth, generated by events like earthquakes, volcanic eruptions, or large explosions.

Seismology The study of seismic waves that move through and around Earth.

Spatial Anything related to the location, arrangement, distribution, and organization of phenomena on Earth's surface.

Strata Distinct layers of rock, soil, or other materials, formed by **deposition** over time.

Stratosphere The second-lowest layer of Earth's atmosphere, located above the **troposphere** and below the mesosphere.

Subduction The geological process where one tectonic plate sinks or is forced beneath another plate at a convergent plate boundary, typically when an oceanic plate collides with a continental plate.

Substrate A rock surface where geological processes such as sediment deposition occur.

Subtropical The climatic and geographic regions located immediately north and south of the tropics and characterized by distinct seasons with hot, humid summers and mild, cool winters.

Sustainability Balancing social, economic, and environmental goals to ensure a long-term, thriving planet and society for everyone.

Plate tectonics A scientific theory that explains how major landforms are created as a result of Earth's subterranean movements.

Tipping point A critical threshold within Earth's system where a small perturbation triggers a sudden, dramatic, and often irreversible change in a system's state or behaviour.

Topographic map A 2-D map representing a 3-D landscape using contour lines that connect points of equal elevation, along with symbols and colours to show natural and artificial features.

Traverse method A technique of establishing a framework of connected lines (of measured distances, angles, and directions) to determine survey points.

Triangulation A surveying technique: by measuring angles from two known points (a baseline) to an unknown point, and using trigonometry, the position of the third point can be calculated precisely.

Tropical The regions surrounding Earth's equator, defined by a warm, humid climate with little seasonal change.

Troposphere The lowest, densest layer of Earth's atmosphere, which contains nearly all atmospheric water vapour and aerosols, making it the region where virtually all weather phenomena, such as clouds and wind, occur.

Tundra A vast, cold, and barren **biome** characterized by frozen ground (permafrost) and a lack of trees. Tundra is found in the Arctic and at high altitudes.

Urbanization The process of making an area more urban through development and population growth.

Vector data Real-world geographic features represented using points, lines, and polygons defined by x, y coordinates.

INDEX

Page numbers in **bold** refer to main entries.

A

Ackerman, Edward 192
Acorn (A Classification of Residential Neighbourhoods) 276, 277
Aeolus mission 45
aereal differentiation 266
aerial photography 15, **40–45**, 58, 292
Agarwal, Anti 247
Agassiz, Louis 64, 86, **87**, 88, 126
ageing demographics 181
Age of Exploration 13, 30, 72
Agricola, Georgius 90
agriculture
　biomes 139
　climate change 237
　development 190
　drainage 107
　floodplains 230, 231
　globalization 252
　greenhouse gases 235
　land degradation and desertification **228–9**, 257, 259
　peripheral countries 199
　pesticide use 240
　population growth 148, 149
　precision 294–5
　remote-sensing **292–5**
　reshaping of landscape 163, 173
　shift to 129
air pollution 288
Airy, George Biddell 64, **94–5**
Alan Turing Institute 306
al-Balkhi, Abu Zayd 13, **316**
albedo effect 238, 239, 289
Albers, Heinrich Christian 31
Alberti, Leon Battista 142
Aleutian Trench 121
Alexander the Great 21
Alfasi, Nurit 208
al-Idrisi, Muhammad 13, **317**
Al-Khwarizmi 20
Alonso, William 150, 184
Alpine-Himalayan orogeny 78, 79
altitude
　climate 100, 101
　plants 136, 137
Altvater, Elmar 208
Alvarez, Walter and Luis 75
Amazon basin 252
Amazon Web Services 280
Amin, Ash **323**
Amin, Samir 198
Anaximander 13, 18, **20–21**, 26
anchor-point theory 195
Andean orogeny 79
Anderson, John 132, 133
angle of slope 105
animals
　biomes and ecological zones 136–9
　changing distribution 101
　climate change 237
　effect of DDT 240
　extinctions 252, 257–8
　monitoring biodiversity 306–7
　palaeoclimatology 127
　soil formation 129, 130
Anselin, Luc 262, 280, 281
Antarctic Bottom Water 141
Antarctic, climate change 239
anthromes 136, 139
Anthropocene epoch 219, **256–9**
anthropogenic biomes 139
Appadurai, Arjun 170, 171
applied geography 12, 262–3
aquifers 69
ArcGIS 54, 56
Arctic, climate change 238–9
Argo programme 140
Aristotle 21, 64, 66, 84, 220, 230, **316**
Arnold, John E. 282
Arons, Arnold 140
Arrhenius, Svante 80, 81, **82–3**, 218, 234, 236
Arrighi, Giovanni 198
artificial intelligence (AI) 48, 61, 214, 268, 277, 281, 306, 307
assimilation 171
associative cultural landscape 175
asthenosphere 95, 120, 122
atmosphere
　and life 241
　carbon dioxide in 218, 234–9
　circulation cells 73, 84, 100
　climate modelling 268, 269, 270, 271
　greenhouse effect 80–83
　insulation of Earth 80, 81, 234
　water in 68
atmospheric circulation, winds **72–3**
atmospheric pressure 141
atolls 125
atomic weapons 258, 259
Aubréville, André 218, 228, **229**, 248
Auerbach, Felix 160
authority constraints 267
Axial Seamount 125
azimuthal projections 26, 29, 31

B

Babylonians 13, 18, 20, 188
"back-to-the-land" movement 242
bacteria 129
Bahuguna, Sundarlal 227
Bailey, Walter 44
Bakker, Karen **323**
Ballard, Robert 120
balloons 42, 43
Baran, Paul 198
Barbari, Jacopo de' 42
Barlow, Maude 66, **322**
Barrows, Harlan H. **162**, 163, 172
basaltic lava 123, 131
Bassett, Edward 265
bathymetry 46, 117
Bauer, Peter 61
Baumeister, Reinhard 262, **264–5**
beacons, radio 48
Beaufort, Francis/Beaufort scale 142
Beck, Ulrich 219, 250, **251**, 253

Bedford, Trevor 313
bedrock 75, 105, 106, 107
Béguyer de Chancourtois, Alexandre-Émile 39
behavioral geography 147, **194–5**
Behnke, Roy 229
BeiDou 51
bematists 21
Benioff, Hugo 124
Berghaus, Heinrich 66
Berg, Peter 219, 241, **243**
Bernhardi, Friedrich von 155
Bid rent theory 150
biocapacity 248, 249
bioclimate classification 98
biodiversity 61, 101, 227, 230, 237
　loss of 243, 252, 257, 259
　monitoring **306–7**, 309
bioenergy 291
biogeochemical cycles 271
biogeography 137
biomass 288
Biomass satellite 308
biomes **136–9**
bioregionalism 218, **242–3**
biosphere 136, 137, 139, 241, 256
biota 241
biotemperature 100
biotic factors 138
Birkeland, Peter 128
birth rates 147, 178, 179, 180, 181
Bjerknes, Vilhelm 60, 268, **320**
Black Americans
　environmental justice 246–7
　social injustice 263, 278
Black, Joseph 80
black smokers 120
Blanche, Paul Vidal de la **157**
Blomley, Nicholas 202
Boas, Franz 146, **166**, 167–8, 169
Bondi, Liz 210
Booth, Charles 37, 274, 278
boreal zones 100, 139
boreholes 112, 127
Boserup, Ester 148, 149, 178, **321**
Bouguer, Pierre 94
Boulineau, Emmanuelle 202
boundaries, tectonic plates 118, 119
Brantingham, Paul and Patricia 194
Braudel, Fernand 182
Bretz, J. Harlen 74
Broecker, Wallace S. 65, 127, **140–41**, 237
Brown, Lester 254
Brundtland Commission/Report 15, 219, **244–5**, 248
Brundtland, Gro Harlem 244, **245**
Brush, Charles F. 290
Bryan, Kirk 270
Brzezinski, Zbigniew 158
Buckland, William 74, 77, 88
buffer analysis 281
Bullard, Edward 14, 118, 123
Bullard, Linda McKeever 246
Bullard, Robert D. 219, **246–7**
Bunda Cliffs (Australia) 75
Bunge, Bill 206
Burgess, Ernest 14, 146, 147, 160, 184, **185**, 186, 204, 274
Bus Riders Union (BRU) (Los Angeles) 207

C

CAESAR landscape evolution model 104
calcification, soil 131
Callendar, Guy Stewart 234–5, 236
Calthorpe, Peter 262, 282, 283
Cambrian Explosion 259
Canada Geographic Information System (CGIS) 55, 304
capability constraints 267
capital
　flows and gentrification 197
　urbanization of 147, **204–5**
capitalism
　capitalist systems 190, 204
　global 198, 250, 251, 252
　unregulated 185
carbon
　deforestation and release of 252
　in lakes 126
　in soil 131
carbon capture and storage 239, 249
carbon cycle 121, 234, 236, 238, 241
carbon dioxide 80
　acidic rain 109
　atmospheric 89, 126, 234–9
　climate change 234–9, 257, 269–70
　greenhouse effect 81–3
carbon footprint 243
carbon storage
　forests 236, 286, 288
　oceans 236
　soil 128, 131
Carlowitz, Hans Carl von 244
Carson, Rachel 15, 218, 219, 227, **240**, 272
cartograms 36–7
cartography 12, 18–19
　aerial photography and remote sensing **40–45**
　coordinates system **22–3**
　crisis mapping **304–5**
　development of **20–21**
　digital twins of Earth **60–61**
　distortions in size 30, 31
　dynamic mapping 51
　geographic information systems (GIS) **52–9**
　geophysical mapping **38–9**
　Global Positioning System (GPS) **48–51**
　infographics 281
　mapping social injustice **278–9**
　mapping the ocean floor **46–7**
　map projections **24–31**
　thematic mapping **34–7**
　tracking disease **310–313**
　triangulation **32–3**
Cascadia project 243
Cassini de Thury, César-François 33
Cassini, Giovanni Domenico 33
Cassini, Jacques 33
Cassini, Jean-Dominique 33
catastrophism 64, **74**
catchment areas 68, 104
caves 108–9
Celsius, Anders 94

INDEX

censuses **188–9**, 274, 275
central business districts 184, 185, 186, 187
central place theory 150, **182–3**, 204
Centre for Ecology & Hydrology 306
centripetal drainage patterns 106
Challenger Deep 46
Challenger HMS 46–7, 124
Charpentier, Jean de 87
Chavis, Benjamin 247
chemical contamination **240**
Chernobyl disaster 251–2
Cheshire, James 280
childcare 210, 212
child labour 179, 180
child mortality 179
Chipko movement 246
cholera 34, 35–6, 54
chorochromatic maps 35, 37
choropleth maps 35, 37
Christaller, Walter 150, 160, **182–3**, 204
chronometers 48
cinder cones 123
circulation cells 73, 84, 100
cities
　and globalization 204
　central place theory 150, **182–3**, 204
　concentric zone model 146, 147, **184–7**, 204
　cultural diffusion 169–70
　garden 242
　gentrification **196–7**
　global 169–70
　heat islands **271**
　migration theory 152, 153
　primate city rule **161**
　rank-size rule 160
　Smart Growth movement **282–3**
　spatial justice **206–7**
　urbanization of capital **204–5**
civil rights, environmental issues 246, 247
Clements, Frederic 136, 137, 138
Cliff, Andrew 280
climate
　and disease 311, 312
　and settlement 156
　biomes 137–9
　extreme event attribution 238
　mapping 36
　migration 219, **254–5**
　modelling 60, 61, 236–7, 263, **268–71**
　ocean currents and 141
　palaeoclimatology **126–7**
climate change 69, 95, 125, 218–19, **232–9**, 243, 250, 253, 257, 259, 283
　Climate Change Adaptation DT 61
　deforestation 286
　drainage 107
　drought 229
　flooding 230
　greenhouse effect **80–3**
　historic 83, 87, 88–9, **126–7**
　migration 254–5
　modelling 100, 269
　natural disasters 298
　societal adaptation 157
　soil and 128, 130, 131
　sustainable development 245
　thermohaline circulation 140
climate mobility 255
climate model cells 270–1
climatic zones **96–101**, 138–9
clouds 67, 68, 69, 101
　high-albedo 239
　imaging through 288, 294

cluster analysis 276, 277
coal 83, 235
Coast, Steve 304
code/space 215
cognitive maps **194–5**
Cold War 55, 158, 159, 191
collisional orogeny 78
colonial expansion 154
Colonna, Fabio 71
Columbus, Christopher 13, 28
Colwell, Robert 292
communication networks
　cultural diffusion 166, 171
　globalization 208
　spatial interaction theory 193
　territorial theory 202, 203
comparative advantage model of development 191
compasses 38
compass roses 28
composite (strato) volcanoes 123
computer modelling 59
concentric zone model 146, 147, **184–7**, 204
Concorcet, Marquis de 148
condensation 67, 68, 69
conic projections 26, 29, 31
coniferous forests 131
conservation **222–7**, 281, 283, 307
constellations 28
construction booms 204, 205
contagious diffusion 168, 169, 170
contaminants 128
continental crust 117, 118, 124
continental drift 46, **47**, 76, 77, 78–9, **110–111**, 116, 122, 124
continental zones 98, 139
continuing landscape 174, 175
contraception 179, 180
contractionism 77
convergent boundaries 78, 79, 119, 121, 122
coordinates system **22–3**, 26
Copernicus, Nicolaus 84
Copernicus programme 45, 263, 308, 309
corals 127, 210, 307
core, Earth's 112, 113
Coriolis deflection 85
Coriolis effect 65, 73, **84–5**
Coriolis, Gaspard-Gustave de 14, 64, 72, 84, **85**
counter-urbanization 187
coupling constraints 267
Covid-19 pandemic 54, 181, 189, 245, 263, 266, 278, 310, 312–13
Cresswell, Tim **323**
crime 194
crisis mapping **304–5**
Croll, James 88
crops 101, 128
　intensive farming 229
　remote-sensing agriculture 292–5, 309
cross-border cooperation zones 202
cross-cutting relationships 71
crowdsourced maps 305
crust, Earth's 75, 94–5, 107, 112, 120
　dating 118–19
　orogeny 77–8
　plate tectonics 116–21
Crutzen, Paul Jozef 219, 256, 257, **258**
Cryogenian period 89
crystals 90, 91, 92, 93
cultural appropriation 171
cultural diffusion 146, **164–71**
cultural landscape 146, **172–5**
cultural relativism 168

currents, ocean 65, 72, 101, **140–41**
Cutter, Susan 298, **303**
Cuvier, Georges 64, **74**, 87
Cvijić, Jovan 108, **109**
cybergeography 214
cyberspace 214
cyclones 142, 143, 302
cylindrical projections 29–30, 31

D

da Gama, Vasco 28
Dalton, Conroy 192
Dalton, John 68
dams 107, 163, 230, 231
Dana, James D. 64, 77, 90–91, **92**
Dangermond, Jack and Laura 54, 56, 312
Darwin, Charles 75, 88, 149, 154, 156
DART system (Deep-ocean Assessment and Reporting of Tsunamis) 300–301
dashboards, interactive 263, 312–13
Dasmann, Raymond 242–3, **321**
data
　collecting personal 313
　geodemographic analysis **274–7**
　linking to maps 54–5, 57
　shared national **59**
　spatial data analysis 15, **280–81**
　visualisation of 19, 34, 37, 54, 59, 281, 311, 312
Davis, Donald 74
Davis, William Morris 64–5, **104**, 105
DDT 219, 240
death rates 147, 178, 179, 180–81
deciduous forests 139
decomposition 129, 131
deforestation 101, 107, 139, 163, 224, 225, 227, 228, 229, 246, 252, 257, 259, 263
　monitoring **286–9**, 309
deltas, subsidence of 94
demographic data
　geodemographic analysis **274–7**, 277, 278
　social injustice 279
　social reform 37
demographic transition 147, 160, **176–81**
dendritic drainage patterns 106, 107
dependency theory 198–9
deposition 70, 71, 75, 104
deranged drainage patterns 106
desertification 101, 218, **228–9**, 254
designed landscape 174
Destination Earth (DestinE) 19, 60, 61, 268
Detroit 187, 206
developed countries 190–91, 198–9
developing countries 190–91, 198, 253, 291
development, theories of **190–91**, 198–9
dew 69
diamond 93
Diamond, Jared 156
diatoms 89
Díaz, Sandra 136
digital culture 171
digital data collection 189
Digital Earth 60
Digital Humanitarian Network 304
digital kinship 215
digital projections 31
digital spaces, as cultural spaces **214–15**
digital twins (DT) 19, **60–61**, 268

diluvian theory 74, 87
dinosaurs, extinction of 75
direct mail 276
disaster relief 263, 304–5
discordant drainage patterns 106
disease
　crops 293
　disaster relief 305
　mapping spread 36
　modelling 313
　population check 149
　tracking **310–313**
disinformation, online 215
displaced persons 254, 255
distance decay 193
divergent boundaries 78, 79, 118–19, 122
Dodge, Martin 214, 215
Dokuchaev, Vasily 128, 130, **319**
domino theory 159
Doppler effect 48, 49
Doppler shift 50
Dorsey, Herbert Groves 46
dot distribution/density 35, 36
drainage basins 104, 106, 107
drainage patterns 106–7
drones 293
droughts 69, 228, 229, 237, 238, 254, 295
drylands 228–9
dry zones 98
Duany, Andres 283
Du Bois, W.E.B. 278
Dupin, Charles 37
Dust Bowl 228, 244
du Toit, Alexander 110, 116, **320**
Dutton, Clarence 94, **319**
dykes 71

E

Early Warnings for All initiative (EW4ALL) 298, 299, 303
early warning systems **296–303**
Earth
　age 75
　axial tilt 88, 89, 100
　circumference 20, 21, 32, 33
　cooling 88
　curvature 26, 30, 43
　digital simulations 60–61
　early concepts of shape 20, 21
　magnetism 38–9
　orbital variation 88, 89
　rotation 72, 84–5, 111, 118, 140
　spherical nature 26
　structure **112–13**
Earth Observing System (EOS) 263, 293, 309
Earth Overshoot Day 248, 249
earthquakes 112, 113, 121, 122, 124
　monitoring/forecasting 51, 125, 298, 299, 304
　mountain-building 76–7, 78
　seismology and Richter Scale **132–5**, 299, 300
　Tohoku 116
　underwater 134–5
Earth Resources Technology Satellite 44
Earth Summit 224
Earth system models 271
Easton, Roger 48, 50
echo-sounding 46, 47
ecological footprint 219, **248–9**
ecological zones **136–9**
ecology
　Anthropocene epoch **256–9**

330 INDEX

human **162–3**
economic activities
  and migration 152, 153
  location 150–51
economic growth
  globalization 250–53
  sustainable development 245
  territoriality theory 202, 203
  theories of development 147, **190–91**, 198–9
  world-systems theory **198–9**
ecoregions 243
ecosystems 65, 69, 138
  capitalism and 252
  integrating with human activity 242, 243, 272
ecozones 139
Egyptians, Ancient 13, 20, 22, 188
Ehrlich, Paul 148
Ekholm, Nils Gustaf 83
Ekman, Vagn Walfrid 140
elections 279, 305
electrical conductivity 38, 39
electricity, renewable sources 290–91
electromagnetic waves 38
elevation, landscape formation 104
El-Hinnawi, Essam 254
Elliot-Smith, Grafton 166
Ellis, Erle 136, 139
El Niño (Southern Oscillation) 141
emigration 153, 180, 181
Engelke, Peter 258, 259
Engels, Friedrich 206, 210, 264
Enhanced Fujita scale 143
Enlightenment 14
environment
  adaptation 156, 162, 163
  Anthropocene epoch 256–9
  cultural variation 167
  globalization 250–53
  human transformation 157, 162
  monitoring change **284–9**
  sustainable development 245
environmental deficit/surplus 248, 249
environmental determinism **156**, 157, 162, 163, 173
environmental DNA (eDNA) 307
environmental geography 12, 15, 218–19
environmental impact assessment (EIA) **273–4**
environmentalism 227
Environmental Justice Movement 219, **246–7**
Environmental Protection Agency (EPA) 227, 262, **272–3**, 283
Environmental Systems Research Institute (Esri) 312
epicentres, earthquake 124, 133, 134, 300
epidemiology, spatial 193, 310, **311**, 312–13
equal-area projections 30–31
equator 20, 72, 73, 100
equirectangular projection method 26
Eratosthenes 13, 18, 20, **21**, 22, 26, 32, 65
erosion 75, 76
  cycle 65, **102–7**
  soil formation 131
erratics 86, 87, 88
Escobar, Arturo 190
eskers 87
Esmark, Jens 87
Essential Biodiversity Variables (EBVs) **306–7**
ethnoscape 170
Eurasia 146, 158, 159
European Commission **60–61**

European Space Agency 19, 42, 45, 60, 286, 293, 308
evacuations 299, 301
evaporation 67, 68, 69, 100, 104, 141
evapotranspiration 100, 101
Evelyn, John 224
evolution, cultural 166–8
Ewing, Maurice 46, 47
expansion diffusion 168
export-led growth 191
extinctions, species 252, 257–8

F

Fainstein, Susan 206
Fairchild, Sherman 43
Fallou, Friedrich 130
famine 149, 178, 228
Faraday, Michael 39
Farr, Dr William 152
fathometers 46
faults/faulting 70, 71, 77, 78, 106
Febvre, Lucien 157
feminist geography 211, 212
Fenster, Tovi 208
Ferguson, Neil 313
Ferrel cells 73, 101
Ferrel, William 73, 84
fertility, human 149, 179, 181
fertilizers, chemical 129, 235, 257, 295
Fisher, Howard 56, 304
flooding 69, 107, 128, 194, 230–31, 237, 238, 239, 254, 289, 301, 305
  catastrophism 74
  early warnings **302–3**
  flood defences 239
floodplains 104, 105, 107
  management 218, **230–31**
flow charts 37
fluorinated gases 235
folding 77, 78
food
  cultural diffusion 168–9
  supply 148–9, 156, 178, 220, 237, 245, 259, 295, 309
Foote, Eunice Newton 82, 234
Ford, Derek 108, 109
forests
  carbon sinks 236
  management and conservation 218, 226
  monitoring change 263, **286–9**
  regeneration 249, 289
  sustainable development 244
  temperatures 101
Forman, Richard 172
fossil fuels 83, 234, 235, 236, 239, 241, 249, 257, 258, 259, 271, 290, 291
fossils 70, 74, 259
  catastrophism 74
  continental drift 110, 111, 116
  marine 70, 71, 74
  palaeoclimatology 126, 127
Fourier, Jean-Baptiste Joseph 64, 80, 81–2, 234
Frank, Andre Gunder 198
Frankfurt Planning Acts 265
free markets 150, 221
Frère de Montizon, Armand Joseph 36
freshwater, in glaciers 68
friction of distance 193
Frisius, Gemma 18–19, 28, 32, 33
Frobenius, Leo 14, 146, 166
Fujita, Tetsuya "Ted"/Fujita scale 143, **321**

G

Gaia hypothesis 219, **241**
Galilei, Galileo 72
Galileo system 51
Gall, James 31
Galpin, Charles 182
gaming 214, **215**
garden cities 242, 265
Gardner, Lauren 310, 311, 312, **313**
Gauss, Carl Friedrich 18, **38–9**, 84
Geddes, Patrick 204, 242
Geiger, Rudolf 98
Geist, Helmut 229
Gemini programme 44
gender
  space theory 147, **210–213**
  spatial justice 207
genetic modifications 259
gentrification **196–7**, 204
geochemical markers 259
geocoding 58
geodemographic analysis 192, 262, **274–7**, 277, 278
geodesy 94
geoengineering 239
GEOGLAM 292
Geographia (Ptolemy) 23, 26–7, 28
geographic information systems (GIS) 15, 19, 34, 37, **52–9**, 192, 263, 267, 278–9, 280, 281, 290, 291, 294, 295, 306, 307, 310, 311, 312
geology
  and biology 241
  Anthropocene epoch 256, 259
  gradual change 75
  mapping 36
  soil formation 130
geomagnetism 39
geophysical mapping **38–9**
geopolitics 18, 155, 159, 245
Geosat 50
geospatial analytics *see* spatial data analysis
Geospatial artificial intelligence (GeoAI) 274, 280
geostatistics 59
geosynclinal theory 77
geotemporal analysis 57
geothermal power 291
gerrymandering 279
geysers 124
Gibson, William 214
Gilbert, Grove Karl 104, **319**
Gilbert, William 38
Gilmartin, Patricia 194
Gilmore, Ruth Wilson 278
GIS *see* geographic information systems
glaciation 64, **86–9**
glaciers 68, 69
  climate change 219, 237
  monitoring **289**, 309
  satellite imagery 45, 51
Glass, Ruth **196**, 197
Glatzmaier, Gary 112
Glendening, Parris N. 283
Global Atlas of Renewable Energy 263
Global Biodiversity Information Facility (GBIF) 306
global cooling 89, 118
global Earth observation programmes **308–9**
Global Earth Observation System of Systems (GEOSS) 309
Global Ecosystem Dynamics Investigation (GEDI) programme 287
Global Footprint Network (GFN) 248, 249
Global Forest Watch (GWF) 263, 286–7
globalization 197, 204, 208, 219, **250–53**
  cultural 170
Global Navigation Satellite Systems (GNSS) 51
Global Positioning System (GPS) 15, 19, **48–51**, 58, 121, 281, 295, 307
global village 166
global warming 69, 89, 219, 234–9, 241, 257, 259, 269, 289
GLONASS 51
Godwin, William 148
Golledge, Reginald 162, 195
Gondwana 110, 116
goods
  export 191
  supply 182–3
Google Earth 57, 286
Google Maps 31
Gordon, H. Scott 221
Gore, Al 60
Gottmann, Jean 161
GPS *see* Global Positioning System
GRACE (Gravity Recovery and Climate Experiment) 45
gravitational anomalies 45
gravity
  isostasy 94–5
  measuring 38, 39
great flood concept 74, 86–7
Great Green Wall projects 228, 229
great ocean conveyor 140, 141
Great Oxygenation Event 259
Great Trigonometrical Survey 94
Greeks, ancient 13, 18, 20–23, 26, 64, 188
green belt **265**
Green Belt Movement (Kenya) 227
greenhouse effect 65, **80–83**, 126, 218, 262
greenhouse gases 80, 89, 101, 125, 126, 127, 229, 234–9, 257, 269, 270, 271, 286, 288, 309
Greenpeace 286
Gregory, Ken 107
grid references 22–3
groundwater 69, 104, 131
Group on Earth Biodiversity Observations Biodiversity Observation Network (GEO BON) 307, 309
Group on Earth Observations (GEO) 309
growing season 100
Guettard, Jean-Étienne 104, **317–18**
Gutenberg, Beno 113, 132, 133, 134
Guterres, António 302, 303
Guyot, Arnold Henry **319**

H

habitability 69, 80, 101
habitat
  evaluation 59
  loss/destruction 237, 249, 252, 257
Hadley cells 73, 101
Hadley, George 14, 64, **72–3**
Haeckel, Ernst 162
Hägerstrand, Torsten 168, 262, 266–7
Haggett, Peter 310, 311
hail 69
Haiti earthquake 263, 304–5
Halley, Edmond 38, 68, 72
Hall, James 77
Hanson, Susan 210, **322**

INDEX

Hardin, Garrett 220, 221
Harris, Chauncy 184
Hartmann, William 74
Hartshorne, Richard 15, 266
Harvey, David 147, 204, **205**, 208, 219, 250, 251, 252–3
Hassabis, Demis 214
Haushofer, Karl 154, 155
Haussmann, Georges-Eugène 205, 264
Haüy, René Just (Abbé) 90
healthcare 179, 180, 267
heartland theory **158**, 159
heat hazes 288
heat islands **271**
heatwaves 237, 238
Hecataeus of Miletus 21
Heezen, Bruce 46, 47
Heidegger, Martin 200
Hess, Harry 65, 76, 110, 111, **116**, 117–18, 124
hierarchical diffusion 168, 169, 170
Hill, Verity 310
hindcasting 271
Hinds, Richard Brinsley 98
Hipparchus 13, 18, **22–3**
historical particularism 168
Hitler, Adolf 155
HIV/AIDS pandemic 181
Holdridge, Leslie 98, 100, **320–21**
Holmes, Arthur 78, 107, 111, 116, 122
Holocene epoch 256, 259
Holocene interglacial period 89
homeostasis 241
Hondius, Jodocus 34
Hothouse Earth 259
hotspots 119, **122–5**
hot springs 124
housing, location of urban 185, 186
Howard, Ebenezer 242, 265
Hoyt, Homer 185, 186–7
Huang-Ho (Yellow) River 230
Hugi, Franz Josef 87
human activity
 Anthropocene epoch **256–9**
 behavioural geography **194–5**
 biomes 136, 137, 139
 climate change 83, 89, 236, 238, 239
 drainage patterns 107
 ecology 146, 162–3
 environmentalism 227, 273
 integrating with ecosystems 242, 243
 land degradation 228–9, 248
 landscape change 157, 162, 163, 172–5
 physical world 162
 satellite imagery 45
 water resources 69
human behaviour, environmental factors **156**, 157, 162, 163
human ecology **162–3**, 172
human geography 12, 146–7
humanistic geography **200–201**
humanitarian response 305
Humboldt, Alexander von 36, 39, 76, 87, 110, 136, 137, 228, 241, **318**
humidity 98, 101
hunter-gatherers 129
Huntington, Ellsworth 156
Huronian ice age 89
hurricanes 142–3, 231, 237, 254, 298, 301, 302, 309
Hutton, James 64, 70, **75**, 77, 241
hydrological cycle **66–9**
hydropower 290, 291
hydrothermal vents **120**
hyperdiffusionism 166
hyperspectral sensors/imaging 292, 306
hypocentre, earthquake 133, 134, 135

I

Ibn Battuta 13, **317**
Ibn Khaldun 156
ice ages 83, **86–9**
ice cores 65, 80, 89, **126–7**, 289
ice sheets 238–9, 289
ideoscape 171
immigration 153, 180, 181
imperialism 198
inclusivity
 gender 213
 spatial justice 206
Indigenous cultures
 and biodiversity 175
 and globalization 252–3
 and landscape 175
 land ownership 243
 marginalization 170
 social injustice 279
industrial capitalism 185
industrialization 147, 151, 154, 160, 178, 190, 199, 224, 234, 248, 251, 253, 264, 265
Industrial Revolution 14, 149, 152, 178, 179, 234, 246, 248, 259, 264
industry 173, 182, 185
infiltration 69
infographics 36, 280, **281**
infrared radiation 81, 82, 218, 234
Inō, Tadataka **318**
insecticides 240
Integrated Digital Earth Analysis System (IDEAS) 61
Intergovernmental Panel on Climate Change (IPCC) 140, 235
interlocking spurs 104, 105
International Commission on Stratigraphy 256
International Energy Agency (IEA) 290
International Geosphere–Biosphere Programme (IGBP) 257
International Renewable Energy Agency (IRENA) 263, 290–91
International Union of Geological Sciences (IUGS) 259
International Year of Glaciers' Preservation 86
Internet 56, 169, 171, 214–15, 276
interpolation 58
invasive species 252
investment 196, 205
Islamic scholars 13, 27, 28
isolated state model 161
isostasy **94–5**, 107
isothermal lines 36

J

Jacobs, Jane 196, **321**
Jefferson, Mark 160, **161**
Jefferson, Thomas 156
Jenny, Hans 65, **128–31**
jet streams 73
Johnson, Robert 226
Joy, Annamma 166
Juan de Fuca Ridge 118, 119, 125

K

Kanamori, Hiroo 135
karst landscape **108–9**
Katz, Cindi **323**

Keeling, Charles David 15, 218, 234, 235, **236**
Kelley, Florence 37
King, Lester 105
Kitchin, Rob **214–15**
Kjellén, Rudolf 155
Kleidon, Axel 241
Klinenberg, Eric 278
Kolbert, Elizabeth 257–8
Kolb, John Harrison 182
Köppen climatic classification 65, **96–101**, 138–9
Köppen-Geiger classification 98, 99, 100
Köppen, Wladimir 65, **98**, 99, 111
Kropotkin, Peter 154
Kyoto Protocol 253

L

labour
 gender 210, 211–12
 low-cost 199
la Cour, Paul 290
lakes, creation of 69
Lambert, Johann Heinrich 26
Lambin, Eric 229
land degradation **228–9**, 248
landfill sites 246–7
land grabbing 253
landmarks 195
Landry, Adolphe 178
Landsat 15, 19, 42, 44–5, 56, 60, 286, 288, 292, 293, 309
landscape
 cultural landscape **172–5**
 erosion cycle **104–7**
 formation 64–5
 glaciation **85–9**
 human activity 157, 163, 172–5
 karst **108–9**
 landscape ecology 172
 preservation 174
landslides 107
land use 139
 aerial photography/satellite imagery 43, 45
 location theory **150–51**, 184
 mapping 55
 unsustainable 229
 urban 184–7
 zoning 147, 262, **264–5**
language, cultural diffusion 166, 169, 171
Laplace, Pierre Simon de 84
lasers, mapping with 287–8, 289
lateral continuity, law of 71
lateral erosion 105
laterization 131
latitude 22, 23, 28, 30, 31
 biomes 136, 138
 climate 100
Laurasia 110, 116
Laussedat, Aimé 43
lava 123
lava domes 123–4
Lawrence, George 42
Lebensraum **154–5**, 158
Lee, Everett S. 152
Lefebvre, Henri 204, 206, **320**
Lehman, Inge **112–13**
Lehmann, Herbert 108
Leonardo da Vinci 70
Levasseur, Pierre Émile 36–7
levees 230, 231
Ley, David 197

LGBTQ+ community 213
Libera satellite 42
LiDAR 45, 58, 287–8, 294, 306, 307
life
 biosphere 136
 Gaia hypothesis **241**
life-cycle assessments (LCAs) 273
life paths 267
life-zones classification 100
limestone 107, 131
 karst landscapes 108–9
Lincoln, Abraham 225
lithosphere 95, 120, 122
Liu Hui 32
living conditions 264, 274–5
Lloyd, William Forster 218, **220–21**, 248
local and global **208–9**
local climate zones (LCZ) 271
location theory 146, **150–51**
logging 227, 244
longitude 23, 26, 29, 30
Longitude Act 48
Loran (Long Range Navigation) system 49
Lorius, Claude 65, 126–7
Lösch, August 182, 183
Lotka, Alfred 178
Lovelock, James 219, **241**
lunar eclipses 21, 23
Lyell, Charles 14, 64, **75**, 76, **77**, 79, 88, 256
Lynch, Kevin 147, 162, 194, **195**, 200

M

Maathai, Wangari 227
McDowell, Linda 147, 210, **211**, 212, 213
McHarg, Ian 56
McKenzie, Dan 110, 119
Mackinder, Halford 15, **158**, 1469
McLuhan, Marshall 166
McNeill, John R. 258, 259
Magellan, Ferdinand 13, 28
magma 79, 117, 119, 120, 123, 124
magnetic anomalies, seafloor 118, 120
magnetic field, Earth's 38–9, 113
 reverse polarity 118
magnetometers 38, 39
Mahan, Alfred Thayer 158
malaria 310, 312
Mallet, Robert 112
Malthusian theory **148–9**
Malthus, Thomas 146, **148–9**, 178, 218, 220
Manabe, Syukuro 237, 262, 263, 268, **269**, 270, 271
mantle, Earth's 94, 95, 112, 113, 120
 continental drift 78
 convection 111, 116, 117, 118
 hotspots 122–3
 viscosity 119, 120
manufacturing 183, 190, 204, 205
mapmaking *see* cartography
Mapping Display and Analysis System (MIDAS) 54
Marbut, Curtis 128
Margulis, Lynn 241
Mariana Trench 117
marine cloud brightening 239
marine (tidal) energy 291
Marinoan glaciation 89
Marinus of Tyre 13, 18, 22, 23, 26–7
maritime power 158
Maritime Silk Road 159
marketing 262, 274–7

332 INDEX

Markusen, Ann 150
Mars
  life 241
  water 90
Marshall Plan 190
Marsh, George Perkins 224, 225
Martel, Édouard-Alfred 108, **320**
Marx, Karl 178, 204, 208
Mason, Ron 118
mass consumption 190, 191
mass extinctions 74, 75, 241, 257–8
Massey, Doreen 14, 147, **208–9**, 210, 211
Massey, Douglas 152
Matthews, Drummond 110, 118
mature landscapes 104–5
Matuyama, Motonori 118
Maury, Matthew Fontaine 72, 117, 140
Mayer, Benoit 254
meandering rivers 105
meaningful place 200, 201
mediascape 171
Medici, Ferdinand II de' 71
Mediterranean zones 139
megacities 160
megalopolises 161
Meier, Patrick 305
meltwater 69
Mendeleev, Dmitri 39
Mercalli, Giuseppe 132
Mercator, Gerardus 14, 18, 22, 26, **28**, 29–31, 32
Mercury programme 44
meridians 20, 21, 22, 23, 26, 27
Mersenne, Marin 84
metamorphosis 78, 79
methane 80, 126, 234, 235, 239
Meydenbauer, Albrecht 43
microplastics **252**, 258
Mid-Atlantic Ridge 19, 46–7, 116, 117, 120
migration patterns 14, 59, 146, 237, 253, 307
  climate migration **254–5**
    cultural diffusion 166, 167, 168
  migration theory **152–3**
Milankovitch, Milutin 88–9
military campaigns 35
military surveillance 43
Millennium Development Goals (MDGs) 244
MIMO (map-in–map-out) 54
Minard, Charles 34, 36
mineralogy **90–3**
minerals, soil 129
mining 258
Mississippi River 231
Mobility as a Service (MaaS) integrated transport apps 266
modernity 251, 253
Modified Mercalli Intensity (MMI) Scale 132
Moho Discontinuity 94, 112
Mohorovičić, Andrija 94, 112
Mohs, Frederick/Mohs scale 93
Moll, Herman 34
moment magnitude (M_w) scale 135
Mongol Empire 158
Montesquieu 157
Moon, formation of 74
moraines 87
Moran, Patrick 280
Morgan, W. Jason 119, 121
Morley, Lawrence 118
Moro, Anton 75
mortality rates 178, 179, 180–81, 206, 238
Mortimore, Michael 228, 229
Mosaic geodemographic system 276, 277

mosquitoes 240, 311
mountains
  marine fossils 70, 74
  orogeny **76–9**
  rainfall 100–101
Muir, John 218, 224, 225–6, **227**
Müller, Paul 240
multi-hazard alert systems 298, 303
Multispectral Scanner System (MSS) 44, 45
Myers, Norman 219, 254–5, 306

N

Napoleon I, Emperor 35
Napoleon III, Emperor 264
NASA 19, 42, 44, 60, 61, 90, 142, 237, 239, 263, 286, 287, 289, 291, 293, 308, 309
National Oceanic and Atmospheric Administration (NOAA) 46, 122, 308
national parks 224–6
National Trust 225
nations, territorial theory 146, **202–3**
NATO 159
natural disasters 254, 263, 308, 309
  early warning systems 263, **296–303**
Natural Earth 59
natural resources
  and capitalism 250
  and population 148–9, 257
  location 93, 151
  management **222–7**
natural selection 149, 154
navigation 18, 192
  at sea 26, 27–9
  GPS **48–51**
Navstar GPS satellites 49, 50–51
neighbourhoods, classification of 262, 276, 277
neocatastrophism 75
neptunism 75
Neumann, Frank 132
Neustadt Project 265
neutral space 200, 201
Newbigin, Marion I. **320**
Newton, Isaac 20
New Urbanism 283
Nickel, Ernest 91
Nickel-Strunz classification 90, 91
Nightingale, Florence 37
Nile, River 230, 231
nitrogen 80
  in soil 131
nitrous oxide 80, 234, 235
Nixon, Richard 226, 227, 272, 273
Noah's Flood 74, 87
nomads 156
non-material culture 169
normalized difference vegetation index (NDVI) 262, 292–3, 294, **295**
North Atlantic Deep Water 141
North Atlantic Oscillation 141
Norwood, Virginia 42, 44, **45**
Notestein, Frank 178, 179, 181
nuclear power 251–2, 258, 259
Nunes, Pedro 28, 29, 30, **317**
Nurske, Ragnar 190
nutrients cycle 129

O

oceanic crust 111, 112, 117–19, 120
  basaltic 117
  recycling 118

oceans
  acidification 259
  carbon sinks 236, 238
  climate modelling 270
  climate zones 101
  mapping 19, 45, **46–7**, 116–17
    monitoring 66
    primeval 75
    protecting 227
    remote sensing and surveying 45
    seafloor spreading 46, 47, 116, 117, 119, 120, 124
    water evaporation 68
    water temperature 219, 307
Odum, Eugene P. 272
Odum, Howard T. 242, 272
Ogunseitan, Oladele A. 200
O'Keefe, Phil 303
Oke, Timothy 271
Oldham, Richard Dixon 113, 133
old landscapes 105
open-source digital mapping 59, 263, 304, 305
optical mineralogy 92–3
Ord, Keith 280
Ordnance Survey 59
Ordnance Survey of Ireland 32
Oregon Experiment 282
organic matter 129, 130
organic theory **154–5**, 159
Organisation for Economic Cooperation and Development (OECD) 258
orogeny **76–9**
Ortelius, Abraham 110
Ostrom, Elinor 220, 221, **322**
Otto, Friederike 238
overaccumulation 252
overgrazing 229
overlays, transparent 34, 54, 55
overpopulation 148
Owen, Robert 178
ownership, territoriality 203
ozone layer 234, 258, 259

P

Paelinck, Jean 281
palaeoclimatology 65, 89, **126–7**
Palissy, Bernard 14, 64, **66–7**, 68, 69
Pangaea 78, 110, 111, 116
parallels 20, 21, 22, 23, 26, 27, 29, 31
Paris Agreement (2015) 234, 250, 253, 283
Parkinson, Bradford 19, 48, **49**, 50
parklands 174
Park, Robert E. 162, 204, 274
Parmenides 98
Pawlyn, Michael 157
pedestrian movements 192
pedology **128–31**
Pei Xiu 22, **316**
Penck, Walther 105
peneplain 105
percolation 69
peripheral countries 198–9
permafrost 239
Perraudin, Jean-Pierre 86
Perrault, Pierre 67–8, 104
personal safety 212–13
Peru-Chile Trench 121
Peschel, Oscar 154
pesticides 219, 240, 272, 294
Peters, Arno 31
Petit, Paul 84
PFAS (forever chemicals) 240
Phillips, Norman 60, 263, 268

photogrammetry 42, **43**, 44
photosynthesis 100, 236
physical geography 12, 64–5
Piaget, Jean 194
Picard, Jean 33
Picquet, Charles 54
Pinchot, Gifford 218, 224, 226, 227, 244
Piri Reis **317**
place
  definition 201
  gendered 212–13
plants
  biomes and ecological zones 136–9
  changing distribution 101
  monitoring biodiversity 306–7
  photosynthesis 236
  soil formation 129, 130
  transpiration 67, 68, 69
Plater-Zyberk, Elizabeth 283
plate tectonics 65, 71, 75, 95, 105, **114–21**, 122
  continental drift 110, 111
  earthquakes 134
  orogeny 76–9
  volcanic activity 122, 123, 124
Plato 64
Platzky, Laurine 209
plutonism 75
podzolization 131
Polanyi, Karl 250
polar and alpine zones 98–9, 139
polar cells 73, 101
poles
  air circulation 72, 73
  cartography 26, 27, 31
  magnetic 39
political power, territoriality theory 202, 203
pollution 44, 219, 227, 246, 257, 263, 308
  air 288
  environmental justice 206, 246, 253, 278
  microplastics **252**, 258
popular culture 169
population
  censuses and population geography **188–9**
  climate change 234, 238
  decline 181
  demographic transition model **176–81**
  displacement 253
  economic development 160
  food supply 220
  forecasting and modelling 178–9, 181
  global increase 146, 245, 257, 258, 259
  human ecology 163
  Malthusian theory **148–9**
  mapping 36–7
portolan charts 27–8, 29, 30
positive feedback loops 239
possibilism 156, **157**, 163
postal (ZIP) codes 275–6
post-colonial nations 191, 198
Pouillet, Claude 80, 82
poverty 37, 245, 247, 275
Pratt, Geraldine 210
Prebsisch, Raúl 198
precipitation 66, 67, 68, 69
  climatic zones 99, 100
preservation vs conservation 226, 227
Priestley, Joseph **318**
primate city rule 160, **161**
Prime Meridian 22, 95
PRIZM 276, 277
probabilism 157

projections, map **24–31**
Project SloMo 112
proportional circles 36, 37
Pruitt, Evelyn 44
Ptolemy, Claudius 13, 18, 22, 23, 26–7, 28, **316**
public health 263, 264, 310, 311–13
public spaces, and gender 212–13
public transport 187, 206, 207, 212, 213
Pulfrich, Carl 43
Pulido, Laura 262, 263, 278, **279**
Pythagoras 13, 21

Q

Quantitative Revolution 15
quantum navigation 48

R

racism
  and environmental determinism 156
  and social injustice 278
  and spatial justice 207
  environmental 246–7
radar 38, 294
Radar for Detecting Deforestation (RADD) 288
radial drainage patterns 106
radioactivity
  dangers of 251–2, 259
  radioactive decay 124
  radioactive heating 118
rainfall
  acidic 108, 109
  biomes 137, 138
  climate change 69, 237
  climatic zones 99, 100–101
  drylands 228
  global distribution 66
  hydrological cycle 66
  water cycle 67, 68–9
rainforests, destruction of 252, 257, 286–9
rain shadows 101
Ramakutty, Navin 136, 139
Ramayana 66
rank-size rule **160**, 161
raster data/layers 57–8
Ratcliffe, Derek 240
Ratzel, Friedrich 146, **154–6**, 159, 162, 172, 173, 202
Ravenstein, Ernst Georg 14, 146, **152–3**
raw materials 151, 199
Reagan, Ronald 48, 51
real estate 204, 205
redevelopment 196–7
Rees, William 219, 248–9
reflexive modernization 251, 253
reforestation 244
regenerative practices 249
relict cultural landscape 174–5
relief agencies 304
religious beliefs
  age of Earth 75
  catastrophism 74
  cultural diffusion 167, 169
  mapping 34
relocation diffusion 168, 169
remote sensing 37, 42, **44–5**, 281, 286, 289, 307
  agriculture **292–5**
renewable energy 224, 253, 263, **290–91**
rent gap theory 205
rent theory 184

reservoirs, underground 67
resource management 221
  ecological footprint 248–9
retailing 182–3, 193
re-urbanization 187
reverse geocoding 58
rhumb lines 27, 28, 30
Ricardo, David 150
Riccioli, Giovanni Battista 84
Richardson, Lewis Fry 84
Richter, Charles F. 65, 132, **133**, 134
Richter Scale **133**, 134, 135
Richthofen, Ferdinand von **319**
ridges and rises, oceanic 116–17, 118, 120
Rift Valley, East African 47
Rimland theory 146, **159**
Rio Earth Summit 244, 245
"risk society" 219, 251–2, 253
Ritter, Carl **318**
rivers
  creation 69
  discharge 104
  landscape formation 65, 104–6, 107
  rainwater 67–8
  shared management 231
Robbin, Jonathan 274, 275
Robinson, Arthur 26
rocks
  erosion cycle 104–7
  folds and faults 70
  formation 70
  mineralogy **90–93**
  rock cycle 75
  soil formation 131
Rockström, Jack 258–9
Romans 23, 188, 264
Roosevelt, Franklin D. 244
Rose, Gillian 147, 210–211
Rostow, Walt 147, **190–91**
Rousseau, Jacques 148, 149
routes 194, 195
run-off 68
rural-urban migration 152, 153

S

Sack, Robert D. 200, 202–3
Saffir, Herbert 142, 143
Saffir-Simpson Hurricane Wind Scale 142, 143
salinity, ocean 140, 141
salinization 131, 237
San Andreas Fault 119, 134
Sanders, Fred 142
sandstone 131
San Francisco earthquake (1906) 133–4
Santos, Milton **321–2**
SARS-CoV-2 virus 310
Sassen, Saskia 204
Satellite laser-ranging (SLR) stations 51
satellites
  biodiversity monitoring 306–7
  climate and weather 98, 101, 142, 208, 237, 269, 294, 301
  conditions for disease 311
  forest monitoring 286–9
  glacier/ice sheet monitoring 289
  global Earth observation programmes **308–9**
  Global Positional System **48–51**
  hydrological cycle 66, 69
  imagery 15, 43–5, 56, 57, 58, 263
  land cover 42
  remote-sensing agriculture 292–5
  solar radiation 42

tsunami warnings 135
Sauer, Carl 146, 156, 157, 162, 172, 173, **173**
Saussure, Horace-Bénédict de 82, 86
scale 22, 32, 33
scarps 105
Scheer, Herman 290, **291**
schemas 194
Schimper, Karl 88
Schlüter, Otto 173
Scientific Revolution 13, 14, 64
scientific/technological development 15, 251–2, 253
Scott, Robert Falcon 86
Scripps, E.W. 178
Seabed 2030 project 46
Seager, Joni **323**
sea ice, retreating 140, 141, 237, 238, 239
sea levels, changes in 94, 95, 107, 237, 259, 289, 309
Seaman, Valentine 36
Seasat 45, 50
seawater 67
sediments
  core sampling 127
  deposition 70, 71, 75, 104, 105, 107
  segmentation 76, 262, 276–7
seismic waves 38, 39, 94, 112–13, 133, 299
seismology 38, 39, 112–13, 119, **132–5**, 299
self-sufficiency, community 242
semi-peripheral countries 199
Semple, Ellen Churchill **156**
Seneca 224
Sentinel satellites 19, 286, 287, 288, 293, 308–9
services
  access to 206–7
  export of 191
  provision of 182
settlements
  central place theory **182–3**
  hierarchy of 160, 182–3
  territoriality theory 202
Seutter, Matthäus 30
sextants 48
shared/common resources **220–21**, 248
Shelford, Victor 137
Shen Kuo 70, 126
Shevky, Eshref 274
shield volcanoes 123
Silent Spring (Carson) 15, 218, 227, **240**, 272
Simon, Julian 148
Simpson, Robert 142
sinkholes 108, 109
Skidmore, Andrew K. 306, **307**
sleet 69
slums 37
Smart Growth movement 262, **282–3**
Smith, Adam 150, 152, 221
Smith, James 70
Smith, Linda Tuhiwai **322–3**
Smith, Neil 196, 197, 204, 205, 250
Snell, Willebrord 32–3
Snider-Pellegrini, Antonio 110
snow 69
snowball Earth 89
Snow, John 19, 34, **35**, 54, 310
social class 184, 186, 187, 207, 251
social injustice, mapping 263, **278–9**
social media 169, 171, 279
social mobility 184
social planning 275
social reform 37

society
  sustainable development 245
  territoriality theory 202–3
sociofacts 169
soil
  and settlement 156
  biomes 138
  classification 128
  degradation 228, 243
  erosion 229, 244
  fertility 128, 129, 130, 131
  formation and pedology 65, **128–31**
  mapping and monitoring 294, 309
Soja, Edward 205, **206–7**
solar power 290, 291
solar radiation 80, 81, 82, 88, 89, 100, 235, 239, 270, 289
Somerville, Mary **318–19**
sonar 46, 47, 116
Soulé, Michael E. 306
space
  and gender **210–213**
  and place 201
space-time **266–7**
Spate, Oskar 157
spatial data analysis 262, **280–81**
spatial econometrics 281
spatial interaction theory **192–3**
spatial justice 205, **206–7**
speleology **108–9**
Spencer, Herbert 154
springs 67, 68
Sputnik 1 43, 48, 49
Spykman, Nicholas 146, **159**
Stahl, Ernst 137
Stahl, Georg 128
statistics
  geostatistics 59
  mapping 19, 35, 36, 37
Steensen, Niels see Steno, Nicolas
Steno, Nicolas 14, 64, **70–71**, 76
step migration 153
stimulus diffusion 168–9
Stoermer, Eugene F. 219, 256, **257**
Stone, Daithí 238
stop and go determinism 156
storm warnings **301–2**
Stott, Peter 238
Strabo 21, 23, 157
Strahler, Arthur Newell 105
Strasbourg 265
strata 70, 71, 74, 76
Strommel, Henry 140
Strunz, Karl Hugo 91
Sturtian glaciation 89
subduction 78, 116, 118, 119, 120–21, 122, 124
submersibles 120
subsidence 77, 105
subtropical zone 99–100, 139
suburbs 186, 187
Suess, Eduard 76–8, 136, 256, **319**
Sun
  hydrological cycle 67, 68, 69
  position 21, 22
  see also solar radiation
supercontinents 76, 78, 110
superposition, Steno's law of **70–71**, 127
super-rotation 112
super-volcanoes 75
sustainability 189, 218, 219, 221, 224, 226, 242, 248–9, 253, 263, 272, 295
sustainable development **244–5**
Sustainable Development Goals (SDGs) 189, 244, 245
SYMAP 56, 304

INDEX

T

Tarde, Gabriel 166
Taylor, Eva **320**
Taylor, Thomas Griffith 156
Teale, Edwin Way 240
technoscape 170–71
tectonic activity *see* plate tectonics
Teilhard de Chardin, Pierre 256
temperate zones 98, 99, 100
temperatures
  biomes 137, 139
  climate change 80, 218–19, 234, 235, 237, 238
  climatic zones 98–9, 100
  greenhouse effect 80–83
  ocean currents 140, 141
Terra Nova expedition 86
territoriality theory 146, **202–3**
territory, control of 154–5
Thaer, Albrecht 130
Thales of Miletus 13
Tharp, Marie 19, 46, **47**, 116–17
thematic mapping 19, **34–7**, 54, 304
thermal contraction 78
thermohaline circulation, global **140–41**
Thompson, Warren 147, 160, **178**, 179
Thrift, Nigel **322**
Thünen, Johann Heinrich von 146, 147, **150–51**, 161, 184
Timation (time navigation) system 49, 50
time-geography 262, **266–7**
tipping points 239, 259
Tobler, Waldo 54, 147, 192–3, 266
Tolman, Edward 194
Tomlinson, Roger 15, 19, 54, 55, **56**, 192, 304
topophilia 147, 172, 200, 201
topsoil 131
tornadoes 143, 237, 302
tourism, cultural diffusion 169–70
Tournachon, Gaspard-Félix (Nadar) 19, **42**, 43
trade
  cultural diffusion 166, 167
  developing countries 198
  development theories 191
  globalization 250–53
trade winds 34, 72
Tragedy of the Commons 218, **220–21**, 248
transform boundaries 119
transition communities 242, 243
Transit system 49, 50
transpiration 67, 68, 69, 101, 104
transport networks 158, 160, 173, 185, 186, 187, 205, 208, 212, 267, 282, 283
tree-planting 239
trellis drainage patterns 106
trenches, oceanic 117, 118, 120–21, 124
Trewartha, Glenn 99–100
triangulation 18, 19, **32–3**
Triassic period 83
trigonometry 21, 23, 42
trilateration 50
Troll, Carl 172
tropical zones 98, 139
  air circulation 72, 73
Truman, Harry S. 159
tsunamis 116, 121, 123, 125, 134–5
  alerts 299–301
Tuan, Yi-Fu 14, 147, 172, 200–201
Tucker, Compton J. 262, **292–3**, 295
Tufte, Edward 281
Turin Papyrus 20
Tylor, Edward Burnett 166–7
Tyndall, John 82, 218, 234
typhoons 143, 302
Tyrwhitt, Jacqueline 34, 54

U

Uberti, Oliver 280
Ukraine, Russian invasion of 158, 178
Ullman, Edward 184
underdeveloped countries 198, 219, 255, 303
uniformitarianism **75**, 77
United Nations
  Brundtland Report 15, 219, **244–5**, 248
  Climate Change Conference (COP21) 283
  Commission on Population and Development 178–9
  Convention to Combat Desertification 229
  environmental justice 246
  Environment Programme (UNEP) 230, 231, 254, 286
  Food and Agriculture Organization (FAO) 286
  Framework Convention on Climate Change (UNFCCC) 253
  Office for Disaster Risk Reduction (UNDRR) 263, 298, 299, 303
  Programme on Reducing Emissions from Deforestation and Forest Degradation (UN-REDD) 286
  Statistics Division (UNSD) 147, 188–9
  Sustainable Development Goals (SDGs) 189, 244, 246
  UNESCO World Heritage Convention 174, 175
  World Population and Housing Census Programme 189
uplift 77, 78
upwelling 141
urbanization 152, 186–7, 199, 264, 275
urban planning 14, 160, 163, 184–7, 206, 207, 242, 262, 264, 265, 281
  discriminatory 278
  gender-sensitive 213
  Smart Growth movement **282–3**
  spatial interaction theory 193
  time-geography 266–7
urban regeneration 187
urban settlements 173
  see also cities
Urdaneta, Andrés de 72
US Geological Survey (USGS) 32, 59
Ushahidi platform 304, 305
U-shaped valleys 105
USSR, collapse of 250

V

Valvasor, Johann Wiekhard von 108
van den Berg, Leo 184, 187
van Dresser, Peter 242
Van Newkirk, Allen 219, 242
vector data/layers 57–8
vegetation type
  aerial photography/satellite imagery 43, 44, 229, 292–3, 306
  biomes 139
  climate 98, 101
  mapping 36
  soil formation 130, 131
Venetz, Ignaz 87
Vernadsky, Vladimir 65, 136, 137, 241, 256
Verne, Jules 42
vertical erosion 104, 105
vertical projections 30
Vespucci, Amerigo 28
Vine, Frederick 110, 118–19
violence, against women 213
virtual reality 59, 192
virtual worlds 214, 215
visualization technology 281, 311
visual simulation 192
volcanic activity
  climate change 89, 239, 270
  hotspots **122–5**
  monitoring and early warnings 51, 301
  orogeny 78
  underwater 124, 125
volcanic islands 119, 123, 124, 125
volcanicity 119
V-shaped valleys 104
vulnerable people, assistance for 303

W

Wackernagel, Mathis 219, **248–9**
Wadati, Kiyoo 122, 124, 133, 135
Waldseemüller, Martin 28
Wallace, Alfred Russel 149
Wallerstein, Immanuel 147, 190, 198, **199**
Walling, Desmond 107
Walter, Heinrich **136–9**
Wang Chong 66
war 149
waste materials 245, 246–7, 249, 255
water
  erosion 104
  hydrological cycle **66–9**
  landscape formation 108
  rainforests 286
  satellite monitoring 309
  soil filtration 128, 131
water cycle 66, **68–9**
waterfalls 105
water supply 69, 237, 259
water vapour 67, 68, 69, 80, 82, 101, 234, 235
Watson-Watt, Robert 298
wayfinding 192
wealth distribution 251
weather
  Coriolis effect 65, 85
  extreme **142–3**, 238, 257, 298
  forecasting 294, 298
  modelling 60, 61, 65
weather bombs 142
weathering 130, 131
Webber, Richard 162, 192, 274, 275, **276**, 278
Weber, Alfred 151
Weber, Wilhelm 18, 39
Web Mercator 31
Wegener, Alfred 14, 65, 76, 78, **110–111**, 116, 124
Werner, Abraham 74, 75, **318**
Western world, as development model 190–91, 198
Westgate, Ken 303
West, Gladys 48, **50**
Wetherald, Richard 236–7, 269
wetlands 231
Whatmore, Sarah **323**
Whewell, William 75
whirlwinds 143
Whiston, William 74
White, Gilbert Fowler 194, 218, **230–31**
Wiechert, Emil 113
Wiechert, Johann 38
wildfires 237, 287, 288, 307, 309
warning systems **301**
Williams, Marilyn 274
Williams, Paul 108, 109
Wilson, John Tuzo 65, 76, 111, 118–19, 122, 124, **125**
Wilson, Woodrow 224
wind power 290, 291
winds
  atmospheric circulation and **72–3**
  circular patterns 72
  climate change 237
  climate zones 100, 101
  early warnings 302
  harnessing 72
  mapping 34
  ocean currents 141
  speeds 142–3
windward side 100–101
Wisner, Ben 303
Wood, Denis **322**
Wood, Harry 132, 133
workforce exploitation 199
working-class, urban living conditions 206
workplaces, gendered 211–12
World Bank 272
World Health Organization (WHO) 310, 311
world map
  first 20–21
  Mercator's 29–30
World Meteorological Organization (WMO) 80, 86, 298, 303
world-systems theory **198–9**
World Trade Organization 250
World War I 43
World War II 146, 190
  navigation 49
World Weather Attribution (WWA) 238
World Wildlife Fund (WWF) 224

X

Xenophanes 74
Xiu Xiake 108

Y

yellow fever 36
youthful landscapes 104

Z

Zedler, Othmar 240
Zhang Heng 112
Zheng He 13, 18, 28, **317**
Zipf, George **160**, 161
zoning regulations 197, **264–5**, 275, 278

QUOTE ATTRIBUTIONS

MAPS AND MAPPING

20 Eratosthenes of Cyrene, Ancient Greek mathematician and geographer

22 Strabo, Ancient Greek geographer

24 Gerardus Mercator, Flemish geographer, cosmographer, and cartographer

32 Gemma Frisius, Dutch physician, mathematician, cartographer, and philosopher

34 John Snow, English physician

38 Carl Friedrich Gauss, German mathematician and scientist

40 Gaspard-Félix "Nadar" Tournachon, French photographer, caricaturist, and balloonist

46 Marie Tharp, American geologist and oceanographic cartographer

48 Bradford Parkinson, American engineer

52 Jack Dangermond, President of Environmental Systems Research Institute (Esri)

60 Peter Bauer, Deputy Director of Research at the European Centre for Medium-Range Weather Forecasts (ECMWF)

PHYSICAL GEOGRAPHY

66 Bernard Palissy, French Huguenot potter and craftsman

70 Nicolas Steno, Danish geologist and anatomist

72 George Hadley, amateur English meteorologist

74 Georges Cuvier, French naturalist

75 James Hutton, Scottish geologist

76 Charles Lyell, Scottish geologist

80 John Tyndall, Irish physicist

84 Gaspard-Gustave de Coriolis, French scientist

86 Louis Agassiz, Swiss-American biologist and geologist

90 James D. Dana, American geologist, mineralogist, volcanologist, and zoologist

94 Clarence Dutton, American geologist

96 Wladimir Köppen, Russian-German geographer, meteorologist, and climatologist

102 William Morris Davis, American geographer, geologist, geomorphologist, and meteorologist

108 Jovan Cvijić, Serbian geographer, ethnologist, and sociologist

110 Alfred Wegener, German climatologist, geologist, geophysicist, and meteorologist

112 Andrija Mohorovičić, Croatian geophysicist

114 Alfred Wegener, German climatologist, geologist, geophysicist, and meteorologist

122 J. Tuzo Wilson, Canadian geophysicist and geologist, and Kevin C. Burke, American geologist

126 Claude Lorius, French glaciologist

128 Hans Jenny, Swiss natural scientist

132 Charles F. Richter, American seismologist

136 Vladimir Vernadsky, Ukrainian and Soviet mineralogist and geochemist

140 Wallace S. Broecker, American geochemist

142 Herbert Saffir, American civil engineer

HUMAN GEOGRAPHY

148 Thomas Malthus, English economist and demographer

150 Johann Heinrich von Thünen, German agronomist and economist

152 Ernst Georg Ravenstein, German-English geographer and cartographer

154 Friedrich Ratzel, German geographer and ethnographer

156 Ellen Churchill Semple, American geographer

157 Lucien Febvre, French historian

158 Halford Mackinder, British geographer, academic, and politician

159 Nicholas J. Spykman, American political scientist

160 George Zipf, American linguist and philologist

161 Mark Jefferson, American geographer and cartographer

162 Harlan H. Barrows, American geographer

164 Franz Boas, German-American anthropologist

172 Carl Sauer, American geographer

176 Warren S. Thompson, American demographer

182 Walter Christaller, German geographer

184 Robert Ezra Park, American urban sociologist

188 United Nations Department of Economic and Social Affairs Statistics Division

190 Walter Rostow, American economist

192 Waldo R. Tobler, American geographer, cartographer, and photogrammetrist

194 Kevin Lynch, American urban planner

196 Ruth Glass, German-British sociologist and urban planner

198 Immanuel Wallerstein, American sociologist

200 Yi-Fu Tuan, Chinese-American geographer

202 Robert D. Sack, American geographer

204 David Harvey, British-American academic

206 David Harvey, British-American academic

208 Doreen Massey, British social scientist and geographer

210 Linda McDowell, British geographer

214 Edward Relph, Canadian geographer

ENVIRONMENTAL GEOGRAPHY

220 William Forster Lloyd, British writer

222 John Muir, Scottish-American conservationist

228 André Aubréville, French botanist

230 Gilbert F. White, American geographer

232 Charles David Keeling, American scientist

240 Rachel Carson, American marine biologist, writer, and conservationist

241 James Lovelock, English independent scientist, environmentalist, and futurist

242 Peter Berg, American environmental writer

244 World Commission on Environment and Development

246 Robert D. Bullard, American sociologist and environmental justice campaigner

248 Mathis Wackernagel, Swiss-born sustainability advocate

250 Ulrich Beck, German sociologist

254 Norman Myers, British environmentalist and biodiversity expert

256 Paul J. Crutzen, Dutch meteorologist, and Eugene F. Stoermer, American limnologist

APPLIED GEOGRAPHY

264 Edward M. Bassett, American lawyer

266 Torsten Hägerstrand, Swedish geographer

268 Syukuro Manabe, Japanese-American meteorologist and climatologist, and Richard T. Wetherald, American climatologist

272 Ian McHarg, Scottish landscape architect

274 Jonathan Robbin, American demographer

278 Edward Soja, American political geographer

280 Luc Anselin, Belgian geographer

282 Jan Gehl, Danish architect

284 Global Forest Watch (GFW)

290 Hermann Scheer, chair of the World Council for Renewable Energy

292 Josef Aschbacher, Director General of the European Space Agency (ESA)

296 Neil Smith, Scottish geographer and anthropologist

304 University of Southampton-based interdisciplinary research group, UK

306 Andrew K. Skidmore, Australian ecologist

308 Joseph Aschbacher, Director General of the European Space Agency

310 Este Geraghty, Chief Medical Officer at the Environmental Systems Research Institute (Esri)

ACKNOWLEDGMENTS

Dorling Kindersley would like to thank Stephanie Farrow for the editorial read; Diana Vowles for proofreading; Helen Peters for indexing; and Vanessa Hamilton for additional illustrations. DK Delhi would like to thank Abhijit Dutta for editorial assistance; Raman Panwar for DTP assistance; and Northwind Content Services LLP for their help with picture research.

PICTURE CREDITS

The publisher would like to thank the following for their kind permission to reproduce their photographs:

(Key: a-above; b-below/bottom; c-centre; f-far; l-left; r-right; t-top)

20 Alamy Stock Photo: North Wind Picture Archives (br). **21 Alamy Stock Photo:** Interfoto / Personalities (bl). **23 Alamy Stock Photo:** Piemags (tl). **26 James Ford Bell Library, University of Minnesota, Minneapolis, Minnesota:** (tr). **27 Getty Images:** Hulton Archive / Heritage Images (b). **28 Alamy Stock Photo:** Chronicle (tl). **30 Alamy Stock Photo:** Art Collection 3 (bl). **31 Engaging Data:** Chris Yang (tl). **33 Alamy Stock Photo:** Science History Images / Photo Researchers (bl). **Getty Images:** De Agostini / Dea Picture Library (tr). **35 Alamy Stock Photo:** Antiqua Print Gallery (bl); IanDagnall Computing (tr). **36 Alamy Stock Photo:** Piemags. **37 Special Collection, University of California Riverside:** Pierre Émile Levasseur (bl). **39 Alamy Stock Photo:** Art Collection (bl). **42 Alamy Stock Photo:** Historic Images (tr); Pictorial Press Ltd (bl). **44 Getty Images:** Hulton Archive / Print Collector (tr). **45 Getty Images:** Sean Gallup (bl). **Naomi Norwood:** (tr). **46 Library of Congress, Washington, D.C.:** (bl). **47 Lamont-Doherty Earth Observatory and the estate of Marie Tharp:** (tl). **49 Bradford Parkinson:** (tr). **50 Alamy Stock Photo:** IanDagnall Computing (bl). **51 Getty Images:** Moment / Witthaya Prasongsin (br). **NASA:** JPL-Caltech (tl). **55 Science Photo Library:** Pasquale Sorrentino (tl). **58 Alamy Stock Photo:** Robert vant Hoenderdaal (br); NG Images (tl). **60 Science Photo Library:** Amazing Aerial Agency / Andrea Simula (br). **67 Alamy Stock Photo:** Old Books Images (tr). **69 Alamy Stock Photo:** Shoults (br). **71 Alamy Stock Photo:** Scott Camazine (tl). **Shutterstock.com:** Shane Myers Photography (br). **72 Getty Images / iStock:** Duncan1890 (bc). **75 Getty Images:** Universal Images Group / Auscape (bc). **77 Alamy Stock Photo:** Recall Pictures (tr). **78 Alamy Stock Photo:** Colin Palmer Photography (tr). **79 Alamy Stock Photo:** Hemis / Escudero Patrick (br). **81 Alamy Stock Photo:** Michael Jones (tl). **83 Alamy Stock Photo:** Science History Images / Photo Researchers (bl). **Depositphotos Inc:** Arch88 (tr). **84 Alamy Stock Photo:** World History Archive (bc). **85 Alamy Stock Photo:** Art Collection (tr). **87 Adobe Stock:** Cliff LeSergent (br). **Alamy Stock Photo:** GL Archive (tl). **88 Alamy Stock Photo:** Minden Pictures / Matthias Breiter (bl). **89 Alamy Stock Photo:** Stocktrek Images, Inc. / Walter Myers (tr). **92 Alamy Stock Photo:** Granger - Historical Picture Archive (bl). **93 Alamy Stock Photo:** Jack Clark Collection / Phil Degginger (br). **94 Alamy Stock Photo:** Jon Arnold Images Ltd (bc). **95 Alamy Stock Photo:** Pictorial Press Ltd (bl). **Science Photo Library:** G. R. Roberts (tr). **98 Alamy Stock Photo:** Historic Images (bl). **100 Alamy Stock Photo:** Robertharding / Peter Groenendijk (br). **101 Alamy Stock Photo:** Adrian Davies (bc). **104 Alamy Stock Photo:** Dipper Historic (bl). **105 Alamy Stock Photo:** Frans Lemmens (bl). **107 Alamy Stock Photo:** Morey Milbradt (tl). **109 Alamy Stock Photo:** Panther Media Global / Casa-Blanca (tr); UtCon Collection (bl). **110 Alamy Stock Photo:** Imagebroker.com / Ingo Schulz (br). **111 Alamy Stock Photo:** IanDagnall Computing (tr). **116 Harry Hammond Hess:** (bl). **117 Dreamstime.com:** Zigaplahutar (tr). **118 Alamy Stock Photo:** Prisma by Dukas Presseagentur GmbH / CCOphotostock_KMN (tr). **120 Science Photo Library:** NOAA (bl); Stan Wayman (tr). **121 Alamy Stock Photo:** Magite Historic (bl). **123 Shutterstock.com:** Deni_Sugandi (tl). **124 Dreamstime.com:** Delstudio (tr). **125 Getty Images:** Mondadori Portfolio (tl). **127 Science Photo Library:** British Antarctic Survey (tl). **129 Alamy Stock Photo:** Rob Walls (tr). **131 Alamy Stock Photo:** David Tipling Photo Library / David Tipling (tl). **133 Alamy Stock Photo:** Granger - Historical Picture Archive (tl, bc). **134 Alamy Stock Photo:** Horizon Images / Motion (tr). **137 Alamy Stock Photo:** Rob Crandall (br). **138 Getty Images:** Moment / Posnov (tr). **139 Alamy Stock Photo:** Dmitrii Melnikov (br). **141 Alamy Stock Photo:** Brian Parker (br). **143 Alamy Stock Photo:** ZUMA Press, Inc. (tr). **149 Alamy Stock Photo:** GL Archive (bl). **Getty Images:** Moment / Pencho Chukov (tr). **151 Alamy Stock Photo:** Piemags (br); Zoonar GmbH / Nando Lardi (tl). **152 Alamy Stock Photo:** SocialHistoryImages (bc). **153 Alamy Stock Photo:** The History Collection (bl). **155 Alamy Stock Photo:** History and Art Collection (bl). **Getty Images:** Archive Photos / FPG (tr). **156 University of California, Berkeley:** Influences of Geographic Environment, on the Basis of Ratzel's System of Anthropo-Geography p.8, GF31 .S47 1911, The Library, University of California, Berkeley (bc). **159 Alamy Stock Photo:** RBM Vintage Images (cr). **161 Alamy Stock Photo:** Roman Sigaev (bc). **163 Alamy Stock Photo:** FLPA (bl). **166 Alamy Stock Photo:** IanDagnall Computing (br); Well / BOT (tr). **167 Dreamstime.com:** Thomas Jenkins (br). **169 Getty Images:** AFP (tl). **170 Alamy Stock Photo:** Wim Wiskerke (t). **171 Alamy Stock Photo:** Shiiko Alexander (tr). **173 Alamy Stock Photo:** Blickwinkel / Woike (tr). **175 Alamy Stock Photo:** Henner Damke (tr). **Getty Images:** The Image Bank / Christian Adams (t). **178 Alamy Stock Photo:** Everett Collection Historical (tr). **179 Alamy Stock Photo:** Jordi Boixareu (tr). **180 Alamy Stock Photo:** Jake Lyell (tr). **181 Alamy Stock Photo:** Shawn.ccf (tr). **183 Alamy Stock Photo:** B.O'Kane (bl). **185 Alamy Stock Photo:** The Keasbury-Gordon Photograph Archive (br). **187 Alamy Stock Photo:** Jerome Labouyrie (bl). **189 Alamy Stock Photo:** Associated Press / National Archives and Records Administration (tl); Sipa USA / Pacific Press / Hussain Ali (br). **190 Getty Images:** Universal Images Group / Pictures from History (bc). **191 Getty Images:** Archive Photos / PhotoQuest (br). **193 Alamy Stock Photo:** Justin Kase Zsixz (br). **195 Alamy Stock Photo:** Roy Johnson (tl). **MIT Museum: Courtesy MIT Museum** (bl). **197 Shutterstock.com:** Creative Lab (bl). **199 Getty Images:** Gamma-Rapho / Louis Monier (tr). **201 Getty Images / iStock:** Jamie McDonald (br). **203 Getty Images:** Bettmann (b). **205 Alamy Stock Photo:** Oriol Clavera (bl); Sean Pavone (cr). **207 Getty Images:** Los Angeles Times / Carolyn Cole (bc). **208 Alamy Stock Photo:** Matthew Ashmore / Stockimo (bc). **209 Alamy Stock Photo:** Chris Batson (tr). The Open University: Doreen Massey (bl). **211 Getty Images:** AFP / STR (tr). **Professor Linda McDowell:** (bl). **213 Alamy Stock Photo:** ZUMA Press, Inc. / Pacific Press / Suraranjan Nandi (tr). **215 Alamy Stock Photo:** Jochen Tack (bl). **221 Shutterstock.com:** Rawpixel.com (tl). **224 Getty Images / iStock:** Zrfphoto (tr). **226 Alamy Stock Photo:** Anthony Dunn (bl). **227 Alamy Stock Photo:** RGB Ventures / SuperStock / Library of Congress (bl). **Shutterstock.com:** Everett Collection (tc). **229 Alamy Stock Photo:** Associated Press / Shigeki Tao (tl). **231 NSIDC:** University of Colorado Boulder / Gilbert Fowler White (bl). **Science Photo Library:** Jim Edds (tr). **234 Depositphotos Inc:** Ranglen (tl). **236 Alamy Stock Photo:** ZUMA Press, Inc. (tl). **237 Adobe Stock:** Knelson20 (tr). **239 Getty Images:** John Moore (tr). **240 Alamy Stock Photo:** Everett Collection Historical (cr). **244 Alamy Stock Photo:** Photo12 / Ann Ronan Picture Library (br). **245 Getty Images:** AFP / Joel Saget (tr). **246 Getty Images:** Bettmann (bc). **247 Getty Images:** Hearst Newspapers / Albany Times Union (tl). **249 Getty Images:** AFP / Evaristo SA (bl). **Mathis Wackernagel:** Nicholas Albrecht (tr). **251 Alamy Stock Photo:** Science History Images / Photo Researchers (br). **Getty Images:** Hulton Archive / Leonardo Cendamo (tl). **252 Alamy Stock Photo:** Melting Spot (bl). **254 Alamy Stock Photo:** Xinhua / Str (br). **257 Alamy Stock Photo:** Emmanuel Lattes (tr). **258 Alamy Stock Photo:** Science History Images / Photo Researchers (br); Sueddeutsche Zeitung Photo / Ingrid von Kruse (tl). **265 Alamy Stock Photo:** EMU History (tl). **267 Alamy Stock Photo:** Brian Overcast (br). **269 © European Union, 2020:** Copernicus Sentinel-X imagery [The Copernicus Marine Environment Monitoring Service (CMEMS)] (bl). **Getty Images:** AFP / Kena Betancur (tr). **271 NASA:** JPL-Caltech (bl). **272 Alamy Stock Photo:** Associated Press / Charles Tasnadi (bc). **273 Alamy Stock Photo:** B Christopher (tr). **275 Alamy Stock Photo:** Mil Image (tl). **276 Getty Images:** ITN (tl). **279 Laura Pulido:** Courtesy of Laura Pulido (tr). **280 GeoDa Center:** Images produced with GeoDa, developed by Luc Anselin and the Center for Spatial Data Science, University of Chicago. https://geodacenter.github.io (cr). **281 Getty Images:** Visual Capitalist (br). **282 Shutterstock.com:** Felix Mizioznikov (br). **283 Alamy Stock Photo:** Jochen Tack (cl). **286 Getty Images:** Gallo Images (bl). **288 Alamy Stock Photo:** Juriaan Wossink (bl). **289 Alamy Stock Photo:** Universal Images Group North America LLC / Planet Observer (tr). **291 Getty Images:** AFP / Rodger Bosch (br); Ullstein Bild (tr). **293 Alamy Stock Photo:** Paul Prestidge (tl). **NASA:** Compton Tucker (br). **299 Getty Images:** Chris McGrath (br). **301 Alamy Stock Photo:** Abaca Press (br); Bob Collet (br). **302 Depositphotos Inc:** Wirestock (tl). **304 Getty Images:** AFP / Simon Maina (br). **306 Alamy Stock Photo:** Rosemary Calvert (bc). **308 ESA:** Copernicus Sentinel data (2024), processed by ESA (bc). **309 NASA:** (tl). **312 Johns Hopkins University:** (tr). **313 Johns Hopkins University:** (bl)